Springer Series in Solid-State Sciences

Volume 202

The Springer Series in Solid-State Sciences features fundamental scientific books prepared by leading and up-and-coming researchers in the field. They strive to communicate, in a systematic and comprehensive way, the basic principles as well as new developments in theoretical and experimental solid-state science.

We welcome submissions for monographs or edited volumes from scholars across this broad domain. Topics of current interest include, but are not limited to:

- Semiconductors and superconductors
- Quantum phenomena
- Spin physics
- Topological insulators
- Multiferroics
- Nano-optics and nanophotonics
- Correlated electron systems and strongly correlated materials
- Vibrational and electronic properties of solids
- Spectroscopy and magnetic resonance

Jiro Kitagawa · Yoshikazu Mizuguchi
Editors

High-Entropy Alloy Superconductors

Exotic Properties, Applications and Materials Design

Springer

Editors
Jiro Kitagawa
Department of Electrical Engineering
Fukuoka Institute of Technology
Fukuoka, Japan

Yoshikazu Mizuguchi
Department of Physics
Tokyo Metropolitan University
Tokyo, Japan

ISSN 0171-1873 ISSN 2197-4179 (electronic)
Springer Series in Solid-State Sciences
ISBN 978-981-97-4128-1 ISBN 978-981-97-4129-8 (eBook)
https://doi.org/10.1007/978-981-97-4129-8

This Springer imprint is published by the registered company Springer Nature Singapore Pte Ltd.
The registered company address is: 152 Beach Road, #21-01/04 Gateway East, Singapore 189721, Singapore

Preface

This book presents a comprehensive survey of the latest research concerning high-entropy alloy (HEA) superconductors, an area of inquiry initiated in 2014 and has since garnered significant attention. Each section is authored by experts who elucidate the exotic properties, applications, and materials design associated with HEA superconductors, drawing upon their original concepts.

HEAs represent a novel class of materials introduced in 2004, renowned for their exceptional mechanical attributes, robust resistance to corrosion, and remarkable thermal stability, among other merits. HEAs are characterized by crystals in which more than four elements randomly occupy a single crystallographic site, thereby augmenting configurational entropy with the increasing number of compositional elements. Presently, the concept of high entropy is extensively applied across various materials, including oxides, chalcogenides, and halides. Moreover, the high-entropy state engenders many enhanced functionalities, such as thermoelectric properties, magnetocaloric effects, catalytic prowess, and more.

Superconductivity has emerged as a particularly prominent subject in this domain following the discovery of a HEA superconductor in 2014. Recent findings have unveiled robust superconductivity under extraordinarily high pressure or ion irradiation. Furthermore, improvement of bulk superconductivity via the high-entropy effect has been documented. Within this realm of research, numerous scholars are endeavoring to discern the distinctions in superconducting properties between HEA superconductors and conventional or unconventional counterparts. Given the expansive compositional space of HEAs, the materials research on HEA superconductors is imperative for identifying unique characteristics associated with the high-entropy state. Additionally, research on the practical applications of HEA superconductors is indispensable for advancing superconducting technology. Consequently, the principal themes of this publication revolve around elucidating the similarities and disparities in superconducting properties between HEA superconductors and other classes of superconductors, as well as exploring applications and materials design for HEA superconductors. From a fundamental perspective, the exploration of the distinctive aspects of HEA superconductors contributes to the establishment of novel principles in superconductivity. Moreover, the research topic holds significant promise for practical applications, such as robust superconducting wires and multifunctional superconductors. Furthermore, the distinctive structural attributes of HEAs render the materials design of HEA superconductors invaluable for achieving a comprehensive understanding of these materials.

The book is organized into 11 chapters. Chapter 1 encompasses the discovery and the current status of HEA superconductors. These HEA superconductors are classified into two primary categories: the first class encompasses alloy systems characterized by body-centered cubic (bcc) and hexagonal close-packed (hcp) structures, while the second class comprises intermetallic types. In each of these classes, we expound upon the exotic properties, applications, and materials design, aligning with the overarching themes of this work. We also elucidate the interdependencies between the topics addressed in this chapter and those explored in the subsequent ten chapters, contributed by leading-edge researchers. Chapters 2–5 delve into bcc HEA superconductors, with Chap. 7 dedicated to hcp HEA superconductors. Chapter 6, on the other hand, offers a comprehensive treatment of both bcc and hcp superconductors. In the domain of high-entropy intermetallic compounds, Chaps. 8–10 delve into NaCl-type, van der Waals-type, and $CuAl_2$-type compounds, respectively. Additionally, the eleventh chapter encompasses both BiS_2-based and YBCO-based HEAs.

This publication addresses the novel facets of superconductivity associated with the high-entropy state, the potential applications under consideration, and the intricacies of materials design. These recent discoveries are poised to captivate a myriad of researchers in the fields of materials science, particularly those engaged in high-entropy alloys and the realm of superconducting properties and technology.

Fukuoka, Japan — Jiro Kitagawa
Hachioji, Japan — Yoshikazu Mizuguchi

The original version of the book has been revised: The affiliation of the second author has been updated. A correction to the book can be found at https://doi.org/10.1007/978-981-97-4129-8_12

Contents

Contributors

W. Abuzaid Department of Mechanical Engineering, American University of Sharjah, Sharjah, United Arab Emirates;
Department of Physics, American University of Sharjah, Sharjah, United Arab Emirates

Alexander J. Browne School of Chemistry, University of St Andrews, St Andrews, UK

M. Egilmez Materials Science and Engineering Program, American University of Sharjah, Sharjah, United Arab Emirates;
Department of Physics, American University of Sharjah, Sharjah, United Arab Emirates

Sylwia Gutowska Faculty of Physics and Applied Computer Science, AGH University of Krakow, Krakow, Poland

Rahmatul Hidayati Department of Applied Physics and Institute of Natural Sciences, Kyung Hee University, Yong-In, Gyong-Gi, Republic of Korea

Sonika Jangid Department of Physics, Indian Institute of Science Education and Research Bhopal, Bhopal, India

Kinga Jasiewicz Faculty of Physics and Applied Computer Science, AGH University of Krakow, Krakow, Poland

Md. Riad Kasem Department of Physics, Tokyo Metropolitan University, Hachioji, Tokyo, Japan

Gareoung Kim Department of Applied Physics and Institute of Natural Sciences, Kyung Hee University, Yong-In, Gyong-Gi, Republic of Korea

Jin Hee Kim Department of Applied Physics and Institute of Natural Sciences, Kyung Hee University, Yong-In, Gyong-Gi, Republic of Korea

Jiro Kitagawa Department of Electrical Engineering, Faculty of Engineering, Fukuoka Institute of Technology, Higashi-ku, Fukuoka, Japan

Saurav Marik School of Physics and Materials Science, Thapar Institute of Engineering and Technology Patiala, Patiala, India

Yoshikazu Mizuguchi Department of Physics, Tokyo Metropolitan University, Hachioji, Tokyo, Japan

Masanori Nagao Center for Crystal Science and Technology, University of Yamanashi, Kofu, Yamanashi, Japan

T. Nawaz Materials Science and Engineering Program, American University of Sharjah, Sharjah, United Arab Emirates;
Department of Mechanical Engineering, American University of Sharjah, Sharjah, United Arab Emirates;
Department of Physics, American University of Sharjah, Sharjah, United Arab Emirates

Terukazu Nishizaki Department of Electrical Engineering, Faculty of Science and Engineering, Kyushu Sangyo University, Fukuoka, Japan

Jong-Soo Rhyee Department of Applied Physics and Institute of Natural Sciences, Kyung Hee University, Yong-In, Gyong-Gi, Republic of Korea

Md. Riad Kasem Department of Physics, Tokyo Metropolitan University, Tokyo, Japan

Ravi Prakash Singh Department of Physics, Indian Institute of Science Education and Research Bhopal, Bhopal, India

Janusz Tobola Faculty of Physics and Applied Computer Science, AGH University of Krakow, Krakow, Poland

Chennan Wang Laboratory for Muon Spin Spectroscopy, Paul Scherrer Institute, Villigen, Switzerland;
Department of Physics, University of Fribourg, Fribourg, Switzerland

Bartlomiej Wiendlocha Faculty of Physics and Applied Computer Science, AGH University of Krakow, Krakow, Poland

Tianping Ying Beijing National Laboratory for Condensed Matter Physics, Institute of Physics, Chinese Academy of Sciences, Beijing, China

Tongxu Yu Suzhou Laboratory, Jiangsu, China

Jae Hyun Yun Department of Applied Physics and Institute of Natural Sciences, Kyung Hee University, Yong-In, Gyong-Gi, Republic of Korea

Chapter 1
Discovery and Current Status of High-Entropy Alloy Superconductors

Jiro Kitagawa and Yoshikazu Mizuguchi

Abstract This chapter reviews the discovery and current status of high-entropy alloy (HEA) superconductors. The categorization of HEA superconductors involves bifurcation into two distinct classes based on their crystallographic features. The first class comprises alloy systems represented by body-centered cubic and hexagonal close-packed structures, while the second class pertains to compound types. Within each of these classes, we expound upon the exotic properties, applications, and materials design in alignment with the themes of this book. Furthermore, we elucidate the interrelationships between the subjects addressed in this chapter and those expounded upon in the subsequent ten chapters.

1.1 Introduction

A high-entropy alloy (HEA) is characterized as an alloy or crystalline system in which at least four principal elements randomly occupy a crystallographic site. HEAs stand in stark contrast to conventional alloys composed predominantly of one or two principal elements. The innovative paradigm of HEAs has garnered substantial attention due to its rich applications, encompassing exceptional mechanical properties, energy storage capabilities, magnetic refrigeration, soft ferromagnetism, catalytic potential, thermoelectricity, and biocompatibility [1, 2]. The thermodynamic entropy state of an alloy is assessed through configurational entropy, denoted as $\Delta S_{\text{mix}} = -R\sum_{i=1}^{n} c_i \ln c_i$, where n is the number of elements, c_i is the atomic fraction, and R is the gas constant. The threshold of ΔS_{mix}, which defines HEAs, is now established at 1.0 R, generally fulfilled by the presence of four constituent ele-

J. Kitagawa (✉)
Department of Electrical Engineering, Faculty of Engineering, Fukuoka Institute of Technology, 3-30-1 Wajiro-higashi, Higashi-ku Fukuoka 811-0295, Japan
e-mail: j-kitagawa@fit.ac.jp

Y. Mizuguchi
Department of Physics, Tokyo Metropolitan University, 1-1, Minami-osawa, Hachioji 192-0397, Japan
e-mail: mizugu@tmu.ac.jp

J. Kitagawa and Y. Mizuguchi (eds.), *High-Entropy Alloy Superconductors*, Springer Series in Solid-State Sciences 202, https://doi.org/10.1007/978-981-97-4129-8_1

ments [3]. HEAs are reputed for their extraordinary attributes, setting them apart from conventional alloy systems. These distinctive properties are collectively referred to as the "four core effects," encompassing the following phenomena:

- high entropy effect, wherein an elevated ΔS_{mix} confers stability to the solid solution phase
- severe lattice distortion, resulting in solid solution strengthening due to high atomic disorder
- sluggish diffusion, characterized by hindered atomic mobility and reduced diffusion coefficients
- cocktail effect, which denotes a synergistic phenomenon leading to physical quantities exceeding the average values predicted by the mixture rule.

Superconductivity has emerged as a prominent area of interest following the discovery of an HEA superconductor in 2014 [4]. Presently, HEA superconductivity is the subject of extensive exploration, extending diverse structures such as body-centered cubic (bcc) [5, 6], hexagonal close-packed (hcp) [7–9], face-centered cubic (fcc) [10], CsCl-type [11], A15 [12], NaCl-type [13], α (or β)-Mn-type [14, 15], σ-phase type [16], $CuAl_2$-type [17], van der Waals-type [18], BiS_2-based [19], and YBCO-based [20] structures. The bcc, hcp, and fcc HEAs are categorized as alloy systems, while the remaining structures are classified as high-entropy compounds. Figure 1.1a and b illustrate the bcc HEA and the $CuAl_2$-type high-entropy compound, respectively. Within the bcc structure, a single crystallographic site, known as the $2a$ site, is randomly populated by multiple atoms. In contrast, the $CuAl_2$-type structure possesses the $4a$ and $8h$ sites. In Fig. 1.1b, the $4a$ site is randomly occupied by five transition metals, while Zr atoms occupy the $8h$ site.

The advent of superconductivity in HEA systems was triggered by the discovery of $Ta_{34}Nb_{33}Hf_8Zr_{14}Ti_{11}$,which has a bcc structure [4]. This alloy exhibits the cocktail effect, as delineated among the four core effects previously mentioned, for the superconducting critical temperature T_c. This discovery has garnered considerable scholarly attention, leading to extensive investigations into alloy-type superconduc-

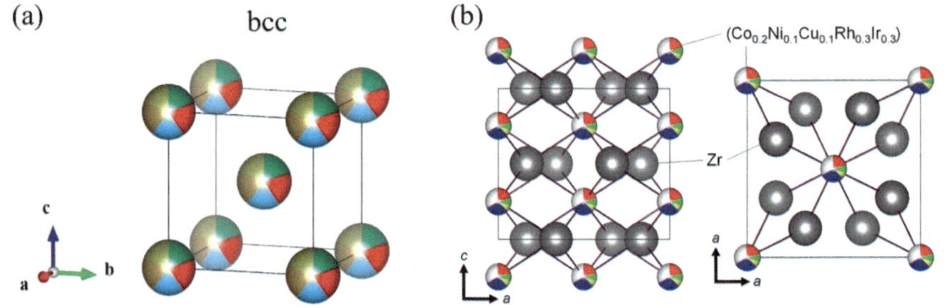

Fig. 1.1 Crystal structures of high-entropy **a** bcc-type and **b** $CuAl_2$-type compounds. The multi-colored balls indicate random site occupation by several atoms. The solid lines represent the unit cells. Reproduced from [6, 17] under the Creative Commons Attribution License

tors. Notable findings include the robustness of superconductivity against high pressures and irradiation, as well as deviations from the Matthias rule [21–23]. Furthermore, the exploration of potential applications for HEA superconductors holds significance for the advancement of superconducting technology. For instance, several bcc HEAs have demonstrated high critical current densities on par with those observed in commercial Nb-Ti superconducting wires, representing a promising avenue for next-generation superconducting wire technology [22, 24, 25]. Additionally, the high hardness or shape memory effect in some HEAs presents promising prospects for developing multifunctional devices [26, 27].

By extending the HEA material variation to compounds having two or more crystallographic sites, we can explore the effects of HEA sites on various superconducting materials, including high-T_c, high-field, unconventional, and low-dimensional superconductors. As shown in Fig. 1.1b, one of the crystallographic sites is the HEA composition in HEA-type compounds. Therefore, various Tr-Zr bonds (Tr: transition metal) are formed in HEA-type $TrZr_2$. The random bond length in HEA-type compounds results in exotic physical properties possibly related to glassy phonons.

Following the brief overview of the discovery of HEA superconductors in both alloy systems and compounds, the subsequent sections of this chapter delve into the current landscape of HEA superconductors. This examination is conducted with a focus on exotic properties, applications, and materials design, all of which are harmonized with the themes of this book. Each class of HEA superconductors will be treated individually in this regard. Furthermore, we elucidate the thematic connections between the topics addressed in this chapter and the subsequent ten chapters contributed by leading-edge researchers.

1.2 Discovery of High-Entropy Alloy Superconductors

The first superconductor of the alloy type, $Ta_{34}Nb_{33}Hf_8Zr_{14}Ti_{11}$, made its debut in 2014 [4]. This compound crystallizes into the bcc structure and manifests itself as a type II superconductor with T_c = 7.3 K. This behavior is comprehensively investigated by the electrical resistivity, magnetization, and specific heat. The specific heat analyses suggest a Bardeen-Cooper-Schrieffer (BCS)-type phonon-mediated superconductor characterized by a weak electron-phonon coupling. The cocktail effect is examined concerning the T_c value, taking into account the T_c values of the constituent elements, which stand at 4.47 K for Ta, 9.25 K for Nb, 0.128 K for Hf, 0.61 K for Zr, and 0.40 K for Ti. Consequently, the composition-weighted T_c is 4.71 K. Remarkably, the experimental T_c significantly surpasses this value of 4.71 K, attributable to the cocktail effect [4].

After the development of HEA superconductors, HEA-type superconducting compounds debuted in 2018. The first example was the layered BiS_2-based superconductor $REO_{0.5}F_{0.5}BiS_2$ (RE: rare earth) [28]. The RE site in the material is the solution of RE = La, Ce, Pr, Nd, Sm. Interestingly, the superconducting properties are improved by an increase in RE-site ΔS_{mix} in the BiS_2-based system [19, 28]. This result suggests that the local HEA site can affect the bulk physical properties of layered compounds.

Subsequently, we explore the present state of HEA superconductors, categorizing them into two distinct classes. The first class includes bcc and hcp HEA superconductors, while the second encompasses the compound type. Chapters 2 through 5 survey bcc HEA superconductors, with Chap. 7 dedicated to hcp HEA superconductors. Chapter 6, on the other hand, undertakes the comprehensive treatment of both bcc and hcp superconductors. In the class of high-entropy compounds, Chaps. 8–10 delve into the NaCl-type, van der Waals-type, and $CuAl_2$-type compounds, respectively. Additionally, the eleventh chapter considers both BiS_2-based and YBCO-based HEAs.

1.3 Body-Centered Cubic and Hexagonal Close-Packed Structures

In this section, we survey the current status of bcc and hcp HEA superconductors, focusing on their exotic properties, applications, and materials design, all harmonized with the themes described within the present book. Within each subsection, we expound upon various topics, many of which will find comprehensive treatment in the ensuing chapters from Chap. 2–7. Table 1.1 summarizes the interrelationships between the subjects addressed in this section and those expounded upon in Chaps. 2 through 7.

Table 1.1 Interrelationships between topics addressed in this section and those expounded upon in Chaps. 2 through 7

Topics	Chapter number
Exotic properties	
Deviation from the Matthias rule	3, 6, 7
Band broadening	3, 5
Strongly correlated electronic state	2
Search for breaking of time-reversal symmetry	4, 6
Applications	
High critical current density	2
High hardness	3
Shape memory and superelasticity	4
Materials design	
Semiempirical parameters	3, 7
First-principles calculations	5

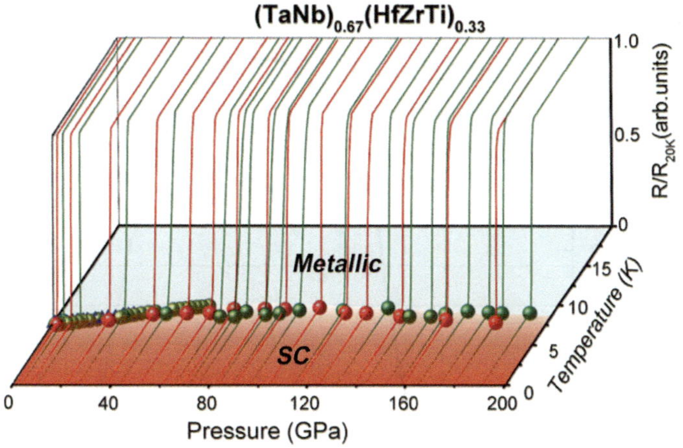

Fig. 1.2 Phase diagram of T_c vs. pressure up to 190.6 GPa for $(TaNb)_{0.67}(HfZrTi)_{0.33}$, combined with resistance vs. temperature plots. Reproduced with permission from [21]

1.3.1 *Exotic Properties*

One of the striking and notable characteristics within this class of materials is the remarkable insensitivity of T_c to extremely high pressure. As depicted in Fig. 1.2, the phase diagram of the bcc $(TaNb)_{0.67}(HfZrTi)_{0.33}$ HEA superconductor offers an insightful perspective [21]. This diagram elegantly combines the relationship between T_c and pressure with the evolution of resistivity with respect to temperature. Notably, the value of T_c, approximately 8 K, exhibits insensitivity to changes over a wide pressure range, from ambient pressure to an astonishing 200 GPa. Furthermore, interestingly, the $TaNb_2HfZrTi$ bcc thin film demonstrated notably robust superconductivity even when subjected to Kr-ion irradiation [22]. This irradiation introduces a heightened level of atomic disorder, typically leading to a reduction in T_c. However, it is noteworthy that the reduction of T_c in $TaNb_2HfZrTi$ remains relatively mild compared to that in famous superconductors such as Nb_3Sn, MgB_2, and YBCO [22]. These findings regarding $(TaNb)_{0.67}(HfZrTi)_{0.33}$ and $TaNb_2HfZrTi$ have profound implications, positioning superconducting HEAs as promising candidates for materials capable of withstanding extreme conditions, such as those encountered in aerospace applications and nuclear fusion. Furthermore, the robustness of superconductivity in the presence of magnetic elements has been observed, underscoring the versatility of HEA materials [29].

The majority of HEAs showcase conventional s-wave BCS-type superconducting behavior. According to the BCS theory, T_c is intrinsically linked to the density of states at the Fermi energy (E_F), expressed as $D(E_F)$, and the strength of the electron-phonon interaction. In the context of bcc and hcp alloy systems, the valence electron concentration per atom (VEC) has a significant relationship with $D(E_F)$. Consequently, the Matthias rule has been established, underscoring the strong correlation

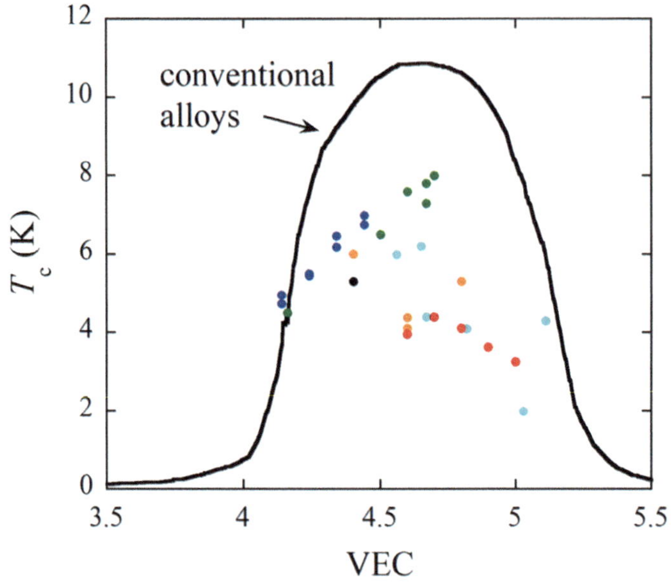

Fig. 1.3 Comparison between conventional Matthias rule (solid curve) and VEC dependence of T_c values of typical quinary bcc HEA superconductors. The correspondence between color and HEA is as follows: green: nonequimolar Hf-Nb-Ta-Ti-Zr; blue: Al-Nb-Ti-V-Zr; black: $Hf_{21}Nb_{25}Ti_{15}V_{15}Zr_{24}$; orange: HfNbTaTiZr, HfNbReTiZr, HfNbTaTiV, and HfMoNbTiZr; light blue: Nb-Ta-Mo-Hf-W, Ti-Zr-Nb-Ta-W, and Ti-Zr-Nb-Ta-V; and red: Ti-Hf-Nb-Ta-Re. Reproduced from [23] under the Creative Commons Attribution License

between VEC and T_c in binary or ternary superconducting transition metal alloys [30]. As demonstrated in Fig. 1.3, the plot of T_c against VEC, represented by the solid curve, distinctly illustrates a pronounced peak near a VEC value of 4.6 [23]. The dataset for typical quinary bcc HEA superconductors is represented by filled circles of various colors, exemplifying different elemental combinations and their respective studies [23, 26, 31–38]. It is worth noting that the results for HEAs lie beneath the solid curve, revealing that pronounced atomic disorder tends to reduce T_c and lead to deviation from the Matthias rule. Recent electronic and phonon structure calculations have revealed band broadening within electronic and phonon band structures, which can be attributed to substantial atomic disorder [39, 40]. This, in turn, contributes to the decrease in the electron and phonon lifetimes. Various research groups have engaged in discussions to comprehend the influence of this band broadening on superconducting properties [26, 40]. The deviation from the conventional Matthias rule within HEAs suggests that more rigorous treatment of band broadening might be required to explain these phenomena. The impact of atomic disorder, aptly assessed by the value of ΔS_{mix}, on superconducting properties has been an important subject, although conclusive findings have remained elusive. The potential significance of band broadening in this context may be a pivotal factor worthy of exploration.

In the early stages of high-entropy alloy superconductor research, macroscopic measurements such as electrical resistivity, magnetization, and specific heat have proven invaluable. More recently, microscopic techniques such as μSR spectroscopy have been employed for a more profound elucidation of the Cooper pair mechanism or to explore potential violations of time-reversal symmetry [27, 41, 42]. As of the current state of research, exotic Cooper pairing remains elusive. It should be noted that certain bcc HEA superconductors have exhibited a strongly correlated electronic state. The renowned Kadowaki-Woods ratio, denoted as A/γ^2, serves as a criterion for assessing the degree of electronic correlation. Here, A is the coefficient of the T^2 (T:temperature) term in the temperature-dependent electrical resistivity, while γ represents the Sommerfeld coefficient. Many heavy fermion compounds containing elements such as Ce or U have established a universal A/γ^2 value of 10 $\mu\Omega{\cdot}cm{\cdot}mol^2{\cdot}K^2/J^2$. Notably, recent findings suggest that bcc HEA superconductors, such as $Ta_{1/6}Nb_{2/6}Hf_{1/6}Zr_{1/6}Ti_{1/6}$ and $Ta_{34}Nb_{33}Hf_8Zr_{14}Ti_{11}$, also align with this trajectory, mirroring the behavior observed in heavy-fermion compounds [43].

1.3.2 Applications

The practical investigations of bcc and hcp HEA superconductors remain nascent. The assessment of the critical current density (J_c) is of paramount importance for the potential application as superconducting wires. Figure 1.4a and b present the field-dependent characteristics of J_c for $HfNb_2TaTiZr$ bcc HEA thin films at temperatures

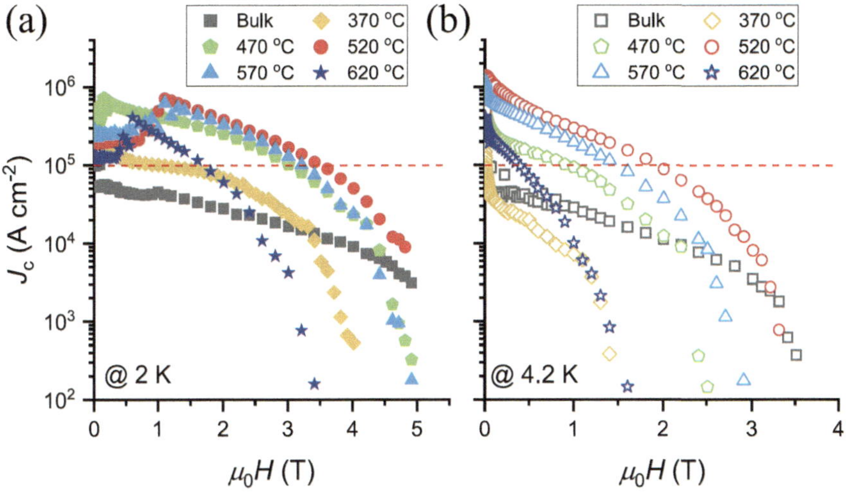

Fig. 1.4 Field-dependent J_c of $HfNb_2TaTiZr$ thin films fabricated at several temperatures denoted in figure at **a** 2 K and **b** 4.2 K. Reproduced from [22] under the Creative Commons Attribution License

of 2 K and 4.2 K, respectively [22]. Notably, the thin-film samples exhibit exceptionally high self-field J_c values, approximating 1 MA/cm^2. The red dashed line in each figure serves as a standard benchmark for large-scale applications, particularly in the context of high-field superconducting magnets. Impressively, the J_c values of $HfNb_2TaTiZr$ thin films surpass the benchmark line up to a magnetic field strength of 2 T at 4.2 K, underscoring their potential as superconducting wires. Another example is $(TaNb)_{0.7}(HfZrTi)_{0.5}$ bulk alloy, which exhibits the unique field-dependent behavior of J_c [24]. In this study [24], the impact of annealing on superconducting properties was explored. Notably, the value of T_c at approximately 7 K remained nearly unaltered after annealing. However, a precipitation phenomenon became evident above an annealing temperature of 500 °C, giving rise to effective flux pinning sites. Although the as-cast sample initially displayed a self-field J_c value less than 0.1 MA/cm^2, thermal annealing at 500 °C significantly enhanced J_c, elevating it by an order of magnitude. Moreover, the bulk alloy exhibited relatively higher J_c values under elevated magnetic fields due to the fishtail effect [24]. It is worth noting that the popular commercial low-temperature superconducting wire, Nb-Ti alloy, relies on the precipitation of α-Ti as flux pinning sites to enhance J_c. This necessitates a meticulous precipitation heat treatment to control the microstructure of the α-Ti precipitates. Consequently, the prospect of discovering an HEA that can deliver larger J_c even in its as-cast state presents a notable advantage over Nb-Ti alloys.

The concurrent realization of superconductivity and high hardness has been the subject of recent investigations within transition metal carbide-based materials [44, 45], representing a multifunctional superconducting prospect. Given that numerous HEAs are characterized by their inherent high hardness, stemming from the pronounced lattice distortion as part of the four core effects, exploring hardness within bcc or hcp HEA superconductors is valuable. Several reports have underscored the relationship between the Vickers microhardness and VEC in bcc and fcc HEAs [26, 46]. For bcc HEAs with VECs ranging from approximately 4 to 5.5, an increase in VEC tends to yield higher hardness because the hard elements are present at VEC values of $\sim$ 5–7. This behavior has been consistently observed in several bcc HEA superconductors [23, 26]. For instance, the Vickers microhardness of $(Ti_{15}Hf_5)(Nb_{35}Ta_{35})Re_{10}$ with VEC=5.0 has been reported to reach as high as 466 HV [23]. Furthermore, the hardness of HEA superconductors has been examined in $(MoReRu)_{(1-2x)/3}(PdPt)_xC_y$ alloys featuring hcp and fcc structures [10]. In this system, C-doping induces a structural transformation from hcp to fcc, facilitated by covalent bonding between metal and carbon atoms, yielding a Vickers microhardness exceeding 1000 HV.

The intriguing duality of the shape memory effect and superelastic behavior has been recently investigated in the bcc HEA superconductor $Ti_{67}Zr_{19}Nb_{11.5}Sn_{2.5}$ [27]. Figure 1.5a shows the temperature-dependent electrical resistivity of this alloy, revealing zero resistivity below T_c = 4.65 K and a martensite-to-austenite transition occurring within the 150-225 K temperature range. This transition is the driving force behind the shape memory effect. Figure 1.5b displays a representative stress-strain curve, indicating the characteristic hysteresis loop of the superelastic behavior.

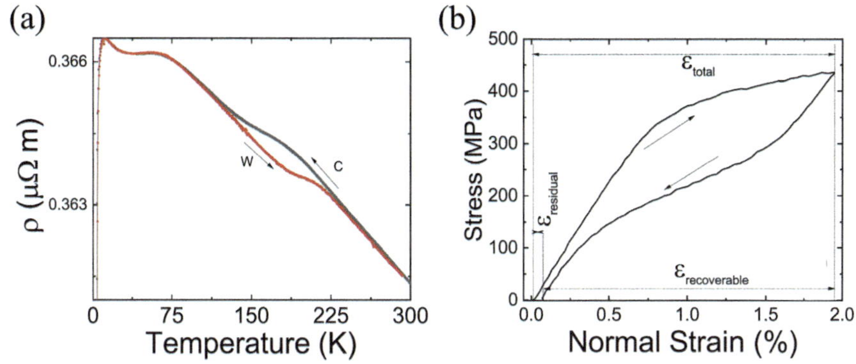

Fig. 1.5 **a** Temperature dependence of electrical resistivity of $Ti_{67}Zr_{19}Nb_{11.5}Sn_{2.5}$. **b** Typical stress-strain curve showing superelasticity in $Ti_{67}Zr_{19}Nb_{11.5}Sn_{2.5}$ measured at 300 K. Reproduced with permission from [27]

This HEA, composed of biocompatible elements, represents an attractive candidate for developing enhanced functional devices.

1.3.3 Materials Design

In general, the prediction of phase formation in HEAs with bcc and fcc structures has been the subject of extensive research, resulting in the proposition of numerous criteria [1, 2, 47]. These criteria are grounded in evaluating physical, chemical, and thermodynamic parameters. Furthermore, user-friendly and cost-free software tools are available for facilitating such predictions [48]. Conversely, the domain of materials design based on semiempirical parameters specific to bcc and hcp HEA superconductors has received relatively limited attention [5, 6, 9]. Currently, the dominant considerations in materials design are VEC and the constituent elements employed.

Using first-principles calculations with the McMillan formula has emerged as a potent avenue for predicting the T_c of HEAs [40]. Notably, this method has demonstrated its ability to accurately reproduce T_c values for bcc Ta-Nb-Hf-Zr-Ti alloys and even the pressure dependence of T_c [40]. In instances where electronic structure calculations necessitate the preservation of translational symmetry, the atomic disorder inherent in HEAs must be modeled using a supercell structure. On the other hand, electronic structure calculations can be conducted using the Korringa-Kohn-Rostoker method with the coherent potential approximation (KKR-CPA) without supercell modeling. Moreover, the phonon band structure can be calculated using supercell-based methodologies [39]. Both the electronic and phonon band structures within HEAs exhibit notable band broadening, exerting a discernible influence on the electron and phonon lifetimes. Consequently, including band-broadening effects is highly desirable for enhancing the precision of superconducting parameter predictions of HEAs.

The third and increasingly prominent avenue for materials design in HEAs is machine learning, a field experiencing rapid application [49, 50]. Machine learning methods fundamentally rely on an expansive student-teacher dataset for effective operation. Consequently, in the domain of bcc and hcp HEA superconductors, an inadequately populated dataset poses a substantial obstacle to advancing materials design through machine learning methodologies.

1.4 High-Entropy Compounds

In this section, we mention the present state of HEA-type superconducting compounds, with a particular emphasis on their unconventional properties, applications, and materials design. Table 1.2 outlines the interconnections between the subjects deliberated in this section and those elucidated in Chaps. 8 through 11.

1.4.1 Exotic Properties

One of the interesting properties of HEA-type compounds was observed in the BiS_2-based layered system $REO_{0.5}F_{0.5}BiS_2$. In the examined samples, the carrier concentration and lattice volume, which affect superconducting properties [51], were fixed by tuning the RE-site composition, and the RE-site ΔS_{mix} was solely varied [19]. With increasing ΔS_{mix}, in-plane structural disorder was suppressed, and the bulkiness of the superconductivity improved. This is one of the cocktail effects of HEA-type compounds.

Another example of exotic superconducting states was observed in HEA-type MTe (M: Ag, In, Sn, Pb, Bi) [13, 52]. As shown in Ref. [21], the robustness of superconductivity to pressure was observed in an HEA superconductor. Similar robustness

Table 1.2 Interrelationships between topics addressed in this section and those expounded upon in Chaps. 8 through 11

Topics	Chapter number
Exotic properties	
Robustness of superconductivity to pressure	8, 9
Glassy phonons	8
Cocktail effect	9, 11
Applications	
High critical temperature	11
Materials design	
Material database	8, 9, 10, 11

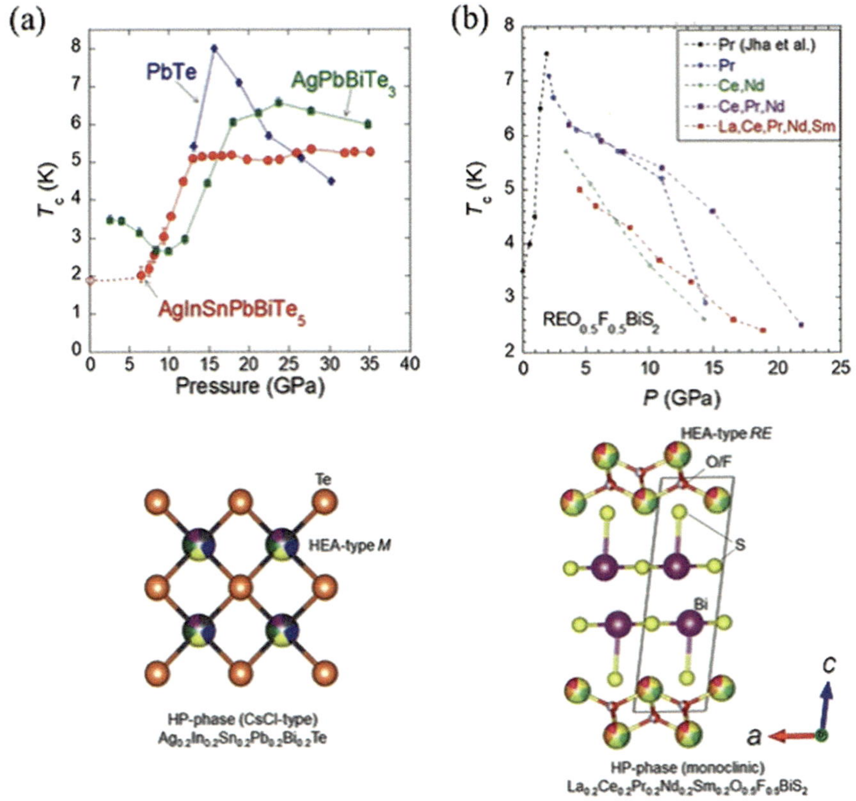

Fig. 1.6 Pressure dependence of T_c and schematic image of high-pressure (HP-phase) crystal structure for **a** MTe and **b** $REO_{0.5}F_{0.5}BiS_2$. Reproduced from [52, 53] under the Creative Commons Attribution License

was observed for HEA-type MTe, $AgInSnPbBiTe_5$ as shown in Fig. 1.6a [52]. High-pressure electrical resistance experiments, X-ray diffraction, and X-ray absorption spectroscopy revealed that the electronic and crystal structures change with pressure, but T_c does not change with pressure [52]. By molecular dynamics simulation of atomic vibrations and electronic band calculation, glassy phonons and blurry electronic band structure were revealed [53]. These results imply that the observed robust superconductivity to pressure is caused by exotic electron-phonon coupling with glassy phonons. In contrast, HEA-type $REO_{0.5}F_{0.5}BiS_2$ does not exhibit robust superconductivity to pressure, as shown in Fig. 1.6b; a similar pressure dependence of T_c is observed for all the samples with different ΔS_{mix} [54]. Because of the RE-site alloying, glassy phonons related to the atomic vibrations of the atoms in the superconducting layers are not induced by RE-site alloying, which would be the reason for the lack of robustness. Therefore, HEA-type superconducting compounds can be a good platform for systematically studying superconductivity with glassy to crystalline phonons by selecting crystal structure and its simplicity.

1.4.2 Applications

As is well known, compound superconductors have high performance because of the variety of superconductors. A high T_c can be achieved in Cu-oxides like YBCO, and a large upper critical field B_{c2} with practical superconducting performance can be achieved in A15-type Nb_3Sn. If we could further improve the performance by the HEA concept, the application of superconductors could be further accelerated. Here, we briefly introduce the improvement in the superconducting performance of HEA-type compounds.

A15-type V_3Ge is one of the high-field superconductors. HEA-type V_3X (X: Al, Si, Ga, Ge, Sn) exhibited a 50%-larger B_{c2} as compared to pure V_3Ge, while the value of T_c was the same [55]. The cause of this enhancement was proposed to be the compositionally-complex-alloy (CCA) state, where the phase separates into different HEA phases [56]. In HEA-type YBCO films, an improvement of the high-field J_c was observed [57]. Improving the J_c performance under high magnetic fields via an HEA could lead to new techniques for superconductivity applications.

1.4.3 Materials Design

As described earlier, various kinds of HEA-type compounds have been synthesized since 2018 [13, 19, 28, 51–57]. However, these materials are basically not new materials from a crystal-structure point of view. Zero- or low-entropy counterparts are known, and HEA-type superconducting compounds have been designed on the basis of previous works on zero- or low-entropy compounds. As shown in Fig. 1.7, $CuAl_2$-type $TrZr_2$ is a known superconductor family with a variation in

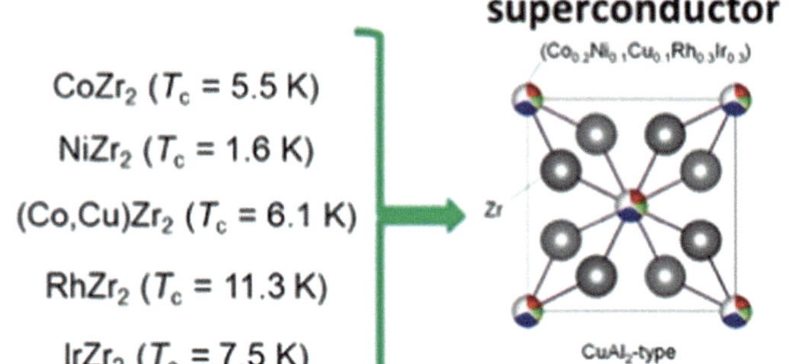

Fig. 1.7 Material design strategy for HEA-type superconducting compounds using material database

T_c. When designing new HEA-type composition, we collect information from material databases or review papers on the system. Then, we determine the HEA composition with a preferable combination for a desired T_c. For various systems, this strategy works, but special synthesis techniques such as high-pressure synthesis (for MTe) and arc melting (for $TrZr_2$ and A15) are required to obtain a single phase.

1.5 Summary

Within this chapter, we describe the discovery and current status of HEA superconductors. The contemporary landscape is characterized by exotic properties, applications, and materials design, all of which are intertwined with the themes of this book. Following the classification of HEA superconductors based on the distinguishing features of crystal structures, this chapter presents the various frontiers of exploration within each subclass. Furthermore, the chapter summarizes the interrelationships between the topics addressed in this chapter and those discussed in the subsequent chapters. In the realm of alloy-type HEAs typified by bcc and hcp structures, we traverse diverse topics, encompassing departure from the Matthias rule, band broadening, the emergence of strongly correlated electronic states, the quest for time-reversal symmetry violation, the critical current density, high hardness, shape memory, and the innovative pursuit of materials design. These topics are detailed on pages spanning Chaps. 2–7. Concerning high-entropy compounds, we surveyed the cocktail effect in HEA-type $REO_{0.5}F_{0.5}BiS_2$, the robustness of superconductivity to pressure, glassy phonons, and a blurred electronic band structure in HEA-type MTe and $AgInSnPbBiTe_5$. The high-T_c HEA-type A15 and YBCO films hold promise for practical applications in enhanced B_{c2} and improved high-field J_c, respectively. At present, a material design strategy for HEA-type superconducting compounds employing a material database proves effective. These subjects are detailed on pages spanning Chaps. 8–11.

References

1. K. Biswas, N.P. Gurao, T. Maiti, R.S. Mishra, *High Entropy Materials* (Springer, Singapore, 2022)
2. M.C. Gao, J.-W. Yeh, P.K. Liaw, Y. Zhang, *High-Entropy Alloys: Fundamentals and Applications* (Springer, Swizerland, 2015)
3. X. Yan, P.K. Liaw, Y. Zhang, Order and disorder in amorphous and high-entropy materials. Metall. Mater. Trans. A **52**, 2111–2122 (2021)
4. P. Koželj, S. Vrtnik, A. Jelen, S. Jazbec, Z. Jagličić, S. Maiti, M. Feuerbacher, W. Steurer, J. Dolinšek, Discovery of a superconducting high-entropy alloy. Phys. Rev. Lett. **113**, 107001 (2014)
5. L. Sun, R.J. Cava, High-entropy alloy superconductors: Status, opportunities, and challenges. Phys. Rev. Mater. **3**, 090301 (2019)

6. J. Kitagawa, S. Hamamoto, N. Ishizu, Cutting edge of high-entropy alloy superconductors from the perspective of materials research. Metals **10**, 1078 (2020)
7. Y.S. Lee, R.J. Cava, Superconductivity in high and medium entropy alloys based on MoReRu. Phys. C **566**, 1353520 (2019)
8. S. Marik, K. Motla, M. Varghese, K.P. Sajilesh, D. Singh, Y. Breard, P. Boullay, R.P. Singh, Superconductivity in a new hexagonal high-entropy alloy. Phys. Rev. Mater. **3**, 060602(R) (2019)
9. A.J. Browne, D.P. Strong, R.J. Cava, Phase stability and possible superconductivity of new 4d and 5d transition metal high-entropy alloys. J. Solid State Chem. **321**, 123881 (2023)
10. Q. Zhu, X. Guorui, Y. Cui, W. Yang, S. Song, G.-H. Cao, Z. Ren, Structural transformation and superconductivity in carbon-added hexagonal high-entropy alloys. J. Alloys Compd. **909**, 164700 (2022)
11. K. Stolze, J. Tao, F.O. von Rohr, T. Kong, R.J. Cava, Sc-Zr-Nb-Rh-Pd and Sc-Zr-Nb-Ta-Rh-Pd high-entropy alloy superconductors on a CsCl-type lattice. Chem. Mater. **30**, 906–914 (2018)
12. A. Yamashita, T.D. Matsuda, Y. Mizuguchi, Synthesis of new high-entropy alloy-type Nb_3(Al, Sn, Ge, Ga, Si) superconductors. J. Alloys Compd. **868**, 159233 (2021)
13. Y. Mizuguchi, Superconductivity in high-entropy-alloy telluride $AgInSnPbBiTe_5$. J. Phys. Soc. Jpn. **88**, 124708 (2019)
14. K. Stolze, F.A. Cevallos, T. Kong, R.J. Cava, High-entropy alloy superconductors on an α-Mn lattice. J. Mater. Chem. C **6**, 10441–10449 (2018)
15. G. Xiao, W. Yang, Q. Zhu, S. Song, G.H. Cao, Z. Ren, Superconductivity with large upper critical field in noncentrosymmetric Cr-bearing high-entropy alloys. Scr. Mater. **223**, 115099 (2023)
16. B. Liu, J. Wu, Y. Cui, Q. Zhu, G. Xiao, H. Wang, S. Wu, G. Cao, Z. Ren, Formation and superconductivity of single-phase high-entropy alloys with a tetragonal structure. ACS Appl. Electron. Mater. **2**, 1130–1137 (2020)
17. Y. Mizuguchi, M. R. Kasem, T. D. Matsuda, Superconductivity in $CuAl_2$-type $Co_{0.2}Ni_{0.1}Cu_{0.1}Rh_{0.3}Ir_{0.3}Zr_2$ with a high-entropy-alloy transition metal site. Mater. Res. Lett. **9**, 141-147 (2020)
18. T. Ying, T. Yu, Y.-S. Shiah, C. Li, J. Li, Y. Qi, H. Hosono, High-entropy van der waals materials formed from mixed metal dichalcogenides, halides, and phosphorus trisulfides. J. Am. Chem. Soc. **143**, 7042–7049 (2021)
19. R. Sogabe, Y. Goto, T. Abe, C. Moriyoshi, Y. Kuroiwa, A. Miura, K. Tadanaga, Y. Mizuguchi, Improvement of superconducting properties by high mixing entropy at blocking layers in BiS_2-based superconductor $REO_{0.5}F_{0.5}BiS_2$. Solid State Commun. **295**, 43-49 (2019)
20. Y. Shukunami, A. Yamashita, Y. Goto, Y. Mizuguchi, Synthesis of RE123 high-T_c superconductors with a high-entropy-alloy-type RE site. Phys. C **572**, 1353623 (2020)
21. J. Guo, H. Wang, F. von Rohr, Z. Wang, S. Cai, Y. Zhou, K. Yang, A. Li, S. Jiang, Q. Wu, R.J. Cava, L. Sun, Robust zero resistance in a superconducting high-entropy alloy at pressures up to 190 GPa. Proc. Natl. Acad. Sci. **114**, 13144–13147 (2017)
22. S.-G. Jung, Y. Han, J.H. Kim, R. Hidayati, J.-S. Rhyee, J.M. Lee, W.N. Kang, W.S. Choi, H.-R. Jeon, J. Suk, T. Park, High critical current density and high-tolerance superconductivity in high-entropy alloy thin films. Nat. Commun. **13**, 3373 (2022)
23. T. Hattori, Y. Watanabe, T. Nishizaki, K. Hiraoka, M. Kakihara, K. Hoshi, Y. Mizuguchi, J. Kitagawa, Metallurgy, superconductivity, and hardness of a new high-entropy alloy superconductor Ti-Hf-Nb-Ta-Re. J. Alloys Metal. Syst. **3**, 100020 (2023)
24. L. Gao, T. Ying, Y. Zhao, W. Cao, C. Li, L. Xiong, Q. Wang, C. Pei, J.-Y. Ge, H. Hosono, Y. Qi, Fishtail effect and the vortex phase diagram of high-entropy alloy superconductor. Appl. Phys. Lett. **120**, 092602 (2022)
25. T. Seki, H. Arima, Y. Kawasaki, T. Nishizaki, Y. Mizuguchi, J. Kitagawa, Effect of annealing in eutectic high-entropy alloy superconductor NbScTiZr. J. Supercond. Nov. Magn. **37**, 1059–1068 (2024)
26. J. Kitagawa, K. Hoshi, Y. Kawasaki, R. Koga, Y. Mizuguchi, T. Nishizaki, Superconductivity and hardness of the equiatomic high-entropy alloy HfMoNbTiZr. J. Alloys Compd. **924**, 166473 (2022)

27. M. Egilmez, O. Batal , W. Abuzaid, Z. Salman, M. Alkhader, H. Akamine, M. Nishida, J. W. A. Robinson, Superconductivity in $Ti_{67}Zr_{19}Nb_{11.5}Sn_{2.5}$ shape memory alloy. Phys. Rev. Mater. **5**, 074802 (2021)
28. R. Sogabe, Y. Goto, Y. Mizuguchi, Superconductivity in $REO_{0.5}F_{0.5}BiS_2$ with high-entropy-alloy-type blocking layers. Appl. Phys. Expr. **11**, 053102 (2018)
29. B. Liu, J. Wu, Y. Cui, Q. Zhu, G. Xiao, S. Wu, G.-H. Cao, Z. Ren, Superconductivity and paramagnetism in Cr-containing tetragonal high-entropy alloys. J. Alloys Compd. **869**, 159293 (2021)
30. B.T. Matthias, Empirical relation between superconductivity and the number of valence electrons per atom. Phys. Rev. **97**, 74–76 (1955)
31. F. von Rohr, M.J. Winiarski, J. Tao, T. Klimczuk, R.J. Cava, Effect of electron count and chemical complexity in the Ta-Nb-Hf-Zr-Ti high-entropy alloy superconductor. Proc. Natl. Acad. Sci. USA **113**, E7144–E7150 (2016)
32. Y. Harayama, J. Kitagawa, Superconductivity in Al-Nb-Ti-V-Zr multicomponent alloy. J. Supercond. Nov. Magn. **34**, 2787–2794 (2021)
33. N. Ishizu, J. Kitagawa, New high-entropy alloy superconductor $Hf_{21}Nb_{25}Ti_{15}V_{15}Zr_{24}$. Res. Phys. **13**, 102275 (2019)
34. S. Vrtnik, P. Koželj, A. Meden, S. Maiti, W. Steurer, M. Feuerbacher, J. Dolinšek, Superconductivity in thermally annealed Ta-Nb-Hf-Zr-Ti high-entropy alloys. J. Alloys Compd. **695**, 3530–3540 (2017)
35. S. Marik, M. Varghese, K.P. Sajilesh, D. Singh, R.P. Singh, Superconductivity in equimolar Nb-Re-Hf-Zr-Ti high entropy alloy. J. Alloys Compd. **769**, 1059–1063 (2018)
36. N.K. Sarkar, C.L. Prajapat, P.S. Ghosh, N. Garg, P.D. Babu, S. Wajhal, P.S.R. Krishna, M.R. Gonal, R. Tewari, P.K. Mishra, Investigations on superconductivity in an equi-atomic disordered Hf-Nb-Ta-Ti-V high entropy alloy. Intermetallics **144**, 107503 (2022)
37. P. Sobota, R. Topolnicki, T. Ossowski, T. Pikula, A. Pikul, R. Idczak, Superconductivity in the high entropy alloy $(NbTa)_{0.67}(MoHfW)_{0.33}$. Phys. Rev. B **106**, 184512 (2022)
38. R. Shu, X. Zhang, S. Rao, A. Febvrier, P. Eklund, Effects of alloying and deposition temperature on phase formation and superconducting properties of TiZrTaNb-based high entropy-alloy films. Appl. Phys. Lett. **120**, 151901 (2022)
39. F. Körmann, Y. Ikeda, B. Grabowski, M.H.F. Sluiter, Phonon broadening in high entropy alloys. Npj Comput. Mater. **3**, 36 (2017)
40. K. Jasiewicz, B. Wiendlocha, K. Górnicka, K. Gofryk, M. Gazda, T. Klimczuk, J. Tobola, Pressure effects on the electronic structure and superconductivity of $(TaNb)_{0.67}(HfZrTi)_{0.33}$ high entropy alloy. Phys. Rev. B **100**, 184503 (2019)
41. K. Motla, P.K. Meena, D.S. Arushi, D. Singh, P.K. Biswas, A.D. Hillier, R.P. Singh, Superconducting and normal-state properties of the high-entropy alloy Nb-Re-Hf-Zr-Ti investigated by muon spin relaxation and rotation. Phys. Rev. B **105**, 144501 (2022)
42. C.K.W. Leung, X. Zhang, F. von Rohr, R. Lortz, B. Jäck, Evidence for isotropic s-wave superconductivity in high-entropy alloys. Sci. Rep. **12**, 12773 (2022)
43. G. Kim, M.-H. Lee, J.H. Yun, P. Rawat, S.-G. Jung, W. Choi, T.-S. You, S.J. Kim, J.-S. Rhyee, Strongly correlated and strongly coupled s-wave superconductivity of the high entropy alloy $Ta_{1/6}Nb_{2/6}Hf_{1/6}Zr_{1/6}Ti_{1/6}$ compound. Acta Mater. **186**, 250–256 (2020)
44. Y. Ge, S. Ma, K. Bao, Q. Tao, X. Zhao, X. Feng, L. Li, B. Liu, P. Zhu, T. Cui, Superconductivity with high hardness in Mo_3C_2. Inorg. Chem. Front. **6**, 1282–1288 (2019)
45. Y. Cui, J. Wu, B. Liu, Q. Zhu, G. Xiao, S. Wu, G. Cao, Z. Ren, Superconductivity and high hardness in metal-rich carbides $MoRe_2C$ and WRe^2C. J. Alloys Compd. **856**, 157314 (2021)
46. F. Tian, L.K. Varga, N. Chen, J. Shen, L. Vitos, Empirical design of single phase high-entropy alloys with high hardness. Intermetallics **58**, 1–6 (2015)
47. D.B. Miracle, O.N. Senkov, A critical review of high entropy alloys and related concepts. Acta Mater. **122**, 448–511 (2017)
48. P. Martina, C.E. Madrid-Cortes, C. Cáceres, N. Araya, C. Aguilar, J.M. Cabrera, HEAPS: A user-friendly tool for the design and exploration of high-entropy alloys based on semi-empirical parameters. Comput. Phys. Commun. **278**, 108398 (2022)

49. X. Liu, J. Zhang, Z. Pei, Machine learning for high-entropy alloys: Progress, challenges and opportunities. Prog. Mater. Sci. **131**, 101018 (2023)
50. Z. Zhou, Y. Zhou, Q. He, Z. Ding, F. Li, Y. Yang, Machine learning guided appraisal and exploration of phase design for high entropy alloys. npj Comput. Mater. **5**, 128 (2019)
51. Y. Mizuguchi, Material development and physical properties of BiS_2-based layered compounds. J. Phys. Soc. Jpn. **88**, 041001 (2019)
52. Md.R. Kasem, Y. Nakahira, H. Yamaoka, R. Matsumoto, A. Yamashita, H. Ishii, N. Hiraoka, Y. Takano, Y. Goto, Y. Mizuguchi, Robustness of superconductivity to external pressure in high-entropy-alloy-type metal telluride $AgInSnPbBiTe_5$. Sci. Rep. **12**, 7789 (2022)
53. Y. Mizuguchi, H. Usui, R. Kurita, K. Takae, Md.R. Kasem, R. Matsumoto, K. Yamane, Y. Takano, Y. Nakahira, A. Yamashita, Y. Goto, A. Miura, C. Moriyoshi, Glassy atomic vibrations and blurry electronic structures created by local structural disorders in high-entropy metal telluride superconductors. Mater. Today Phys. **32**, 101019 (2023)
54. Y. Mizuguchi, K. Yamane, R. Matsumoto, K. Hoshi, A. Yamashita, A. Miura, Y. Nakahira, S. I Kawaguchi, Y. Takano, Absence of T_c-Pinning phenomenon under high pressure in high-entropy $REO_{0.5}F_{0.5}BiS_2$ layered superconductor. J. Phys. Soc. Jpn. **92**, 114702 (2023)
55. Y. Nakahira, R. Kiyama, A. Yamashita, H. Itou, A. Miura, C. Moriyoshi, Y. Goto, Y. Mizuguchi, Tuning of upper critical field in a vanadium-based A15 superconductor by the compositionally-complex-alloy concept. J. Mater. Sci. **33**, 15990 (2022)
56. A.M. Manzoni, U. Glatzel, New multiphase compositionally complex alloys driven by the high entropy alloy approach. Mater. Charact. **147**, 512 (2019)
57. A. Yamashita, K. Hashimoto, S. Suzuki, Y. Nakanishi, Y. Miyata, T. Maeda, Y. Mizuguchi, Fabrication of high-entropy $REBa_2Cu_3O_{7-\delta}$ thin films by pulsed laser deposition. Jpn. J. Appl. Phys. **61**, 050905 (2022)

Part I
Body-Centered Cubic and Hexagonal Close-Packing Structures

Chapter 2
Strongly Correlated High-Entropy Alloy Superconducting Properties and Vortex Dynamics of $Ta_{1/6}Nb_{2/6}Hf_{1/6}Zr_{1/6}Ti_{1/6}$

Rahmatul Hidayati, Jin Hee Kim, Gareoung Kim, Jae Hyun Yun, and Jong-Soo Rhyee

Abstract High-entropy alloys (HEAs) are generally synthesized using conventional alloying techniques, such as arc melting, induction melting, and powder metallurgy. Here we report the superconducting properties of the representative high-entropy alloys $Ta_{1/6}Nb_{2/6}Hf_{1/6}Zr_{1/6}Ti_{1/6}$. The superconducting properties of HEA superconductors are significantly different between arc melting and powder metallurgy methods, such as spark plasma sintering (SPS) and hot press (HP) sintering. The fundamental mechanisms underlying the properties of HEAs are actively being studied, and new compositions and processing methods are being developed to further enhance HEA performance. Here, we introduce the strongly correlated and strongly coupled s-wave superconducting properties of the arc-melted $Ta_{1/6}Nb_{2/6}Hf_{1/6}Zr_{1/6}Ti_{1/6}$ compound. The SPS mixture of $Ta_{1/6}Nb_{2/6}Hf_{1/6}Zr_{1/6}Ti_{1/6}$ exhibited significantly increased vortex pinning, resulting in an increase in the critical current density. The vortex dynamics exhibit unconventional behavior, such as the magnetic vortex avalanche effect. We compare magnetic vortex avalanches and the vortex pinning effect in terms of the composition and synthesis methods.

2.1 Introduction

High-entropy alloys (HEAs) are a class of advanced metallic materials that typically consist of five or more principal elements in near-equal atomic concentrations (usually between 5 and 35 at.%) [1–5]. This phenomenon refers to the high configurational entropy of these alloys, which arises from the large number of possible ways to distribute the different elements in a lattice. The unique composition of HEAs leads to a complex microstructure and results in exceptional mechanical properties, such as high strength, ductility, toughness, and corrosion resistance. These alloys

R. Hidayati · J. H. Kim · G. Kim · J. H. Yun · J.-S. Rhyee (✉)
Department of Applied Physics and Institute of Natural Sciences, Kyung Hee University, Yong-In, Gyong-Gi 17104, Republic of Korea
e-mail: jsrhyee@khu.ac.kr

J. Kitagawa and Y. Mizuguchi (eds.), *High-Entropy Alloy Superconductors*, Springer Series in Solid-State Sciences 202, https://doi.org/10.1007/978-981-97-4129-8_2

also exhibit good thermal stability and wear resistance, making them suitable for a wide range of applications in aerospace, automotive, energy, and other industries [6–8].

Several recent studies have explored the superconducting properties of high-entropy alloys (HEAs). For instance, type II bulk superconducting characteristics were identified in the Ta-Nb-Hf-Zr-Ti system, in which the critical temperature (T_c) was 7.3 K [9], and the superconducting state remained stable under high pressures up to 190 GPa [10]. Additional research has reported various HEA superconductors, including compositionally modified $(TaNb)_{1-x}(HfZrTi)_x$ [11], Nb-Re-Hf-Zr-Ti (T_c = 5.3 K) [12], $(ScZrNb)_{1-x}(RhPd)_x$ [13], and Sc-Zr–Nb-Ta-Rh-Pd (T_c = 9.3 K) [13]. These findings indicate that these HEAs function as conventional s-wave phonon-mediated superconductors. Furthermore, the superconducting state is resistant to disorder, as the transition temperature remains unaffected by disorder in both the simple crystalline and amorphous metallic states.

Among intermetallic superconductors, Nb_3Ge boasts the highest critical temperature (T_c) of 23.2 K and a high critical field (H_c) of 37 T. This thin film is created on a sapphire substrate using a chemical transport reaction method [14]. Nb_3Sn, a commercially available superconductor with a T_c of 18.3 K and H_c of 30 T, is utilized in high-field magnet applications such as thermonuclear fusion reactors, particle accelerator magnets, magnetic resonance imaging (MRI), and nuclear magnetic resonance (NMR) [15, 16]. Currently, the critical temperature and critical current density of high-entropy alloys (HEAs) are inferior to those of Nb_3Sn superconductors, which have a high critical current density (J_c) of 3000 A/mm^2.

Despite their current limitations in terms of superconducting properties, high-entropy alloys (HEAs) possess immense potential for practical superconducting applications due to their exceptional mechanical strength; resistance to fracture, wear, corrosion, and oxidation; and high thermal and structural stability. By improving their superconducting properties through material design and process engineering, HEAs could become increasingly relevant in this domain. However, as this area of research is still in its nascent stage, investigations of the physical properties of HEA superconductors are limited.

It has been recently proposed that the critical temperature of high-entropy alloy (HEA) superconductors could be improved by increasing the valence electron count in both crystalline and amorphous transition metal alloys [17]. However, this method is controversial, as it has not been conclusively shown that the critical temperature depends on the valence electron count [18]. Research into the superconducting properties of HEAs has focused primarily on examining how these properties vary with changes in elemental composition.

We present distinctive properties of the superconducting compound $Ta_{1/6}Nb_{2/6}Hf_{1/6}Zr_{1/6}Ti_{1/6}$, which is nearly optimally doped based on valence electron count measurements in both the normal and superconducting states. The temperature-dependent electrical resistivity ($\rho(T)$) exhibited a parabolic dependency near the critical temperature (T_c) in the normal state, transitioning to linear behavior at higher temperatures. Despite being within the dirty limit, the compound displayed a relatively large specific heat jump ($\Delta C/\gamma T_c$). We also observed that the Kadowaki–Woods ratios (A/γ^2) of the compound closely resemble those of heavy-fermion states, suggesting a strongly correlated system with a substantial effective mass (29 m_e). The presence of strongly correlated electron behavior in s-wave superconductivity is significant, as many superconductors in such systems are heavy-fermion and high-T_c cuprate superconductors, which are nodal superconductors. This study contributes to the growing understanding of the unconventional superconducting and physical properties of the HEA $Ta_{1/6}Nb_{2/6}Hf_{1/6}Zr_{1/6}Ti_{1/6}$ compound and its potential use as a metallic superconductor.

Superconductivity in high-entropy alloys (HEAs) can be classified into four types based on the crystal structure and valence electron count (VEC) of each atom. The superconducting transition temperature (T_c) of HEA superconductors is closely linked to the VEC [3]. Type A HEA superconductors have a body-centered cubic (BCC) structure and a VEC range of 4.3–4.8 [19]. In contrast, type-B HEA superconductors exhibit a more complex structure and a different VEC range of 6.5–6.8 [13]. Type-C HEA superconductors feature a CsCl-type structure with a VEC range of 5.9–6.3. Recently, type D HEA superconductors characterized by a hexagonal close-packed (HCP) structure and a VEC value of 5.8 have been reported [12].

While the $Ta_{1/6}Nb_{2/6}Hf_{1/6}Zr_{1/6}Ti_{1/6}$ HEA superconductor, synthesized by the arc-melting method, exhibited strongly correlated and strongly coupled s-wave superconducting properties according to various superconducting parameters, such as a high Kadowaki-Woods ratio, a sizeable specific heat jump, and a high effective mass, the $Ta_{1/6}Nb_{2/6}Hf_{1/6}Zr_{1/6}Ti_{1/6}$ compound obtained using planetary ball milling and spark plasma sintering methods exhibited significantly different superconducting properties. The valence electron count (VEC) of the compound was 4.5. The high mechanical strength of HEA compounds is crucial for practical applications; however, their extreme strength makes it challenging to manipulate their shape through cutting and grinding. Powder metallurgy processes such as ball milling and spark plasma sintering (SPS) can broaden the range of HEA applications by enabling the formation of samples with various shapes, including those suitable for 3D printing applications.

Upon comparing the spark plasma sintering (SPS)-processed $Ta_{1/6}Nb_{2/6}Hf_{1/6}Zr_{1/6}Ti_{1/6}$ HEA with the arc-melted sample, we observed a significant increase in the critical current density and pinning force in the sample synthesized using ball milling and SPS sintering. The strong vortex pinning force in the SPS-sintered sample resulted in stable remanent magnetization over time after the applied magnetic field was turned off. Moreover, we detected magnetic flux avalanches during magnetic hysteresis loop measurements at low temperatures ($T \leq 4$ K) and low magnetic field regions ($H \leq 1$ T). The vortex avalanche effect can be explained by vortex quantum creep or flux jump at weakly coupled superconducting grain boundaries.

2.2 Strongly Correlated S-Wave Superconductivity of the High-Entropy Alloy $Ta_{1/6}Nb_{2/6}Hf_{1/6}Zr_{1/6}Ti_{1/6}$ Synthesized by the Arc Melting Method

The primary XRD pattern (Fig. 2.1) for $Ta_{1/6}Nb_{2/6}Hf_{1/6}Zr_{1/6}Ti_{1/6}$ can be identified as the body-centered cubic (BCC) phase (space group $Im\overline{3}m$), with a lattice parameter of a = 3.38 Å. This parameter is comparable to that of other HEAs with similar compositions [9, 11, 12]. The sample exhibited a single phase and was devoid of impurity peaks. A slight widening of the peak signals is attributed to the significant level of HEA disorder [9].

Figure 2.2a displays the electrical resistivity $\rho(T)$ as a function of temperature, ranging from 2 to 300 K without a magnetic field. A sharp superconducting transition is observed at $T_c = 7.85$ K, with a 50% reduction in resistivity from the transition point (shown as a dashed line in the inset). Above T_c, the $\rho(T)$ temperature dependence demonstrates a linearly increasing metallic behavior as the temperature increases. The linear characteristic of $\rho(T)$ may stem from the compound's high disorder [11]. The superconducting transition temperature of this compound is similar to that of other compounds, such as $Ta_{34}Nb_{33}Hf_8Zr_{14}Ti_{11}$ ($T_c = 7.3$ K), $[TaNb]_{0.67}(HfZrTi)_{0.33}$ ($T_c = 7.7$ K), and $[TaNb]_{0.7}(ZrHfTi)_{0.3}$ ($T_c = 8$ K) [9, 11]. The room temperature electrical resistivity is 67.6 μΩ-cm, with a residual resistivity ratio (RRR) of ρ(300 K)/ρ(8 K) ≈ 1.17, comparable to that of previously reported HEAs, with values ranging from ≈

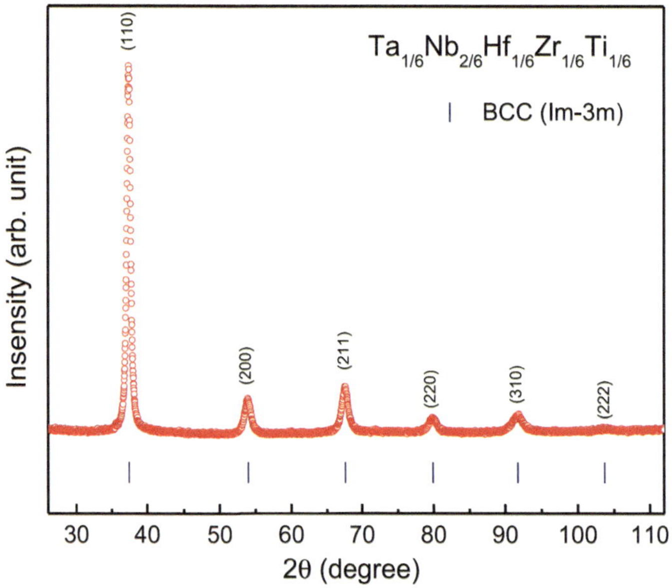

Fig. 2.1 X-ray diffraction pattern of the bulk $Ta_{1/6}Nb_{2/6}Hf_{1/6}Zr_{1/6}Ti_{1/6}$ compound. Reproduced from Ref. [45] with permission from Elsevier

1.07 to 1.33 [9, 11]. Extremely low RRR values could result from highly disordered intermetallic compounds, such as atomic-scale disorder and small domain sizes in HEAs [9, 11, 12, 19].

The inset of Fig. 2.2a illustrates the temperature-dependent electrical resistivity $\rho(T)$ under different static magnetic fields, ranging from 0 to 11 T in increments of 1 T. As the magnetic field increases, the T_c value consistently decreases. The onset of superconductivity is still detectable under an 11 T magnetic field, suggesting that the compound possesses a relatively high upper critical magnetic field $\mu_0 H_{c2}(0)$.

In Fig. 2.2b, the temperature dependence of the upper critical magnetic field $\mu_0 H_{c2}(T)$ is depicted. $\mu_0 H_{c2}(T)$ can be accurately characterized by the Ginzburg–Landau (GL) expression provided in Eq. (2.1). The upper critical magnetic field at the zero-temperature limit, $\mu_0 H_{c2}(0)$, is estimated to be 12.05 T, which surpasses the values of other HEA superconductors (7 ~ 9 T) as shown in Table 2.1 [11, 12].

$$H_{c2}(T) = H_{c2}(0)\frac{(1 - (T/T_c)^2)}{(1 + (T/T_c)^2)} \tag{2.1}$$

Based on the GL theory, with $H_{c2}(0) = \Phi_0/2\pi\xi^2$, the zero-temperature coherence length ξ_GL(0) for the compound is estimated to be 5.23 nm (where $\phi_0 = 2.07 \times 10^{-15}$ Wb represents the flux quanta) [20].

Figure 2.3a displays the low-temperature magnetic susceptibility $\chi(T)$ under zero-field cooling (ZFC, closed circles) and field cooling (FC, open circles) conditions with a static magnetic field of $\mu_0 H = 1$ mT. A diamagnetic superconducting transition is observed at T = 7.82 K in the ZFC measurement, which is close to the critical temperature determined by the electrical resistivity $\rho(T)$. While an ideal superconductor should have a dimensionless magnetic susceptibility of $\chi = -1$, the χ value of the compound should be greater than -1.6 due to the demagnetization effect [9, 11]. Figure 2.3b presents magnetic hysteresis (*M-H*) loops at various isothermal temperatures, with a sweeping magnetic field ranging from -2 T to 2 T (-5 T ~ 5 T only for T = 2 K) and a sweep rate of 60 mT/min. These magnetic hysteresis loops indicate that this compound is a typical type II superconductor, and the temperature-dependent lower critical magnetic field $\mu_0 H_{c1}(T)$ [inset of Fig. 2.3b] is determined from the *M(H)* minimum point. $\mu_0 H_{c1}(T)$ is also fitted using the GL theory [red line in the inset of Fig. 2.3b]. The zero-temperature lower critical magnetic field $\mu_0 H_{c1}{}^{min}(0)$ reaches 23 mT.

Figure 2.4a displays the specific heat C_p divided by temperature (C_p/T) in the absence of a magnetic field plotted against the square temperature (T^2). The Sommerfeld coefficient $\gamma = 7.42$ mJ mol^{-1} K^{-2} and $\beta = 0.27$ mJ mol^{-1} K^{-4} were derived from linear fitting using the $C_p/T = \gamma + \beta T^2$ result (solid red line) above the T_c. The Debye temperature $\Theta_D = 194$ K is determined from $\beta = 12\pi^4 nR/5{\Theta_D}^3$, , where n represents the number of atoms per formula unit (in this case, n = 1) and R = 8.31 J mol^{-1} K^{-1} is the ideal gas constant.

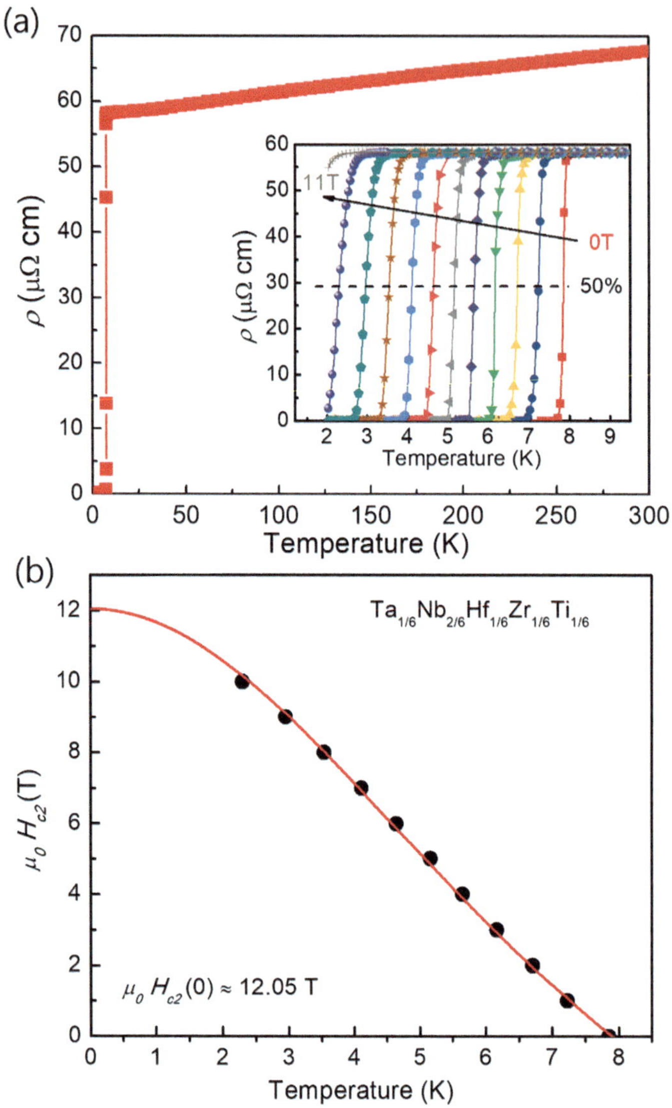

Fig. 2.2 **a** Electrical resistivity $\rho(T)$ of the $Ta_{1/6}Nb_{2/6}Hf_{1/6}Zr_{1/6}Ti_{1/6}$ compound between 2 and 300 K. The inset shows the electrical resistivity $\rho(T)$ under static magnetic fields of $\mu_0 H = 1$ to 11 T in step intervals of 1 T between 2 and 9 K. The dotted line represents 50% of the normal-state electrical resistivity. **b** Temperature-dependent upper critical field $\mu_0 H_{c2}(T)$ obtained from the electrical resistivity. The solid line is the Ginzburg–Landau fit of $\mu_0 H_{c2}(T)$. Reproduced from Ref. [45] with permission from Elsevier

Table 2.1 Superconducting parameters for $Ta_{1/6}Nb_{2/6}Hf_{1/6}Zr_{1/6}Ti_{1/6}$ and certain other HEAs (Refs. [11, 12])

Parameters	$Ta_{1/6}Nb_{2/6}Hf_{1/6}Zr_{1/6}Ti_{1/6}$ [a]	$[TaNb]_{0.67}[HfZrTi]_{0.33}$ [b] [11]	$[TaNb]_{0.16}[HfZrTi]_{0.84}$ [c] [11]	$Nb_{21}Re_{16}Zr_{20}Hf_{23}Ti_{20}$ [12]
T_c [K]	7.85	7.70	4.59	9.3
γ [mJ/molK^2]	7.451	9.97(5)	6.45(8)	5.7
β [mJ/molK^4]	0.274	0.170	0.311(9)	0.1178
Θ_D [K]	192.3	225	184	254.6
$\lambda_{el\text{-}ph}$	0.9	0.83	0.73	0.69
$V_{el\text{-}ph}$ [molK^2/mJ]	0.147	0.104*	0.113*	0.121*
$D(E_F)$ [st./eV]	1.6	1.9	1.6	1.4*
$\Delta C/\lambda T_c$	1.97	1.89	1.98	1.81
$\Delta(0)$ [meV]	1.596	1.21(2)	0.71	1.57*
$2\Delta(0)/k_B T_c$	4.72	3.7	3.6	3.92
$\mu_0 H_{c2}(0)$ [T]	12.05	7.75	9.02	9.7
ξ [nm]	5.59	4.85*	5.23*	5.42
J_c [A/cm^2]	10,655	N/A	N/A	N/A
Valence El. count	4.5	4.67	4.16	4.69

The value marked in star(*) is obtained by our estimation. Reproduced from Ref. [45] with permission from Elsevier

[a] $Ta_{0.17}Nb_{0.33}Hf_{0.17}Zr_{0.17}Ti_{0.17}$

[b] $Ta_{0.335}Nb_{0.335}Hf_{0.11}Zr_{0.11}Ti_{0.11}$

[c] $Ta_{0.08}Nb_{0.08}Hf_{0.28}Zr_{0.28}Ti_{0.28}$

[d] $Nb_{0.21}Re_{0.16}Zr_{0.2}Hf_{0.23}Ti_{0.2}$

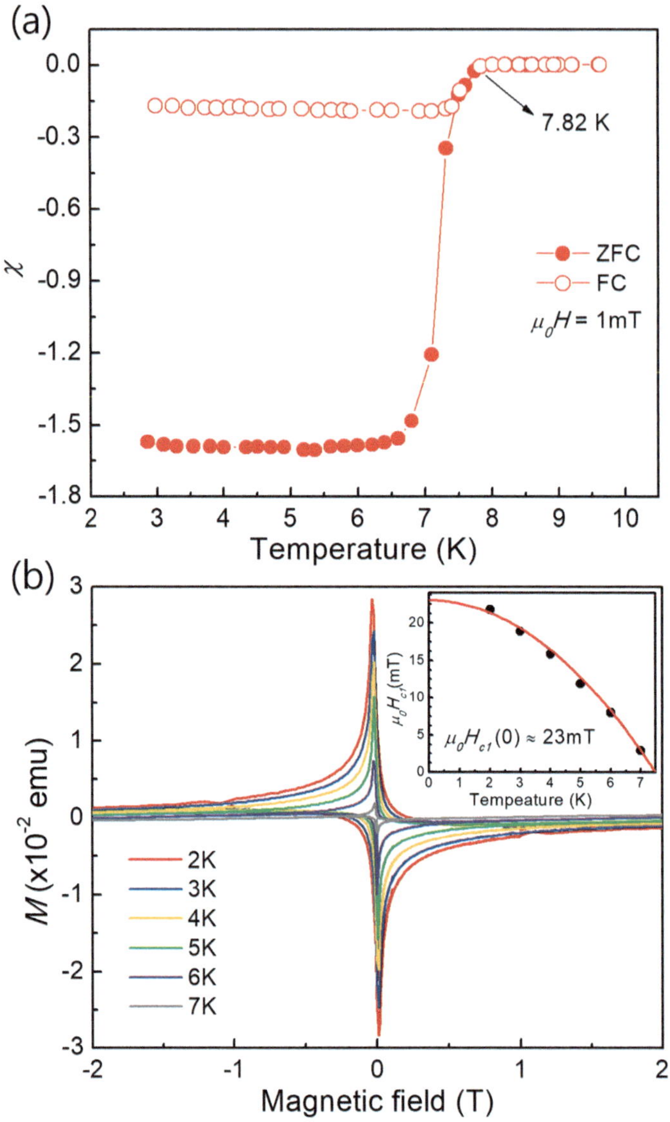

Fig. 2.3 **a** Temperature-dependent field cooling (FC) and zero-field cooling (ZFC) magnetic susceptibility of the $Ta_{1/6}Nb_{2/6}Hf_{1/6}Zr_{1/6}Ti_{1/6}$ compound under a static magnetic field of $\mu_0 H =$ 1 mT. **b** Isothermal magnetic hysteresis (M-H) loops at various temperatures. The inset shows the temperature-dependent lower critical fields $\mu_0 H_{c1}$ (mT) with Ginzburg–Landau fitting. Reproduced from Ref. [45] with permission from Elsevier

$$\lambda_{el-boson} = \frac{\mu^*\ln\left(\frac{1.45T_c}{\Theta_D}\right) - 1.04}{1.04 + \ln(\frac{1.45T_c}{\Theta_D})(1 - 0.62\mu^*)} \tag{2.2}$$

The electron–boson coupling constant $\lambda_{el\text{-}boson} \simeq 0.9$ is determined from the McMillan formula in Eq. (2.2), assuming a Coulomb pseudopotential $\mu^* = 0.13$ [21]. The electron–phonon potential V_{el-ph} is calculated as $V_{el-ph} = \lambda_{el-ph}/D(E_F)$, where $D(E_F)$ represents the density of states at the Fermi level. These findings suggested that the compound is an intermediate coupled superconductor. The obtained values for γ, β, Θ_D, and $\lambda_{el\text{-}boson}$, as well as various superconducting parameters of the $Ta_{1/6}Nb_{2/6}Hf_{1/6}Zr_{1/6}Ti_{1/6}$ compound, are compared with those of other HEA superconductors in Table 2.1. The electron-boson (phonon) coupling constant and electron–phonon potential of the compound are greater than those found in other HEA superconductors.

The inset of Fig. 2.4a displays the temperature dependence of $(C–C_n)/T$ at low temperatures. The normal-state specific heat C_n is subtracted from the specific heat C_p obtained from extrapolation with $C_p = \gamma T + \beta T^3$ [red line in Fig. 2.4a]. Two types of specific heat jumps are extracted, $\Delta C/\gamma T_c^{mid} = 2.22$, using a linear extrapolation of $(C–C_n)/T$ at $T = T_c$ and $\Delta C/\gamma T_c^{zero} = 1.56$ from the maximum value of the $(C–C_n)/T$ data. Both $C/\gamma T_c$ values are greater than the standard weak-coupling Bardeen-Cooper-Schrieffer (BCS) value of $\Delta C/\gamma T_c = 1.43$, which is typically observed in other HEAs as shown in Table 2.1 [9, 11, 12, 19]. A specific heat jump greater than the theoretical BCS weak coupling suggests that the superconducting state of the $Ta_{1/6}Nb_{2/6}Hf_{1/6}Zr_{1/6}Ti_{1/6}$ compound lies within the intermediate to strong coupling regime.

In Fig. 2.4b, the temperature-dependent electronic specific heat divided by the temperature (C_{el}/T) is plotted after subtracting the lattice contribution of the specific heat. The zero-temperature superconducting gap ($\Delta(0)$) can be estimated from the low-temperature electronic specific heat data. Padamsee et al. [22] developed a formula to calculate a single $\Delta(0)$ gap using the semiempirical α-model in a strongly coupled superconductor system. The thermodynamic properties of the normalized entropy S and electronic specific heat C_{el} as a function of temperature in the superconducting state can be described by the following equations:

$$\frac{S}{\gamma_n T_c} = -\frac{6}{\pi^2}\left(\frac{\Delta(0)}{k_B T_c}\right)\int_0^\infty [f In(f) + (1-f)In(1-f)]dy,\ \frac{C_{el}}{\gamma_n T_c} = t\frac{d\left(\frac{S}{\gamma_n T_c}\right)}{dt}, \tag{2.3}$$

where $f(\xi) = \left[\exp(E(\xi)/k_B T) + 1\right]^{-1}$ is the Fermi function, $E(\xi) = [\xi^2 + \Delta^2(t)]^{0.5}$ is the energy of the quasiparticles, ξ is the energy of normal electrons relative to the chemical potential, $y = \xi/\Delta(0)$ is a dimensionless integration variable, $t = T/T_C$ is the reduced temperature, and $\Delta(t)/\Delta(0) = \tanh[1.82(1.018((1/t) - 1))^{0.51}]$ is a phenomenological formula for the temperature dependence of the energy gap. Using Eq. (2.3), fitting of the specific heat data of the compound produces $\Delta(0)/k_B T_c =$

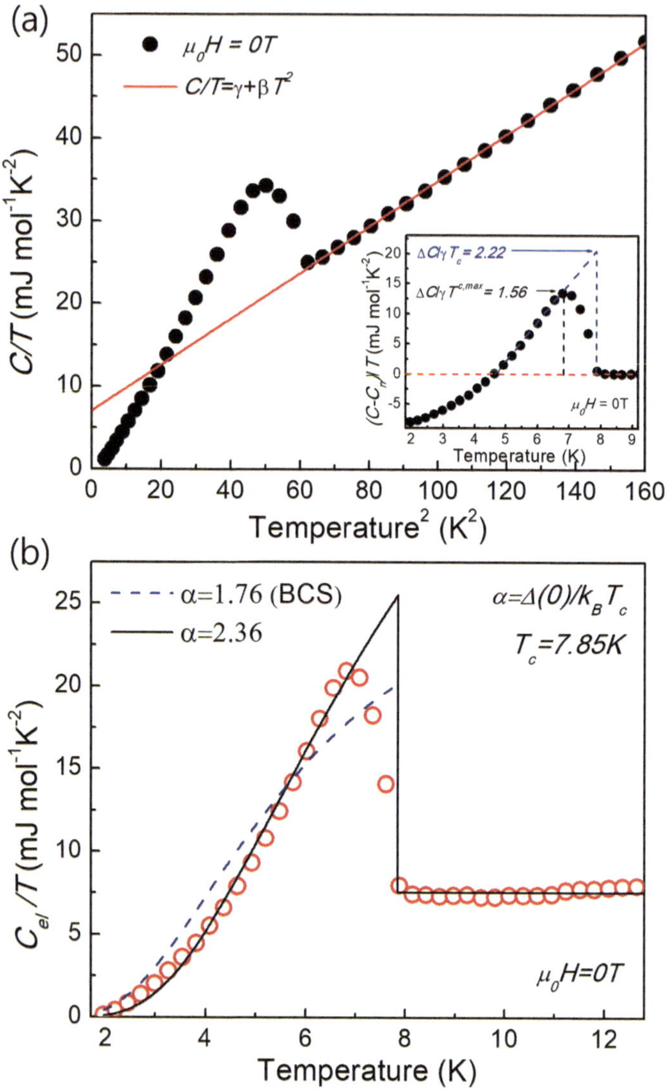

Fig. 2.4 **a** Specific heat divided by temperature C_p/T versus the square temperature T^2 for the $Ta_{1/6}Nb_{2/6}Hf_{1/6}Zr_{1/6}Ti_{1/6}$ compound between 2 and 160 K. The red line is the fit of the Debye model $C_p/T = \gamma + \beta T^2$. The inset shows a temperature-dependent specific heat jump in the vicinity of the superconducting phase transition obtained by subtracting the specific heat value in the normal state. **b** Temperature-dependent electronic specific heat divided by the temperature C_{el}/T in the vicinity of the superconducting phase transition. The solid black line is the fit to C_{el}/T based on the calculated $\alpha = 2.36$, and the red dotted line is the fit using the $\alpha = 1.76$ value from the weak-coupling BCS model, where $\alpha = \Delta(0)/k_B T_c$. Reproduced from Ref. [45] with permission from Elsevier

2.36 ± 0.01 (solid black line in Fig. 2.4b), which is higher than the weak coupling limit BCS $\Delta(0)/k_B T_c = 1.76$. We also confirmed the superconducting gap obtained by $C_{el}(T) = A\exp(-\Delta(0)/k_B T)$, where k_B is the Boltzmann constant. The BCS weak coupling limit of $\Delta(0)/k_B T_c = 1.764$ could not reproduce the $C_{el}(T)$ data, as shown in Fig. 2.4b (blue dashed line).

To assess dirtiness, we estimated the mean free path, which is given by

$$l = \frac{3\pi^2\hbar^3}{e^2\rho_0 m^{*2} v_F^2}, \tag{2.5}$$

where $\hbar$ is Plank's constant divided by 2π, e is the elementary charge of the electron, m^* is the effective mass, v_F is the Fermi velocity, and ρ_0 is the residual resistivity, assumed by static impurity scattering. The value of ρ_0 was obtained by fitting the low-temperature electrical resistivity (see the inset of Fig. 2.7, red line). The Fermi velocity is related to the coherence length $\xi(0) = \hbar v_f/\pi\Delta(0)$ and is estimated by $v_f \approx 43000 m/s$ [23].

$$\mathrm{n} = \frac{1}{3\pi^2}\left(\frac{m^* v_f}{\hbar}\right)^3 \tag{2.6}$$

We obtained the carrier density, $\mathrm{n} = 4 \times 10^{22}\ \mathrm{cm}^{-3}$, by measuring the Hall resistivity. The effective mass is estimated by $m^* = 2.6205 \times 10^{-29}\mathrm{kg} = 29m_e$, corresponding to 29 times the free electron mass, which is a very high value in an intermetallic compound. From Eq. (2.5), the mean free path is estimated by $l = 1.87$ nm, and the scattering time is obtained by the relation $\tau = l/v_f = 0.439$ ps. Therefore, the compound had a dirty limit of $\Delta(0) < \hbar/2\tau = 1.596$ meV < 7.501 meV.

The critical current density (J_c) is estimated from the *M-H* hysteresis curves by the Bean model given as $J_c = 20\Delta M/[a(1 - \frac{a}{3b})]$, where ΔM is the vertical width of the *M-H* loop [24, 25]. Constants a and b are the sample widths, where $a < b$. A semilogarithmic plot of the isothermal $J_c(H)$ with a magnetic field below 5 K is shown in Fig. 2.5a. The zero-field superconducting critical current density is approximately 10,655 A cm^{-2} at 2 K; in comparison, the $J_c(0)$ of a conventional binary alloy NbTi superconductor exceeds ~ 10^5 A cm^{-2} at 4.2 K [26–28].

To understand the pinning mechanism of the $Ta_{1/6}Nb_{2/6}Hf_{1/6}Zr_{1/6}Ti_{1/6}$ compound, we plotted the $J_c(H)$ graph on a double-logarithmic scale, as shown in Fig. 2.5b. The critical exponent α of $J_c(H) \propto H^{-\alpha}$ is 0.85 at 2 K and increases to 0.93 at 5 K with increasing temperature, which is associated with collective pinning. The Kramer plot $J_c^{0.5}H^{0.25} \propto (H^* - H)$ was used to estimate the empirical irreversible field H^*, as shown in the inset of Fig. 2.5a [29–32]. The irreversible field systematically decreased from 2.6 T at 2 K to 1.4 T at 5 K, as indicated by the arrow in the inset.

Figure 2.6 shows the dependence of the normalized flux pinning force density $f_P(= F_P/F_{P,\max})$ on temperature, as indicated, on the reduced field $h(= H/H^*)$, where F_P is the flux pinning force density and $F_{P,max}$ is the maximum value of F_P. To obtain a reduced field $h(= \frac{H}{H^*})$, we used the empirical irreversible field H^* from

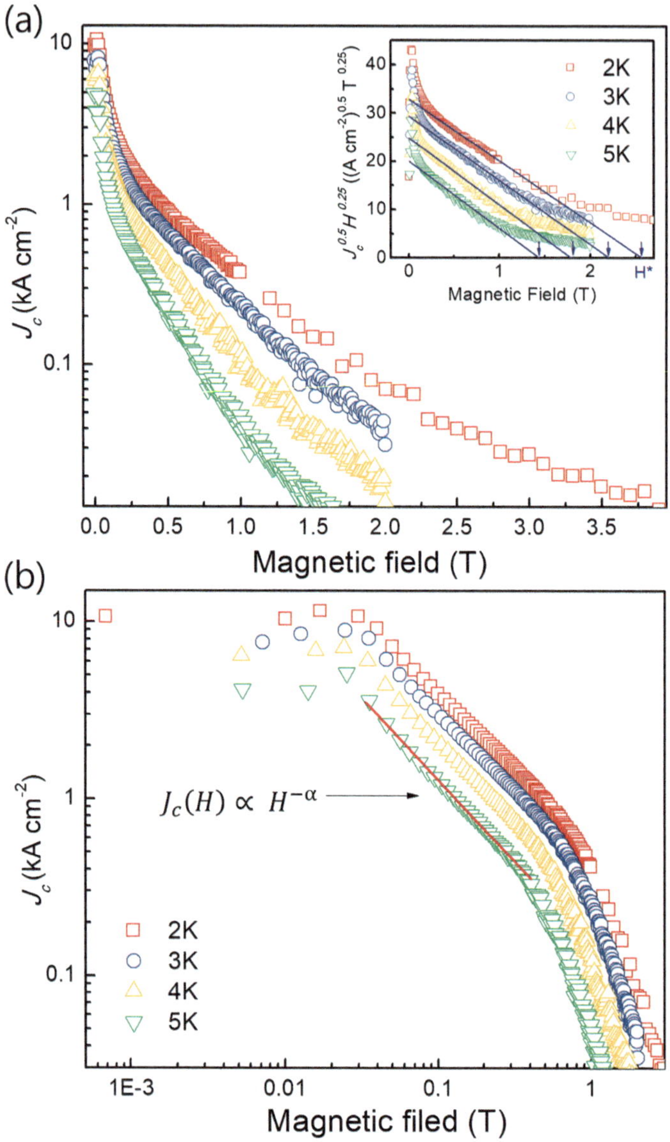

Fig. 2.5 **a** Semilog plot of the magnetic field dependence of the critical current density J_c of $Ta_{1/6}Nb_{2/6}Hf_{1/6}Zr_{1/6}Ti_{1/6}$ at various isothermal temperatures. The inset shows the Kramer plots (solid lines) used to determine the irreversible field H^* (as indicated by the arrow). **b** Log–log plot of the magnetic field dependence of the critical current density J_c at various temperatures. The solid line is the power-law field dependence of the critical current density $J_c(H) \propto H^{-\alpha}$. Reproduced from Ref. [45] with permission from Elsevier

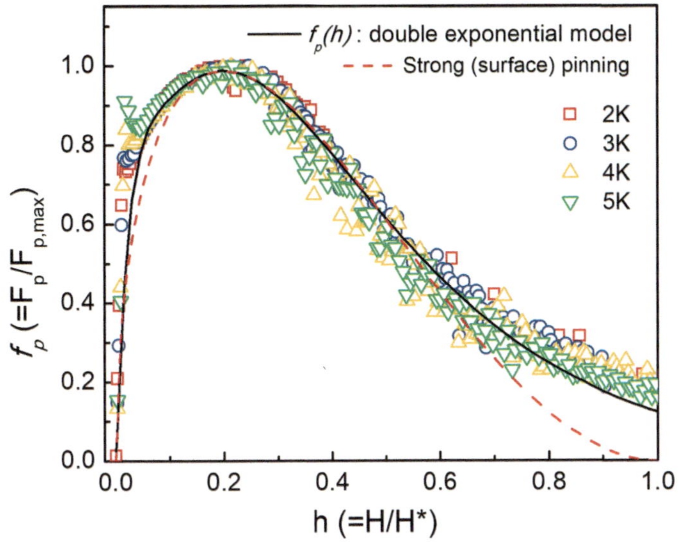

Fig. 2.6 Normalized flux pinning force density $f_P(=F_P/F_{P,\max})$ with decreasing field $h(=H/H^*)$ for the $Ta_{1/6}Nb_{2/6}Hf_{1/6}Zr_{1/6}Ti_{1/6}$ compound at various isothermal temperatures. The solid line and the dashed line are fitting curves obtained for the double exponential model and surface pinning mechanism model, respectively. Reproduced from Ref. [45] with permission from Elsevier

the inset of Fig. 2.5a. The $f_P(h)$ graphs are well scaled using the strong (surface) pinning model of Dew-Huges, $f_P(h) \propto h^{0.5}(1-h)^2$, and the double exponential model using $f_p(h) = J_1\exp(-B_1h)h + J_2\exp(-B_2h)h$ (black line) [34]. These pinning mechanisms result in a two-gap superconductor, such as MgB_2 [35–38]. It is not clear whether HEA compounds have a two-gap structure, but we can surmise that there are two types of pinning mechanisms involved.

According to the Fermi liquid theory, electrons contribute to the linear temperature dependence of heat capacity $C_{el}(T) = \gamma T$ and the quadratic temperature dependence of electrical resistivity $\rho(T) = \rho_0 + AT^2$. Various transition metals exhibit a constant Kadowaki–Woods ratio (KWR) of $\frac{A}{\gamma^2} \approx a_{TM} = 0.4\mu\Omega$ cm mol^{2}K^{2}J^{-2}, while heavy-fermion compounds display more pronounced values according to $A/\gamma^2 \approx a_{HF} = 10\mu\Omega$ cm mol^{2}K^{2}J^{-2} [39–41]. The temperature-dependent electrical resistivity of the $Ta_{1/6}Nb_{2/6}Hf_{1/6}Zr_{1/6}Ti_{1/6}$ compound fits well with the Fermi-liquid behavior $\rho(T) = \rho_0 + AT^2$ near or above T_c, as illustrated in the inset of Fig. 2.7. When plotting the KWR (A versus γ) for various transition metals and heavy-fermions alongside the $Ta_{1/6}Nb_{2/6}Hf_{1/6}Zr_{1/6}Ti_{1/6}$ compound, as shown in Fig. 2.7, both the $Ta_{1/6}Nb_{2/6}Hf_{1/6}Zr_{1/6}Ti_{1/6}$ compound and $Ta_{34}Nb_{33}Hf_8Zr_{14}Ti_{11}$ lie on the heavy-fermion system line. The compound's high KWR compared to that of conventional transition metals suggests that $Ta_{1/6}Nb_{2/6}Hf_{1/6}Zr_{1/6}Ti_{1/6}$ and other HEAs possess more dynamic properties [41], which could stem from impurity scattering, proximity to a phase transition, and/or electron–phonon interactions [42–44]. The heavy-fermionic character of the KWR of $Ta_{1/6}Nb_{2/6}Hf_{1/6}Zr_{1/6}Ti_{1/6}$ is consistent

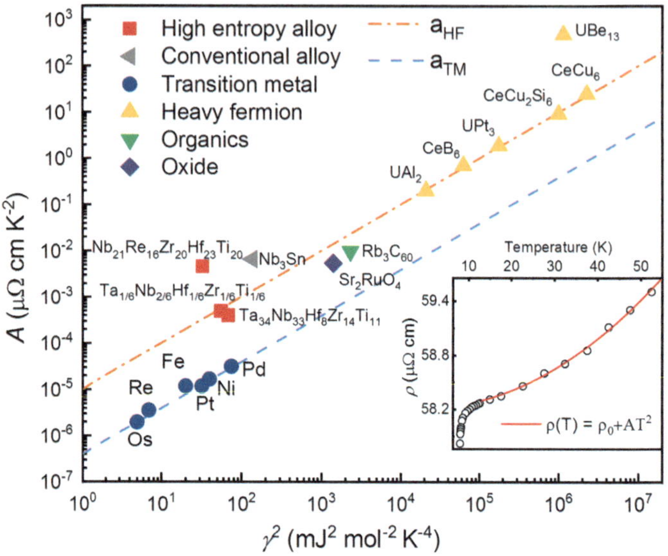

Fig. 2.7 Kadowaki–Woods plot for transition metals ($a_{TM} = 0.4\ \mu\Omega$ cm mol^2 K^2 J^{-2}, dark blue dashed line) and for the heavy-fermion system ($a_{HF} = 10\ \mu\Omega$ cm mol^2 K^2 J^{-2}, orange dash-dotted line). The inset shows the temperature-dependent electrical resistivity above T_c, fitted by the formula $\rho(T) = \rho_0 + AT^2$. Reproduced from Ref. [45] with permission from Elsevier

with the high effective mass of the carrier (29 m_e), as discussed earlier. A high KWR combined with a high effective mass in a compound has significant implications for strongly correlated superconductivity in HEA superconductors.

2.3 Strong Vortex Pinning and Enhancement of the Critical Current Density in $Ta_{1/6}Nb_{2/6}Hf_{1/6}Zr_{1/6}Ti_{1/6}$ Alloys Synthesized by the Spark Plasma Sintering Method

In this section, the $Ta_{1/6}Nb_{2/6}Hf_{1/6}Zr_{1/6}Ti_{1/6}$ compounds are synthesized by the powder metallurgy method using planetary ball milling and the spark plasma sintering method. The $Ta_{1/6}Nb_{2/6}Hf_{1/6}Zr_{1/6}Ti_{1/6}$ HEA, which has a BCC structure (type-A HEA superconductor), exhibited type-II superconducting behavior with T_c = 7.85 K and a VEC value of 4.5 [45]. In a previous section, it was proposed that the $Ta_{1/6}Nb_{2/6}Hf_{1/6}Zr_{1/6}Ti_{1/6}$ HEA superconductor, synthesized by the arc-melting method, displayed strongly correlated and strongly coupled s-wave superconducting properties based on various superconducting parameters, such as a high Kadowaki–Woods ratio, significant specific heat jump, and high effective mass [45]. A comparison of the SPS-sintered HEA and the arc-melted sample revealed a substantial increase in the critical current density and pinning force for the sample synthesized by ball milling and SPS sintering. The strong vortex pinning force on the SPS-sintered sample resulted in stable remanent magnetization with time after the applied magnetic field was turned off. Furthermore, flux avalanches were observed in a magnetic hysteresis loop measurement in the low magnetic field region ($H < 1$ T). The vortex avalanche effect can be understood by vortex quantum creep or flux jump at weakly coupled superconducting grain boundaries [46].

Figure 2.8 shows the X-ray diffraction (XRD) patterns for both the ball-milled powder (represented by a black line) and the SPS-sintered (depicted with a red line) $Ta_{1/6}Nb_{2/6}Hf_{1/6}Zr_{1/6}Ti_{1/6}$ compound. The XRD peaks correspond to the body-centered cubic (BCC) structure ($IM\overline{3}m$, , space group No. 229). The lattice parameters of the ball-milled powder and SPS-sintered samples ($a = 3.37$ Å) were nearly identical to the previously reported lattice parameter ($a = 3.38$ Å) for the arc-melted sample [45]. Despite the ball-milled powder exhibiting a single phase, the peak profile was broad, suggesting an extremely small particle size on the submicron scale. Upon spark plasma sintering, minor phases were observed, identified as Hf_2Fe ($Fd\overline{3}m$, space group No. 227, $a = 12.07$ Å) [47], Zr_2Fe (I4/mcm, No. 140, $a = 6.320$ Å, $c = 5.740$ Å) [48], and β-Ta (P 42/mnm, No. 136, $a = 10.194$ Å, $c = 5.313$ Å) [49]. Fe-based minor phases such as $(Zr,Hf)_2Fe$ were also detected in HfTaTiNbZr-based alloys produced through mechanical alloying and spark plasma sintering [50]. The presence of Zr_2Fe and Hf_2Fe impurity phases can be attributed to the reaction between the metal powder and the stainless steel jar during the mechanical alloying process.

Scanning electron microscopy (SEM) images and energy dispersive X-ray spectroscopy (EDX) elemental mapping of the SPS-sintered $Ta_{1/6}Nb_{2/6}Hf_{1/6}Zr_{1/6}Ti_{1/6}$ compounds are displayed in Fig. 2.9 and reveal no evidence of secondary phase agglomeration at the micron scale. The SEM/EDX findings suggest that the Fe-based secondary phases are dispersed randomly throughout the $Ta_{1/6}Nb_{2/6}Hf_{1/6}Zr_{1/6}Ti_{1/6}$ matrix, as Fe is consistently observed across the sample. By averaging the elemental concentrations at five distinct locations in the EDX profile, the composition was determined to be $Ta_{0.156}Nb_{0.316}Hf_{0.163}Zr_{0.163}Ti_{0.158}$, featuring a minor Fe impurity (4.31 at.%). This composition closely resembles the original compound $Ta_{0.166}Nb_{0.333}Hf_{0.166}Zr_{0.166}Ti_{0.166}$.

The temperature-dependent electrical resistivity, $\rho(T)$, of the sintered $Ta_{1/6}Nb_{2/6}Hf_{1/6}Zr_{1/6}Ti_{1/6}$ compound reveals a superconducting transition at $T_c = 7.83$ K, as depicted in Fig. 2.10a. The superconducting transition temperature for the sintered $Ta_{1/6}Nb_{2/6}Hf_{1/6}Zr_{1/6}Ti_{1/6}$ compound is nearly identical to that of the arc-melted sample ($T_c = 7.85$ K) [45], suggesting that the superconducting transition temperature is not significantly affected by minor impurity phases. The residual resistivity ratio (RRR) in the metallic region above T_c for the SPS-sintered sample exhibits an RRR = $\rho(300\text{ K})/\rho(10\text{ K})$ of 1.2, which is similar to the RRR of 1.17 for the arc-melted sample [45]. Conversely, the $\rho(T)$ values (65 μΩ-cm at 10 K, 79 μΩ-cm at 300 K) for the SPS-sintered sample are greater than those (58 μΩ-cm at 10 K, 68 μΩ-cm at 300 K) for the arc-melted sample [45]. The increased electrical resistivity can be ascribed to impurities and grain boundaries in the sintered sample.

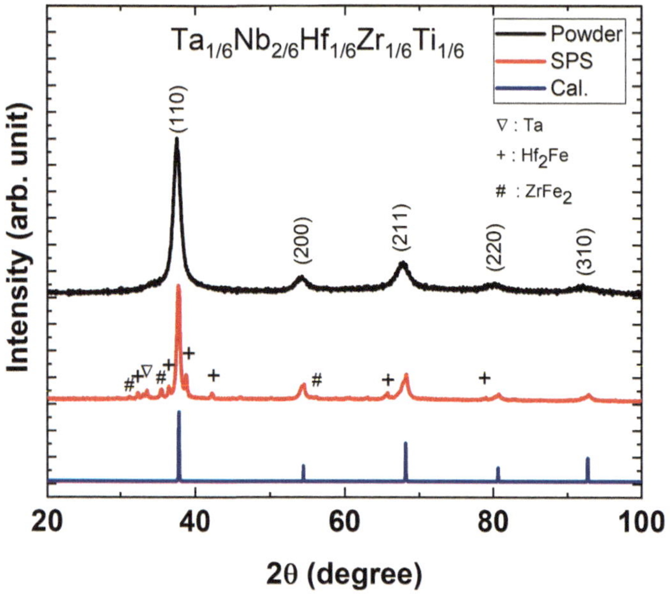

Fig. 2.8 X-ray diffraction (XRD) patterns of the ball-milled (black), SPS-sintered (red), and calculated (blue) $Ta_{1/6}Nb_{2/6}Hf_{1/6}Zr_{1/6}Ti_{1/6}$ compounds. Reproduced from Ref. [61] with permission from Elsevier

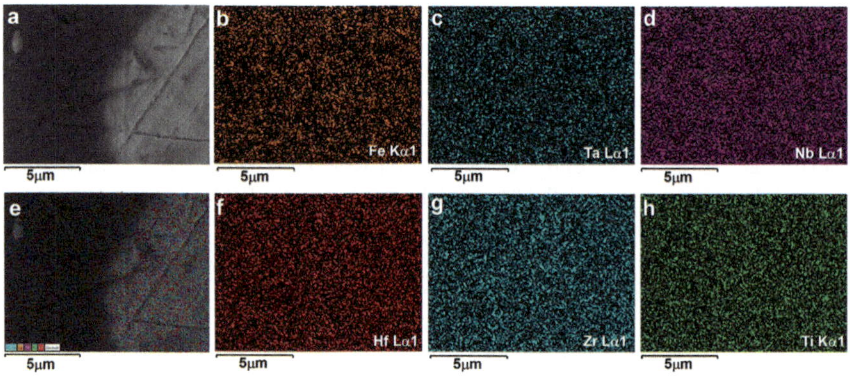

Fig. 2.9 Scanning electron microscopy (SEM) images (**a**) and their elemental map (**e**). Elemental mapping of Fe (**b**), Ta (**c**), Nb (**d**), Hf (**f**), Zr (**g**), and Ti (**h**) in the SPS-sintered $Ta_{1/6}Nb_{2/6}Hf_{1/6}Zr_{1/6}Ti_{1/6}$ compound. Reproduced from Ref. [61] with permission from Elsevier

The inset in Fig. 2.10a displays the temperature-dependent electrical resistivities under varying static magnetic fields, ranging from 0 to 9 T. The superconducting transition temperatures consistently decrease as the magnetic fields increase.

The upper critical magnetic fields, $\mu_0 H_{c2}(T)$, are derived from the superconducting transition temperatures under different magnetic fields, as illustrated

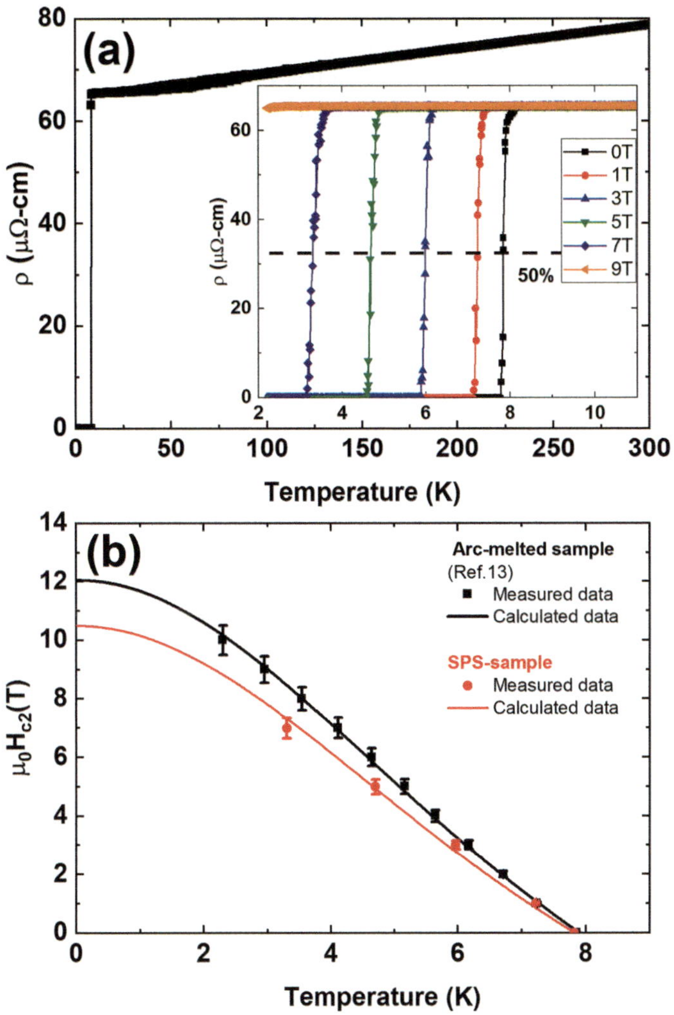

Fig. 2.10 **a** Temperature-dependent electrical resistivity $\rho(T)$ of the SPS-sintered $Ta_{1/6}Nb_{2/6}Hf_{1/6}Zr_{1/6}Ti_{1/6}$ compound. The inset shows the electrical resistivity under various static magnetic fields from 0 to 9 T near T_c. The dotted line represents 50% of the normal-state electrical resistivity. **b** The upper critical magnetic fields $H_{c2}(T)$ of the SPS-sintered and arc-melted $Ta_{1/6}Nb_{2/6}Hf_{1/6}Zr_{1/6}Ti_{1/6}$ [45]. The solid line is the Ginzburg–Landau fitting result of $H_{c2}(T)$. Reproduced from Ref. [61] with permission from Elsevier

in Fig. 2.10b. The $\mu_0 H_{c2}(T)$ values asymptotically increase as the temperature decreases. The $\mu_0 H_{c2}(T)$ of the arc-melted samples are slightly greater than those of the samples produced by spark plasma sintering. The experimental lines fitted to the data are based on the Ginzburg–Landau equations; Eq. (2.1).

The zero-temperature limit of the upper critical magnetic field, $\mu_0 H_{c2}(0)$, obtained from the Ginzburg–Landau (GL) equations reveals that the $H_{c2}(0)$ of the sintered sample (10.5 T) is marginally lower than that of the arc-melted sample (12.05 T) [45]. The coherence length, ξ, is calculated using the Ginzburg–Landau theory and the relation $H_{c2}(0) = \Phi_0/2\pi\xi^2$, where Φ_0 (= 2.07×10^{-15} Wb) represents the flux quanta.

The coherence length ξ obtained for the SPS sample is 5.66 nm, which is longer than that of the arc-melted sample (5.23 nm) [45]. This observation is notable because powder metallurgically sintered samples containing magnetic Fe impurities and incoherent grain boundaries might be expected to have shorter coherence lengths and greater penetration depths than arc-melted samples. Magnetic impurities can degrade the superconducting spin singlet state, leading to diminished superconducting properties, such as reduced critical temperature and coherence length. Furthermore, grain boundaries can limit coherence length, as incoherent grain boundaries can scatter superelectrons. The similar or longer coherence length in the SPS sample relative to that in the arc-melted sample suggested that the magnetic Fe impurities and incoherent grain boundaries in the sintered $Ta_{1/6}Nb_{2/6}Hf_{1/6}Zr_{1/6}Ti_{1/6}$ compound did not negatively impact the superconducting properties.

Figure 2.11a displays the temperature-dependent magnetic susceptibility, $\chi(T)$, under a low magnetic field (H = 50 Oe) for both zero-field cooling (ZFC, black square) and field cooling (FC, red circle) processes near T_c. A sharp diamagnetic superconducting transition is observed at $T_c = 7.6$ K in the ZFC measurements. The magnetic susceptibility, $\chi(T)$, is corrected by the demagnetization factor N using $\chi = \chi_0/(1 - N\chi_0)$, where χ_0 represents the measured magnetic susceptibility. The demagnetization factor N for a rectangular sample is calculated using $N = 4ab/[4ab + 3c(a + b)]$, where $2a$, $2b$, and $2c$ are the sample dimensions and c is parallel to the applied magnetic field direction [11]. The zero-temperature magnetic susceptibility is approximately $\chi(0) = 0.94$, indicating that the superconducting volume fraction is nearly 100%.

Figure 2.11b displays the field-dependent isothermal magnetic hysteresis *M(H)* at various temperatures, ranging from −5 T to +5 T. The *M(H)* hysteresis loops exhibit typical type II superconducting behavior. It is important to note the occurrence of flux avalanches at low temperatures ($T \leq 5$ K) and low magnetic fields ($H \leq 2$ T), where magnetization is suddenly suppressed at specific irregular magnetic fields. Flux avalanches or magnetization quenching are more pronounced at low temperatures than at high temperatures. For instance, magnetization quenching is more frequently observed over a broad magnetic field range ($H \leq 2$ T) at T = 2 K, while the magnetic field range for magnetization quenching narrows with increasing temperature ($H \leq 0.5$ T for $T = 5$ K). Flux quenching is related to the flux jump, which has also been observed in polycrystalline MgB_2 [51, 52]. A flux jump is not observed for the arc-melted $Ta_{1/6}Nb_{2/6}Hf_{1/6}Zr_{1/6}Ti_{1/6}$ compound [45], suggesting that the flux avalanche in the sintered sample results from unstable vortex dynamics near impurities and grain boundaries.

Interestingly, the flux avalanche effect might be connected to the critical current density J_c, as studies have shown that a higher frequency of flux jumps leads to an

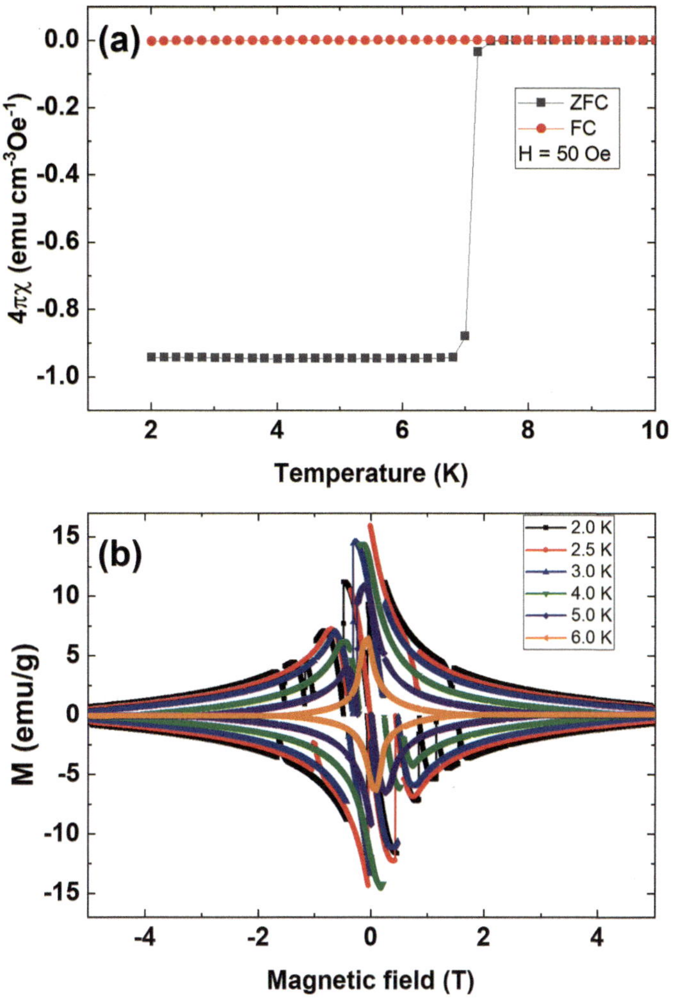

Fig. 2.11 **a** Temperature-dependent magnetic susceptibility $\chi(T)$ of the sintered $Ta_{1/6}Nb_{2/6}Hf_{1/6}Zr_{1/6}Ti_{1/6}$ compound under a static magnetic field (50 Oe) with field cooling (FC) and zero-field cooling (ZFC) thermal hysteresis. **b** Magnetic field-dependent isothermal magnetic hysteresis *M(H)* loops of the sintered $Ta_{1/6}Nb_{2/6}Hf_{1/6}Zr_{1/6}Ti_{1/6}$ compound at various temperatures, as indicated. Reproduced from Ref. [61] with permission from Elsevier

increased critical current density (J_c) [51]. The critical current densities J_c, displayed in Fig. 2.12a, are derived using the Bean model. The critical current densities J_c of the sintered $Ta_{1/6}Nb_{2/6}Hf_{1/6}Zr_{1/6}Ti_{1/6}$ compound are approximately 30,500 A cm^{-2} (at T = 2 K, H = 0.01 T) and 73,200 A cm^{-2} (at T = 4 K, H = 0.01 T), showing a remarkable increase of approximately 286% (at T = 2 K, H = 0.01 T) and 687% (at T = 4 K, H = 0.01 T) compared to the 10,655 A cm^{-2} (at T = 2 K, H = 0.01 T) for the arc-melted sample [45]. The improvement in the critical current density can be

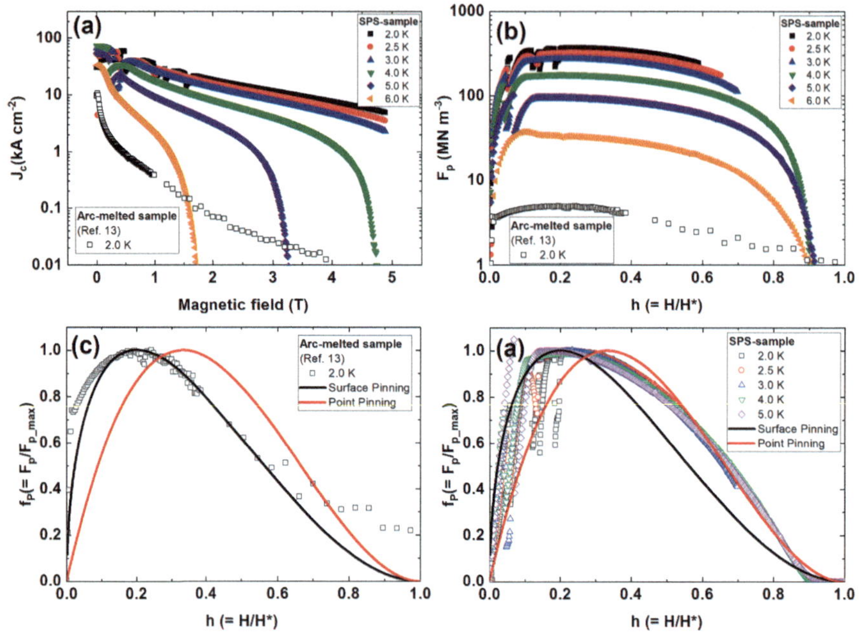

Fig. 2.12 **a** Magnetic field-dependent critical current density $J_c(H)$ of the sintered and arc-melted $Ta_{1/6}Nb_{2/6}Hf_{1/6}Zr_{1/6}Ti_{1/6}$ compounds. **b** Pinning force density F_p with a reduced magnetic field h = H/H* for the sintered and arc-melted $Ta_{1/6}Nb_{2/6}Hf_{1/6}Zr_{1/6}Ti_{1/6}$ compounds. **c** The normalized flux pinning density f_p of the arc-melted sample and **d** f_p of the sintered $Ta_{1/6}Nb_{2/6}Hf_{1/6}Zr_{1/6}Ti_{1/6}$ compound. Reproduced from Ref. [61] with permission from Elsevier

attributed to flux pinning by impurities. A similar enhancement in J_c was observed for nanoparticle-dispersed $YBa_2Cu_3O_7$ films [53].

Figure 2.12b displays the pinning force densities derived from $\boldsymbol{F}_p = \boldsymbol{J}_c \times \boldsymbol{B}$. The pinning force densities of the sintered HEA sample are evidently greater than those of the arc-melted sample (open square) [45]. The normalized flux pinning density f_p (= $F_p/F_p(max)$) with a reduced magnetic field h ($h = H/H^*$, where H^* is the empirical irreversible magnetic field) is presented in Fig. 2.12c, d for the arc-melted and SPS-sintered $Ta_{1/6}Nb_{2/6}Hf_{1/6}Zr_{1/6}Ti_{1/6}$ compounds, respectively. The black and red lines in Fig. 2.12c, d represent the calculated results for the surface pinning $f_p = h^{0.5}(1–h)^2$ and point pinning model $f_p = h(1–h)^2$, respectively [54, 55]. While the normalized flux pinning density of the arc-melted sample fits relatively well with the surface pinning model, the f_p of the sintered sample contributes to both the surface and point pinning models. These findings suggest that the impurities or grain boundaries in the sintered $Ta_{1/6}Nb_{2/6}Hf_{1/6}Zr_{1/6}Ti_{1/6}$ compound enhance the critical current density by increasing both the point pinning density and surface pinning strength.

Figure 2.13a displays the time-dependent magnetization $M(t)$ of the sintered $Ta_{1/6}Nb_{2/6}Hf_{1/6}Zr_{1/6}Ti_{1/6}$ compound. The red circle (left axis) represents the $M(t)$ response to the applied magnetic field (black line, right axis). When magnetic fields

($H = 0.5$, 1.0, 2.0, or 3.0 T) are introduced at specific times, a negative superconducting magnetization arises due to the Meissner effect. This negative magnetization persists for 5 min under the applied magnetic fields. Upon removing the magnetic field, the magnetization rapidly becomes positive as a result of the pinned vortices. Generally, the remanent magnetizations M_{rem} in the absence of a magnetic field exhibit exponential behavior due to dynamic vortex relaxation [56], as described in the following equation:

$$M_{rem}(t) = M_0 e^{-\left(\frac{t}{\tau}\right)^b} \tag{2.7}$$

In the equation, M_0, τ, and b represent the remanent magnetization at $t = 0$, the relaxation time, and the dynamic exponent, respectively. However, it is important to note that the remanent magnetization does not decay over time, as shown in Fig. 2.13a. Expanding the regions where the magnetic field is turned on and off in Fig. 2.13b reveals that the magnetization decreases and increases almost linearly, corresponding to the linear increase and decrease in the applied magnetic fields, respectively. When

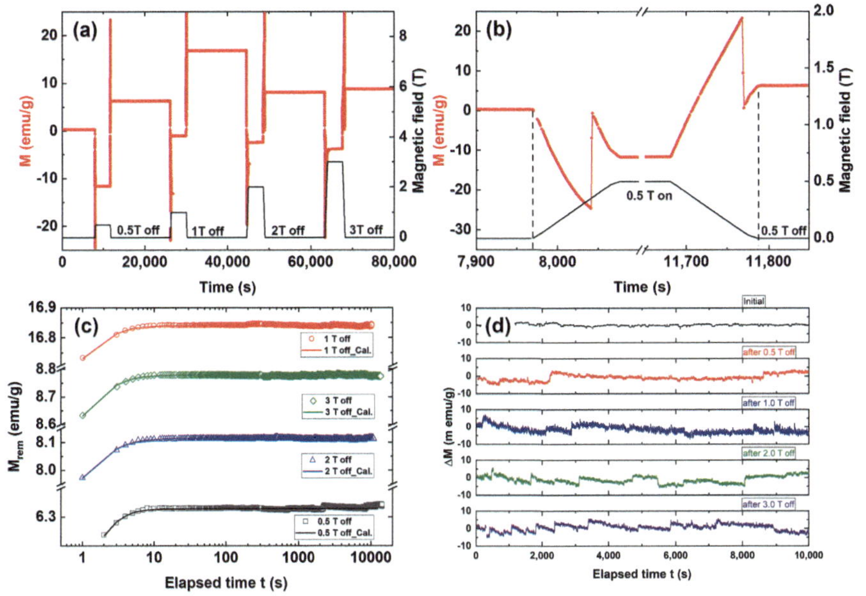

Fig. 2.13 **a** Time-dependent magnetization *M(t)* of the sintered $Ta_{1/6}Nb_{2/6}Hf_{1/6}Zr_{1/6}Ti_{1/6}$ compound for various magnetic field switches in the on and off states. **b** Time-dependent magnetization *M(t)* (red circle, left axis) for a magnetic field profile up to 0.5 T (black line, right axis). **c** Remanent magnetization with time $M_{rem}(t)$ after turning off magnetic fields, as indicated at $T = 2$ K. **d** Fluctuation in the remanent magnetization with time $\Delta M(t) = M_{rem}(t)/M_{rem}(0)$, where $M_{rem}(0)$ is the magnetization when M_{rem} saturates after turning off magnetic fields. Reproduced from Ref. [61] with permission from Elsevier

a static magnetic field is present, the magnetization remains consistently negative due to the Meissner effect.

The remanent magnetization of the sintered $Ta_{1/6}Nb_{2/6}Hf_{1/6}Zr_{1/6}Ti_{1/6}$ compound for various magnetic fields, as indicated, is displayed in Fig. 2.13c. M_{rem} fits the aforementioned equation, and relaxation times are obtained at 0.25 s (0.5 T off), 0.17 s (1 T off), 0.25 s (2 T off), and 0.28 s (3 T off). Once the magnetic field is turned off, the remanent magnetizations of the sintered $Ta_{1/6}Nb_{2/6}Hf_{1/6}Zr_{1/6}Ti_{1/6}$ compound remain remarkably stable for more than 10,000 s without exponential decay, as shown in Fig. 2.13d. The Anderson–Kim model describes this magnetic relaxation with exponential decay in superconductors. Many superconductors, such as UPt_3[57], $YBa_2Cu_3O_7$[58], Li_2Pt_3B[59], and $HgBa_2CuO_{4+\delta}$ [60], exhibit exponential decay over time. The stable remanent magnetization of the sintered $Ta_{1/6}Nb_{2/6}Hf_{1/6}Zr_{1/6}Ti_{1/6}$ compound without exponential decay is highly unusual and can be attributed to the effect of strong vortex pinning.

When plotting the change in magnetization $\Delta M = M(t) - M(0)$, as shown in Fig. 2.13d, where $M(0)$ is defined as the magnetization at the initial stable point, the ΔM fluctuates over time without decaying. The magnetization fluctuations become more significant as the magnetic field increases. The fluctuation in magnetization can be explained by the variation in the trapped vortex flux due to the strong pinning force. The strong pinning force in the SPS sintered sample is associated with the increase in the critical current density in the $Ta_{1/6}Nb_{2/6}Hf_{1/6}Zr_{1/6}Ti_{1/6}$ compound.

2.4 Summary

We synthesized HEA $Ta_{1/6}Nb_{2/6}Hf_{1/6}Zr_{1/6}Ti_{1/6}$ by arc melting as a single phase with a simple BCC crystal structure ($Im\bar{3}m$). We observed a superconducting phase transition at $T_c \approx 7.85$ K and confirmed that the resulting material was an s-wave type II superconductor with relatively high upper and lower critical fields of $\mu_0 H_{c2}^{onset}(0) \approx 12.05$ T and $\mu_0 H_{c1}(0) \approx 23$ mT, respectively. From the isothermal magnetization hysteresis loop, we found that the flux pinning mechanism corresponded with the double exponential model from the Dew–Huges formula, including strong (surface) pinning, indicating that there were two types of pinning mechanisms.

The superconducting gap obtained by the semiempirical α-model from the specific heat exceeded the BCS weak coupling limit, indicating the presence of a strongly coupled superconductor in the compound. Even though the superconducting properties of the $Ta_{1/6}Nb_{2/6}Hf_{1/6}Zr_{1/6}Ti_{1/6}$ compound were within the dirty limit, the relatively significant specific heat jumps ($\Delta C/\gamma T_c^{mid} = 2.22$ and $\Delta C/\gamma T_c^{zero} = 1.56$) showed that strongly correlated behavior existed beyond the BCS theory. Notably, the plot of the Kadowaki–Woods ratio for the compound was consistent with that of a heavy-fermion system with a heavy effective mass (29 m_e), which implied a strongly correlated system. The strongly correlated s-wave metallic superconductivity of the compound raises very interesting questions: (1) Is it a heavy-fermionic-like superconductor? (2) What is the superconducting mechanism of the compound? Does the electron–phonon interaction mediate the superconductivity under such a strong electron–electron interaction? (3) Does the poor metal behavior (i.e., a resistance that exceeds the Mott–Ioffe–Regel limit) originate from the phase fluctuations between transitional metal alloys and transitional amorphous materials? To answer these questions, a strategy guided by further research on the HEA $Ta_{1/6}Nb_{2/6}Hf_{1/6}Zr_{1/6}Ti_{1/6}$ compound, such as quantum critical behavior under an external magnetic field and pressure and gap-symmetry investigations, such as point contact spectroscopy, is encouraged as a pathway toward investigating single-crystal growth and the various physical properties of the compound.

Despite the presence of minor impurities in the sintered sample, when we synthesized the high-entropy alloy (HEA) $Ta_{1/6}Nb_{2/6}Hf_{1/6}Zr_{1/6}Ti_{1/6}$ compound using powder ball milling and spark plasma sintering (SPS), the superconducting phase transitions in terms of electrical resistivity (T_c ~ 7.8 K) and magnetic susceptibility (T_c ~ 7.6 K) were comparable to those of the arc-melted sample. The zero-temperature upper critical field $H_{c2}(0)$ was estimated to be 10.5 T, which was lower than that of the arc-melted sample (12.05 T). The obtained coherence length (5.66 nm) was greater than that of the arc-melted sample (5.23 nm). Isothermal magnetic hysteresis measurements *M(H)* revealed a flux avalanche effect in the low magnetic field region. The critical current densities J_c were significantly enhanced to 30,500 A cm^{-2} (at T = 2 K, H = 0.01 T) and 73,200 A cm^{-2} (at T = 4 K, H = 0.01 T), which were approximately 286% (at 2 K, 0.01 T) and 687% (at 4 K, 0.01 T) greater than the J_c values of the arc-melted sample. The normalized flux pinning density f_p with decreasing magnetic field h in the arc-melted sample could be described by the surface pinning model, while the f_p of the sintered sample exhibited both point pinning and surface pinning. The remanent magnetization M_{rem}(t) displayed unusual nondecaying behavior over time, unlike conventional superconducting vortex relaxation. Stable vortex pinning increases the vortex pinning force, leading to an increase in the critical current density in the SPS-sintered $Ta_{1/6}Nb_{2/6}Hf_{1/6}Zr_{1/6}Ti_{1/6}$ compound.

References

1. A.J.B. Vincent, *A Study of Three Multicomponent Alloys: BSc Part II Thesis* (University of Sussex, UK, 1981)
2. Y.F. Ye, Q. Wang, J. Lu, C.T. Liu, Y. Yang, High-entropy alloy: challenges and prospects. Mater. Today **19**, 349–362 (2016)
3. L. Sun, R.J. Cava, High entropy alloy superconductors: status, opportunities, and challenges. Phys. Rev. Mater. **3**, 090301 (2019)
4. J.-W. Yeh, S.-K. Chen, S.-J. Lin, J.-Y. Gan, T.-S. Chin, T.-T. Shun, C.-H. Tsau, S.-Y. Chang, Nanostructured high-entropy alloys with multiple principal elements: novel alloy design concepts and outcomes. Adv. Eng. Mater. **6**, 299–303 (2004)
5. J.-W. Yeh, Alloy design strategies and future trends in high-entropy alloys. JOM **65**, 1759–1771 (2013)
6. M.-H. Tsai, J.-W. Yeh, High-entropy alloys: a critical review. Mater. Res. Lett. **2**, 107–123 (2014)
7. B. Cantor, Multicomponent and high entropy alloys. Entropy **16**, 4749–4768 (2014)

8. K.M. Youssef, A.J. Zaddach, C. Niu, D.L. Irving, C.C. Koch, A novel low-density, high-hardness, high-entropy alloy with close-packed single-phase nanocrystalline structures. Mater. Res. Lett. **3**, 95–99 (2015)
9. P. Koželj, S. Vrtnik, A. Jelen, S. Jazbec, Z. Jagličić, S. Maiti, M. Feuerbacher, W. Steurer, J. Dolinšek, Discovery of a superconducting high-entropy alloy. Phys. Rev. Lett. **113**, 107001 (2014)
10. J. Guo, H. Wang, F. von Rohr, Z. Wang, S. Cai, Y. Zhou, K. Yang, A. Li, S. Jiang, Q. Wu, R.J. Cava, L. Sun, Robust zero resistance in a superconducting high-entropy alloy at pressures up to 190 GPa. Proc. Natl. Acad. Sci. U.S.A. **114**, 13144–13147 (2017)
11. F. von Rohr, M.J. Winiarski, J. Tao, T. Klimczuk, R.J. Cava, Effect of electron count and chemical complexity in the Ta-Nb-Hf-Zr-Ti high-entropy alloy superconductor. Proc. Natl. Acad. Sci. U.S.A. **113**, E7144–E7150 (2016)
12. S. Marik, M. Varghese, K.P. Sajilesh, D. Singh, R.P. Singh, Superconductivity in a new hexagonal high entropy alloy. J. Alloys Compd. **769**, 1059–1063 (2018)
13. K. Stolze, J. Tao, F. von Rohr, T. Kong, R.J. Cava, Sc−Zr−Nb−Rh−Pd and Sc−Zr−Nb−Ta−Rh−Pd high-entropy alloy superconductors on a CsCl-type lattice. Chem. Mater. **30**, 906–914 (2018)
14. G.I. Oya, E.J. Saur, Preparation of Nb_3Ge films by chemical transport reaction and their critical properties. J. Low Temp. Phys. **34**, 569–583 (1979)
15. X. Xu, A review and prospects for Nb3Sn superconductor development. Supercond. Sci. Technol. **30**, 093001 (2017)
16. T. Baig, Z. Yao, D. Doll, M. Tomsic, M. Martens, Conduction cooled magnet design for 1.5 T, 3.0 T and 7.0 T MRI systems. Supercond. Sci. Technol. **27**, 125012 (2014)
17. K. Stolze, F.A. Cevallos, T. Kong, R.J. Cava, High-entropy alloy superconductors on an α-Mn lattice. J. Mater. Chem. C **6**, 10441 (2018)
18. F.O. Rohr, R.J. Cava, Isoelectronic substitutions and aluminum alloying in the Ta-Nb-Hf-Zr-Ti high-entropy alloy superconductor. Phys. Rev. Mater. **2**, 034801 (2018)
19. K.-Y. Wu, S.-K. Chen, J.-M. Wu, Superconducting in equal molar NbTaTiZr-based high-entropy alloys. Nat. Sci. **10**, 110–124 (2018)
20. M. Tinkam, *Introduction to Superconductivity*, 2nd chap. 4 (Dover Publication, New York, 2004)
21. W.L. McMillan, Transition temperature of strong-coupled superconductors. Phys. Rev. **167**, 331 (1968)
22. H. Padamsee, J.E. Neighbor, C.A. Shiffman, Quasiparticle phenomenology for thermodynamics of strong-coupling superconductors. J. Low Temp. Phys. **12**, 387–411 (1973)
23. J.C. Loudon, S. Yazdi, T. Kasama, N.D. Zhigadlo, J. Karpinski, Measurement of the penetration depth and coherence length of MgB_2 in all directions using transmission electron microscopy. Phys. Rev. B **91**, 054505 (2015)
24. C.P. Bean, Magnetization of high-field superconductors. Rev. Mod. Phys. **36**, 31–39 (1964)
25. P.C. Poole, Handbook of superconductivity. Academic Press, new York, Ch. **9**, 478–483 (1999)
26. C.R. Li, X.-Z. Wu, N. Zhou, An analysis of fluxons in long Josephson junctions. IEEE Trans. Magn. **19**, 1196–1200 (1983)
27. K. Matsumoto, H. Takewaki, Y. Tanaka, O. Miura, K. Yamafuji, K. Funaki, M. Iwakuma, T. Matsushita, Enhanced J_c properties in superconducting NbTi composites by introducing Nb artificial pins with a layered structure. Appl. Phys. Lett. **64**, 115 (1994)
28. K. Yamafuji, T. Fujiyoshi, K. Toko, T. Matsuno, T. Kobayashi, K. Kishio, On the magnetic field dependence of critical current density in single crystals of high-T_c superconductors. Physica C **226**, 133–142 (1994)
29. B. Dam, J.M. Huijbregtse, F.C. Klaassen, R.C.F. van der Geest, G. Doornbos, J.H. Rector, A.M. Testa, S. Freisem, J.C. Martinez, B. Stäuble-Pümpin, R. Griessen, Origin of high critical currents in $YBa_2Cu_3O_{7-\delta}$ superconducting thin films. Nature (London) **399**, 439–442 (1999)
30. L. Civale, B. Maiorov, A. Serquis, J.O. Willis, J.Y. Coulter, H. Wang, Q.X. Jia, P.N. Arendt, J.L. MacManus-Driscoll, M.P. Maley, S.R. Foltyn, Angular-dependent vortex pinning mechanisms in $YBa_2Cu_3O_7$ coated conductors and thin films. Appl. Phys. Lett. **84**, 2121–2123 (2004)

31. J.P. Rodriguez, M.P. Maley, Collective pinning of imperfect vortex lattices by material line defects in extreme type-II superconductors. Phys. Rev. B **73**, 094502 (2006)
32. S.-G. Jung, J. Vanacken, V. V. Moshchalkov, S. T. Renosto, C. A. M. dos Santos, A. J. S. Machado, Z. Fisk, J. A. Aguiar, Critical current density and flux pinning in $Zr_{0.96}V_{0.04}B_2$ superconductor with AlB_2 structure. J. Appl. Phys. **114**, 133905 (2013)
33. D.C. Larbalestier, L.D. Cooley, M.O. Rikel, A.A. Polyanskii, J. Jiang, S. Patnaik, X.Y. Cai, D.M. Feldmann, A. Gurevich, A.A. Squitieri, M.T. Naus, C.B. Eom, E.E. Hellstrom, R.J. Cava, K.A. Regan, N. Rogado, M.A. Hayward, T. He, J.S. Slusky, P. Khalifah, K. Inumaru, M. Haas, Strongly linked current flow in polycrystalline forms of the superconductor MgB_2. Nature (London) **410**, 186–189 (2001)
34. D. Dew-Hughes, Flux pinning mechanisms in type II superconductors. J. Theor. Exp. Appl. Phys. **30**, 293–305 (1974)
35. J. Wang, Z.X. Shi, H. Lv, T. Tamegai, Effect of two-gap structure on flux pinning in MgB_2. Physica C **445–448**, 462–465 (2006)
36. J. Wang, Z.X. Shi, H. Lv, T. Tamegai, Effect of anisotropic superconductivity on flux pinning in polycrystalline MgB_2. Physica C **449**, 104–108 (2006)
37. S. Keshavarzi, M.J. Qin, S. Soltanian, H.K. Liu, S.X. Dou, Vortex dynamics in pure and SiC-doped MgB_2. Physica C **408–410**, 601–602 (2004)
38. D. Tripathi, T.K. Dey, A new scaling relation for n-AlN doped superconducting MgB_2. J. Appl. Phys. **114**, 093906 (2013)
39. M.J. Rice, Electron–electron scattering in transition metals. Phys. Rev. Lett. **20**, 1439–1441 (1968)
40. K. Kadowaki, S.B. Woods, Universal relationship of the resistivity and specific heat in heavy-fermion compounds. Solid State Commun. **58**, 507–509 (1986)
41. A.C. Jacko, J.O. Fjærestad, B.J. Powell, A unified explanation of the Kadowaki-Woods ratio in strongly correlated metals. Nat. Phys. **5**, 422–425 (2009)
42. K. Miyake, T. Matsuura, C.M. Varma, Relation between resistivity and effective mass in heavy-fermion and A15 compounds. Solid State Commun. **71**, 1149–1153 (1989)
43. S.Y. Li, L. Taillefer, D.G. Hawthorn, M.A. Tanatar, J. Paglione, M. Sutherland, R.W. Hill, C.H. Wang, X.H. Chen, Giant electron-electron scattering in the fermi-liquid state of $Na_{0.7}CoO_2$, Phys. Rev. Lett. **93**, 056401 (2004)
44. C. Strack, C. Akinci, V. Pashchenko, B. Wolf, E. Uhrig, W. Assmus, M. Lang, J. Schreuer, L. Wiehl, J.A. Schlueter, J. Wosnitza, D. Schweitzer, J. Müller, J. Wykhoff, Resistivity studies under hydrostatic pressure on a low-resistance variant of the quasitwo-dimensional organic superconductor $\kappa-(BEDT-TTF)_2Cu[N(CN)_2]$ Br : search for intrinsic scattering contributions. Phys. Rev. B **72**, 054511 (2005)
45. G. Kim, M.-H. Lee, J.H. Yun, P. Rawat, S.-G. Jung, W. Choi, T.-S. You, S.J. Kim, J.-S. Rhyee, Strongly correlated and strongly coupled s-wave superconductivity of the high entropy alloy $Ta_{1/6}Nb_{2/6}Hf_{1/6}Zr_{1/6}Ti_{1/6}$ compound. Acta Mater. **186**, 250–256 (2020)
46. V. Sandu, G. Aldica, S. Popa, P. Badica, E. Cimpoiasu, F. Dumitrache, E. Sandu, Transport properties of superconducting MgB_2 composites with carbon-encapsulated fe nanospheres. J. Appl. Phys. **110**, 123921 (2011)
47. V.D. Levytskyi, V.S. Babizhetskyi, D.R. Myakush, I.V. Kovalchuk, D.B. Ryabov, B.Y. Kotur, Phase equilibria in the Y-Hf-Fe system and hydrogen sorption properties of alloys. Mater. Sci. **47**, 822–829 (2012)
48. D. Chattaraj, C. Majumder, S. Dash, Structural, electronic, elastic and thermodynamic properties of Zr_2Fe and Zr_2FeH_5: a comprehensive study using first principles approach. J. Alloy. Compd. **615**, 234–242 (2014)
49. A. Jiang, T.A. Tyson, L. Axe, L. Gladczuk, M. Sosnowski, P. Cote, The structure and stability of β-ta thin films. Thin Solid Films **479**, 166–173 (2005)
50. B. Xu, M.B. Toffolo, L. Regev, E. Boaretto, K.M. Poduska, Structural differences in archaeologically relevant calcite. Anal. Methods (2015)
51. R. Prozorov, V.G. Kogan, Effective demagnetizing factors of diamagnetic samples of various shapes. Phys. Rev. Appl. **10**, 014030 (2018)

52. K. Murai, J. Hori, Y. Fujii, J. Shaver, G. Kozlowski, Magnetic flux pinning and flux jumps in polycrystalline MgB_2. Cryogenics **45**, 415–420 (2005)
53. T. Haugan, P.N. Barnes, R. Wheeler, F. Meisenkothen, M. Sumption, Addition of nanoparticle dispersions to enhance flux pinning of the $YBa_2Cu_3O_{7-x}$ superconductor. Nature **430**, 867–870 (2004)
54. T. Matsushita, *Flux Pinning in Superconductors*, 2nd edn. (Springer, New York, 2014). (Chapter 1)
55. S. Arumugam, M. Krishnan, K. Ishigaki, J. Gouchi, R. Pervin, G.K. Selvan, P.M. Shirage, Y. Uwatoko, Enhancement of superconducting properties and flux pinning mechanism on $Cr_{0.0005}NbSe_2$ single crystal under hydrostatic pressure. Sci. Rep. **9**, 347 (2019)
56. D. Gajda, A.J. Zaleski, A. Morawski, T. Cetner, M. Maecka, M. Rindfleisch, The optimization of J_c, Birr, n value, wires diameter, and research of the dominant pinning mechanism for applications of undoped MgB_2 in superconducting multisection coils. J. Supercond. Nov. Magn. **33**, 3717–3725 (2020)
57. A. Amann, P. Visani, K. Aupke, A.C. Mota, M.B. Maple, Y. Dalichaouch, P.E. Armstrong, Z. Fisk, Unconventional vortex dynamics in the low-field superconducting phases of UPt3. Europhys. Lett. **33**, 303–308 (1996)
58. V.Y. Monarkha, V.A. Paschenko, V.P. Timofeev, Pinning features of the magnetic flux trapped by YBCO single crystals in weak constant magnetic fields. Low Temp. Phys. **39** (2013)
59. C.F. Miclea, A.C. Mota, M. Sigrist, F. Steglich, T.A. Sayles, B.J. Taylor, C.A. McElroy, M.B. Maple, Vortex avalanches in the noncentrosymmetric superconductor Li_2Pt_3B. Phys. Rev. B **80**, 132502 (2009)
60. S. Eley, R. Willa, M.K. Chan, E.D. Bauer, L. Civale, Vortex phases and glassy dynamics in the highly anisotropic superconductor $HgBa_2CuO_{4+\delta}$. Sci. Rep. **10**, 10239 (2020)
61. J.H. Kim, R. Hidayatia, S.-G. Jung, Y.A. Salawu, H.-J. Kim, J.H. Yun, J.-S. Rhyee, Enhancement of critical current density and strong vortex pinning in high entropy alloy superconductor $Ta_{1/6}Nb_{2/6}Hf_{1/6}Zr_{1/6}Ti_{1/6}$ synthesized by spark plasma sintering. Acta Mater. **232**, 117971 (2022)

Chapter 3
Superconductivity, Hardness, and Materials Design of Body-Centered Cubic High-Entropy Alloys

Jiro Kitagawa, Yoshikazu Mizuguchi, and Terukazu Nishizaki

Abstract This chapter discusses the superconductivity and hardness of Hf–Nb–Ti–V–Zr, Al–Nb–Ti–V–Zr, and HfMoNbTiZr body-centered cubic (bcc) high-entropy alloys (HEAs). In particular, owing to the discovery in superconductivity of equiatomic quinary HfMoNbTiZr, superconducting properties are compared under a fixed maximum configurational entropy. This comparison reveals a negative correlation between the Debye temperature, θ_D, and the superconducting critical temperature, T_c, which is in contrast with the positive correlation expected for Bardeen–Cooper–Schrieffer superconductors. A negative correlation is a new feature of superconductivity in a high-entropy state, and can be explained by the uncertainty principle. The hardness of bcc HEA superconductors depends on their valence electron count. Furthermore, T_c decreases with increasing hardness, which is also related to the negative correlation between θ_D and T_c. Finally, in this chapter, a simple materials design for bcc and hexagonal close-packed HEA superconductors is presented.

3.1 Introduction

High-entropy alloys (HEAs) exhibit rich functionalities such as high strength, corrosion resistance, energy storage, radiation protection, superconductivity, soft ferromagnetism, and biocompatibility [1–9]. In addition, HEAs possess a large com-

J. Kitagawa (✉)
Department of Electrical Engineering, Faculty of Engineering, Fukuoka Institute of Technology, 3-30-1 Wajiro-higashi, Higashi-ku Fukuoka 811-0295, Japan
e-mail: j-kitagawa@fit.ac.jp

Y. Mizuguchi
Department of Physics, Tokyo Metropolitan University, 1-1, Minami-osawa, Hachioji 192-0397, Japan
e-mail: mizugu@tmu.ac.jp

T. Nishizaki
Department of Electrical Engineering, Faculty of Science and Engineering, Kyushu Sangyo University, 2-3-1 Matsukadai, Higashi-ku, Fukuoka 813-8503, Japan
e-mail: terukazu@ip.kyusan-u.ac.jp

J. Kitagawa and Y. Mizuguchi (eds.), *High-Entropy Alloy Superconductors*, Springer Series in Solid-State Sciences 202, https://doi.org/10.1007/978-981-97-4129-8_3

positional space, and materials research on HEAs showing novel phenomena has generated significant interest [10, 11]. Materials studies on HEA superconductors have yielded remarkable results. One example is the robust superconductivity of $(TaNb)_{0.67}(HfZrTi)_{0.33}$ under extremely high pressures [12]. This HEA shows zero resistance at 8 K under ambient pressure. The superconducting critical temperature, T_c, is almost unchanged up to a high pressure of 190.6 GPa. Therefore, HEA superconductors are promising candidates as superconducting materials that can operate under extreme conditions. The robustness of superconductivity under high pressure has also been reported for high-entropy intermetallic superconductors [13]. The cocktail effect, which is the enhancement of physical properties beyond those of simple mixtures of constituent elements, is also an unconventional phenomenon and a central issue of HEAs. HEA materials with catalytic [14], thermoelectric [15], magnetic [16], and capacitive energy storage [17] functions frequently exhibit cocktail effects. This effect has also been reported for the BiS_2-based superconductor [18] $REO_{0.5}F_{0.5}BiS_2$ (RE = rare earth). $PrO_{0.5}F_{0.5}BiS_2$ shows a weak diamagnetic signal, which is enhanced by the introduction of a high-entropy state at the rare-earth site, as in $(La_{0.2}Ce_{0.2}Pr_{0.2}Nd_{0.2}Sm_{0.2})O_{0.5}F_{0.5}BiS_2$. Another exotic feature is robust superconductivity in the presence of a magnetic element [19]. σ-phase $Ta_{10}Mo_{35-x}Cr_xRe_{35}Ru_{20}$ shows a systematic reduction in T_c with increasing Cr content. The Cr atoms exhibit an antiferromagnetic correlation. The reduction in T_c as a function of the Cr content is relatively gradual compared to that in conventional Cr-containing superconductors, which suggests that the superconductivity in a high-entropy state is robust against magnetism. Materials research plays a vital role in exploring other remarkable phenomena in HEA superconductors. In particular, materials research toward understanding the high-entropy effect on superconductivity is highly desirable.

Hardness is conveniently used to describe the mechanical properties of metallic materials, and many HEAs exhibit hardness values higher than those of conventional alloys. Superconductivity combined with high hardness has recently been reported for transition metal carbide materials [20, 21], which have generated significant interest as multifunctional superconductors. Considering that many HEAs have high hardness, elucidating the relationship between the hardness and superconductivity of HEA superconductors is required for designing a multifunctional HEA superconductor. Moreover, the hardness of these materials is related to their phonon properties. Because many HEA superconductors follow the Bardeen–Cooper–Schrieffer (BCS) theory based on electron–phonon coupling, a hardness study is essential to obtain deep insight into superconductivity in a high-entropy state.

We studied body-centered cubic (bcc) HEA superconductors [22–24] because the bcc structure is simple and its materials design is more straightforward than that of other structures. Table 3.1 summarizes the T_c, upper critical field H_{c2}, and valence electron count (VEC) values of single-phase bcc HEA superconductors reported in the literature. This chapter discusses the results for $Hf_{21}Nb_{25}Ti_{15}V_{15}Zr_{24}$, Al–Nb–Ti–V–Zr, and HfMoNbTiZr. In the study on HfMoNbTiZr, the significant finding was that T_c decreases with increasing Debye temperature θ_D owing to weakened electron–phonon coupling. This is contrary to the prediction based on the BCS theory, i.e.,

Table 3.1 Superconducting critical temperature T_c, upper critical field H_{c2}, valence electron count (VEC), and δ-parameter of single-phase body-centered cubic HEA superconductors reported

HEA	T_c (K)	H_{c2} (kOe)	VEC	δ	Ref.
$Ta_{34}Nb_{33}Hf_8Zr_{14}Ti_{11}$	7.3	82	4.67	4.56	[25]
TaNbHfZrTi	6.0	105	4.4	4.98	[26]
$(TaNb)_{0.7}(HfZrTi)_{0.3}$	8.0	67	4.7	4.34	[27]
$(TaNb)_{0.67}(HfZrTi)_{0.33}$	7.8	78	4.67	4.47	[27]
$(TaNb)_{0.6}(HfZrTi)_{0.4}$	7.6	84	4.6	4.71	[27]
$(TaNb)_{0.5}(HfZrTi)_{0.5}$	6.5	117	4.5	4.91	[27]
$(TaNb)_{0.16}(HfZrTi)_{0.84}$	4.5	90	4.16	4.64	[27]
$(TaNbV)_{0.67}(HfZrTi)_{0.33}$	4.3	–	4.67	6.39	[27]
$(NbV)_{0.67}(HfZrTi)_{0.33}$	7.2	–	4.67	7.04	[27]
$(TaV)_{0.67}(HfZrTi)_{0.33}$	4.0	–	4.67	7.04	[27]
$(Sc_{0.33}Cr_{0.67}Nb)_{0.67}(HfZrTi)_{0.33}$	5.6	–	4.67	9.14	[28]
$(TaSc_{0.33}Cr_{0.67})_{0.67}(HfZrTi)_{0.33}$	4.4	–	4.67	9.13	[28]
$(TaNb)_{0.67}(Sc_{0.67}Cr_{0.33}ZrTi)_{0.33}$	7.5	–	4.67	5.68	[28]
$(TaNb)_{0.67}(HfSc_{0.67}Cr_{0.33}Ti)_{0.33}$	7.4	–	4.67	5.39	[28]
$(TaNb)_{0.67}(HfZrSc_{0.67}Cr_{0.33})_{0.33}$	7.6	–	4.67	6.15	[28]
$(Y_{0.33}Mo_{0.67}Nb)_{0.67}(HfZrTi)_{0.33}$	4.7	–	4.67	8.98	[28]
$(TaY_{0.33}Mo_{0.67})_{0.67}(HfZrTi)_{0.33}$	3.5	–	4.67	8.96	[28]
$(TaNb)_{0.67}(Y_{0.67}Mo_{0.33}ZrTi)_{0.33}$	7.6	–	4.67	7.25	[28]
$(TaNb)_{0.67}(HfY_{0.67}Mo_{0.33}Ti)_{0.33}$	6.7	–	4.67	7.06	[28]
$(TaNb)_{0.67}(HfZrY_{0.67}Mo_{0.33})_{0.33}$	7.5	–	4.67	7.47	[28]
$(Sc_{0.33}Mo_{0.67}Nb)_{0.67}(HfZrTi)_{0.33}$	4.4	–	4.67	6.62	[28]
$(TaSc_{0.33}Mo_{0.67})_{0.67}(HfZrTi)_{0.33}$	2.9	–	4.67	6.61	[28]
$(TaNb)_{0.67}(Sc_{0.67}Mo_{0.33}ZrTi)_{0.33}$	7.5	–	4.67	5.10	[28]
$(TaNb)_{0.67}(HfSc_{0.67}Mo_{0.33}Ti)_{0.33}$	6.6	–	4.67	4.78	[28]
$(TaNb)_{0.67}(HfZrSc_{0.67}Mo_{0.33})_{0.33}$	7.5	–	4.67	5.58	[28]
$(TaNb)_{0.67}(HfZrTi)_{0.33}Al_{0.15}$	6.6	–	4.45	4.27	[28]
$(TaNb)_{0.67}(HfZrTi)_{0.33}Al_{0.3}$	3.0	–	4.29	4.08	[28]
$Ta_{1/6}Nb_{2/6}Hf_{1/6}Zr_{1/6}Ti_{1/6}$	7.9	120	4.5	4.92	[29]
$(TaNb)_{0.31}(HfUTi)_{0.69}$	3.2	64	4.31	4.17	[30]
NbReZrHfTi	5.3	89	4.8	5.86	[31]
$Hf_{21}Nb_{25}Ti_{15}V_{15}Zr_{24}$	5.3	–	4.4	6.80	[22]
$Al_5Nb_{24}Ti_{40}V_5Zr_{26}$	5.5	–	4.24	5.28	[23]
$Al_5Nb_{14}Ti_{35}V_5Zr_{41}$	4.73	–	4.14	5.72	[23]
$Al_5Nb_{24}Ti_{35}V_5Zr_{31}$	5.45	–	4.24	5.53	[23]
$Al_5Nb_{34}Ti_{35}V_5Zr_{21}$	6.47	–	4.34	5.06	[23]
$Al_5Nb_{44}Ti_{35}V_5Zr_{11}$	6.75	–	4.44	4.18	[23]
$Al_5Nb_{14}Ti_{45}V_5Zr_{31}$	4.96	–	4.14	5.40	[23]
$Al_5Nb_{34}Ti_{25}V_5Zr_{31}$	6.18	–	4.34	5.65	[23]
$Al_5Nb_{44}Ti_{15}V_5Zr_{31}$	6.98	–	4.44	5.77	[23]
HfNbTaTiV	5.0	66.3	4.6	5.79	[32]
HfMoNbTiZr	4.1	55.5	4.6	6.09	[24]

(continued)

Table 3.1 (continued)

HEA	T_c (K)	H_{c2} (kOe)	VEC	δ	Ref.
$(NbTa)_{0.67}(MoHfW)_{0.33}$	4.3	14.5	5.11	4.03	[33]
$(TaNb)_{0.79}(HfZrTi)_{0.21}$ [a]	6.2	58	4.79	3.82	[34]
$(TaNb)_{0.67}(HfZrTi)_{0.33}$ [a]	6.4	71	4.67	4.47	[34]
$(TaNb)_{0.57}(HfZrTi)_{0.43}$ [a]	6.8	88	4.57	4.79	[34]
$(TaNb)_{0.46}(HfZrTi)_{0.54}$ [a]	6.1	99	4.46	4.96	[34]
$(TaNb)_{0.35}(HfZrTi)_{0.65}$ [a]	5.6	110	4.35	4.97	[34]
$(TaNb)_{0.24}(HfZrTi)_{0.76}$ [a]	4.6	104	4.24	4.84	[34]
$(TaNb)_{0.12}(HfZrTi)_{0.88}$ [a]	2.8	62	4.12	4.51	[34]
$(TiZrNbTa)_{0.9}W_{0.1}$ [a]	6.2	–	4.65	5.17	[35]
$(TiZrNbTa)_{0.79}W_{0.21}$ [a]	4.1	–	4.82	5.40	[35]
$(TiZrNbTa)_{0.65}W_{0.35}$ [a]	2.0	–	5.03	5.49	[35]
$(TiZrNbTa)_{0.89}V_{0.11}$ [a]	6.0	–	4.56	5.80	[35]
$(TiZrNbTa)_{0.67}V_{0.33}$ [a]	4.4	–	4.67	6.81	[35]

[a] Thin-film sample

generally, T_c increases as θ_D increases. In this chapter, this new finding is explained by the uncertainty principle between the phonon energy uncertainty and the phonon lifetime. We also present the relationship between the hardness and T_c of bcc HEA superconductors. Reflecting the negative correlation between T_c and θ_D, increasing the hardness tends to reduce T_c. Finally, the simple materials design of a bcc HEA superconductor is described and compared with the design rule for hexagonal close-packed (hcp) HEA superconductors.

3.2 Hf–Nb–Ti–V–Zr

In 2019, Pacheco et al. reported the thermal stability of equiatomic HfNbTiVZr [36]. An as-cast sample showed a single phase with a bcc structure, whereas heat treatment at 600 ~ 1200 °C yielded a secondary phase with a C14 or hcp structure. We searched for a nonequiatomic Hf–Nb–Ti–V–Zr alloy that maintained a single bcc phase after heat treatment [22].

The arc-melted samples were annealed at 800 °C for four days in evacuated quartz tubes. Figure 3.1a shows the X-ray diffraction (XRD) patterns of several samples and their starting chemical compositions. The equiatomic sample possesses a secondary C14 phase, which is consistent with the findings of Pacheco et al. [36]. $Hf_{21}Nb_{22}Ti_{23}V_{11}Zr_{23}$ exhibited a weak XRD peak stemming from the C14 phase, whereas $Hf_{21}Nb_{25}Ti_{15}V_{15}Zr_{24}$ maintained a single bcc phase. The temperature dependence of the real parts of the ac magnetic susceptibility, $\chi'_{ac}(T)$, of these samples is presented in Fig. 3.1b. All the samples show diamagnetic signals owing to superconductivity. Moreover, T_c slightly depends on the chemical composition.

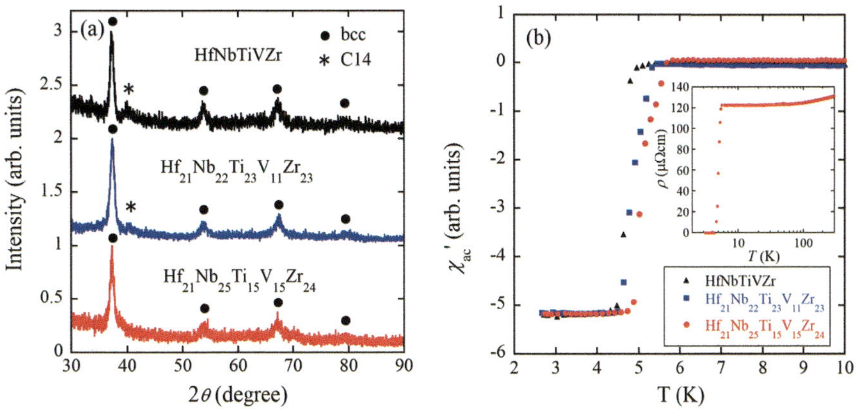

Fig. 3.1 **a** X-ray diffraction patterns and **b** temperature dependence of ac magnetic susceptibility of three Hf–Nb–Ti–V–Zr alloys. The inset shows the temperature dependence of the electrical resistivity of $Hf_{21}Nb_{25}Ti_{15}V_{15}Zr_{24}$. Reproduced from [22] under the Creative Commons Attribution License

The inset of Fig. 3.1b shows the temperature dependence of the electrical resistivity, $\rho(T)$, of $Hf_{21}Nb_{25}Ti_{15}V_{15}Zr_{24}$, which exhibits zero resistivity below approximately $T_c = 5.3$ K.

3.3 Al–Nb–Ti–V–Zr

A specific class of β-Ti alloys has generated considerable interest because of the coexistence of high strength and low Young's modulus after cold rolling. These alloys are called gum metals [37], and a representative alloy is Ti_3(Nb,Ta,V)+(Zr,Hf)+O. This chemical composition is similar to that of a bcc HEA superconductor, and the VEC is 4.24, which is appropriate for superconductivity according to the Matthias rule. Therefore, gum metals have potential for application as next-generation superconducting wires if they can exhibit superconductivity [7]. In 2020, Zherebtsov et al. reported that $Al_5Nb_{24}Ti_{40}V_5Zr_{26}$ exhibits gum-metal-like mechanical behavior after cold rolling [10]. We investigated the superconducting properties of Al–Nb–Ti–V–Zr systems [23], including $Al_5Nb_{24}Ti_{40}V_5Zr_{26}$. Each sample possessing a single bcc phase was fabricated using an arc furnace, without heat treatment or deformation such as cold rolling.

Figure 3.2 shows the temperature-dependent χ'_{ac} and ρ of the gum metal, $Al_5Nb_{24}Ti_{40}V_5Zr_{26}$. A diamagnetic signal and zero resistivity are observed below $T_c = 5.5$ K. $Al_5Nb_xTi_{35}V_5Zr_{55-x}$ and $Al_5Nb_xTi_{59-x}V_5Zr_{31}$ ($x = 14, 24, 34, 44$) were also investigated. For each alloy, the VEC shifts from 4.14 to 4.44 with increasing x. Figure 3.3 shows the $\chi'_{ac}(T)$ and $\rho(T)$ of $Al_5Nb_xTi_{59-x}V_5Zr_{31}$, which exhibit superconductivity and an increase in T_c with increasing VEC. Similar results are obtained

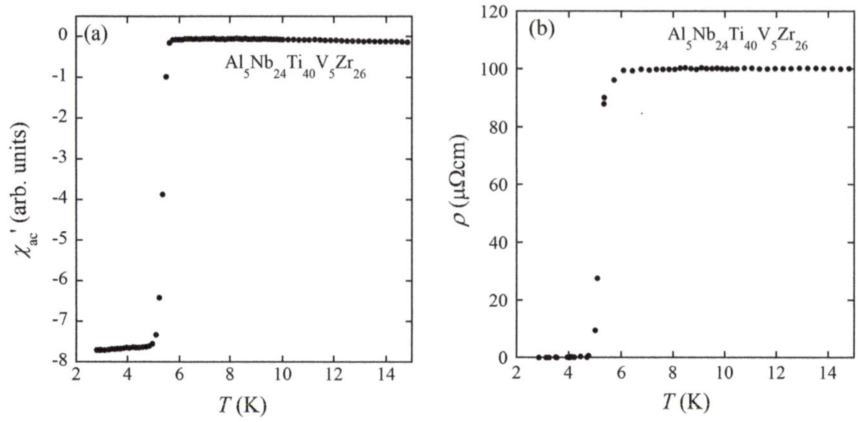

Fig. 3.2 Temperature dependence of **a** ac magnetic susceptibility and **b** electrical resistivity of $Al_5Nb_{24}Ti_{40}V_5Zr_{26}$. Reproduced with permission from [23]

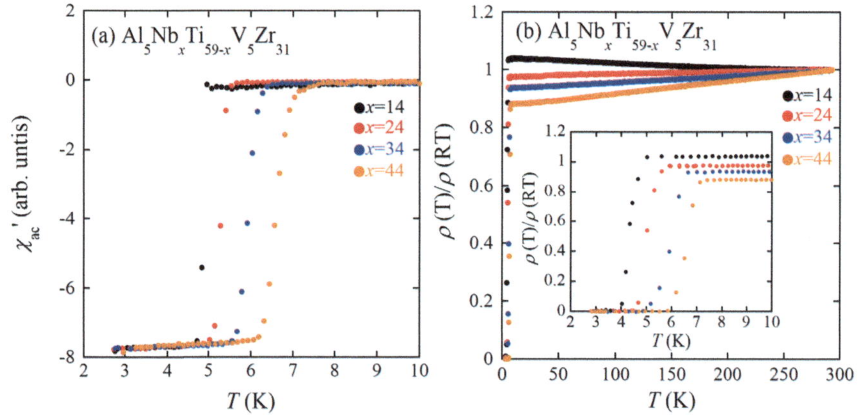

Fig. 3.3 Temperature dependence of **a** ac magnetic susceptibility and **b** electrical resistivity of $Al_5Nb_xTi_{59-x}V_5Zr_{31}$ (x = 14, 24, 34, 44). The inset of **b** shows the low-temperature electrical resistivity. Reproduced with permission from [23]

for $Al_5Nb_xTi_{35}V_5Zr_{55-x}$. Figure 3.4 shows the VEC dependence of the T_c of Al–Nb–Ti–V–Zr and typical bcc HEA superconductors (Ta–Nb–Hf–Zr–Ti, NbReZrHfTi, and $Hf_{21}Nb_{25}Ti_{15}V_{15}Zr_{24}$). The dotted curve represents the Matthias rule for binary and ternary d-electron metallic alloys. Remarkably, almost all the HEAs, including the Al–Nb–Ti–V–Zr HEA, are located in a straight line, suggesting universal behavior. At a small VEC, the T_c values of the HEAs are comparable to those of conventional alloys. However, with increasing VEC, the former becomes lower than the latter. The configurational entropy, ΔS_{mix}, characterizes a high-entropy state and is expressed as

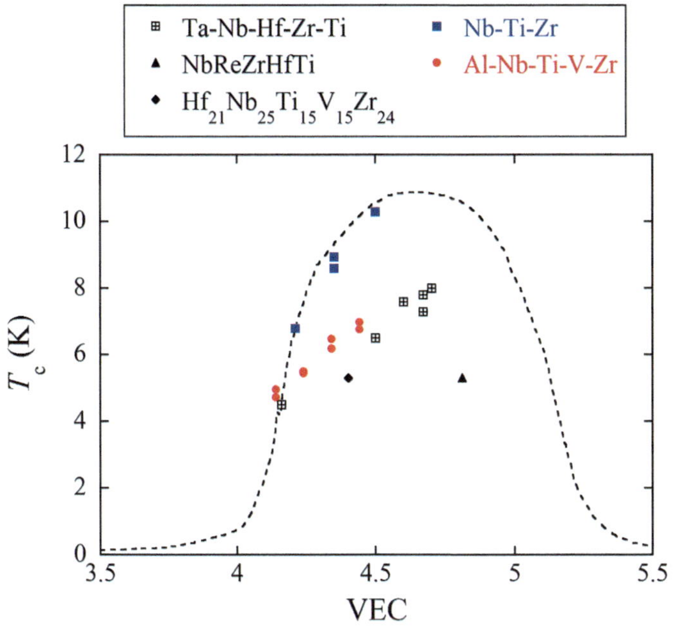

Fig. 3.4 VEC dependence of T_c of different alloys. The dotted line is the VEC dependence of T_c of conventional d-electron metallic alloys. Reproduced with permission from [23]

$$\Delta S_{\text{mix}} = -R\sum_{i=1}^{n} c_i ln c_i, \tag{3.1}$$

where c_i is the atomic fraction of the i-th element and R is the gas constant. Based on this equation, we classified solid solutions into low-entropy alloys with a ΔS_{mix} less than $1.00R$, medium-entropy alloys with a ΔS_{mix} between $1.00R$ and $1.50R$, and HEAs with a ΔS_{mix} greater than $1.50R$ [38]. Generally, the ΔS_{mix} of typical bcc HEA superconductors exceeds $1.50R$ (Ta–Nb–Hf–Zr–Ti: $1.43{\sim}1.59R$, NbReZrHfTi: $1.61R$, $Hf_{21}Nb_{25}Ti_{15}V_{15}Zr_{24}$: $1.59R$). Because the ΔS_{mix} of Al–Nb–Ti–V–Zr alloys ranges from $1.271R$ to $1.376R$, this alloy system is categorized as a medium-entropy alloy according to the definition of an HEA by ΔS_{mix}. However, the VEC dependence of T_c of Al–Nb–Ti–V–Zr is similar to that of a typical HEA superconductor. Furthermore, the ternary Nb–Ti–Zr alloys follow the Matthias rule of conventional alloys ($Nb_{50}Ti_{10}Zr_{40}$: T_c =10.3 K, $Nb_{35}Ti_{15}Zr_{50}$: T_c =8.95 K, $Nb_{35}Ti_{60}Zr_5$: T_c =8.6 K, and $Nb_{21}Ti_{61}Zr_{18}$: T_c =6.8 K [39]), as shown in Fig. 3.4. Therefore, even in the case of medium-entropy alloys, increasing the number of elements up to five results in HEA superconductors.

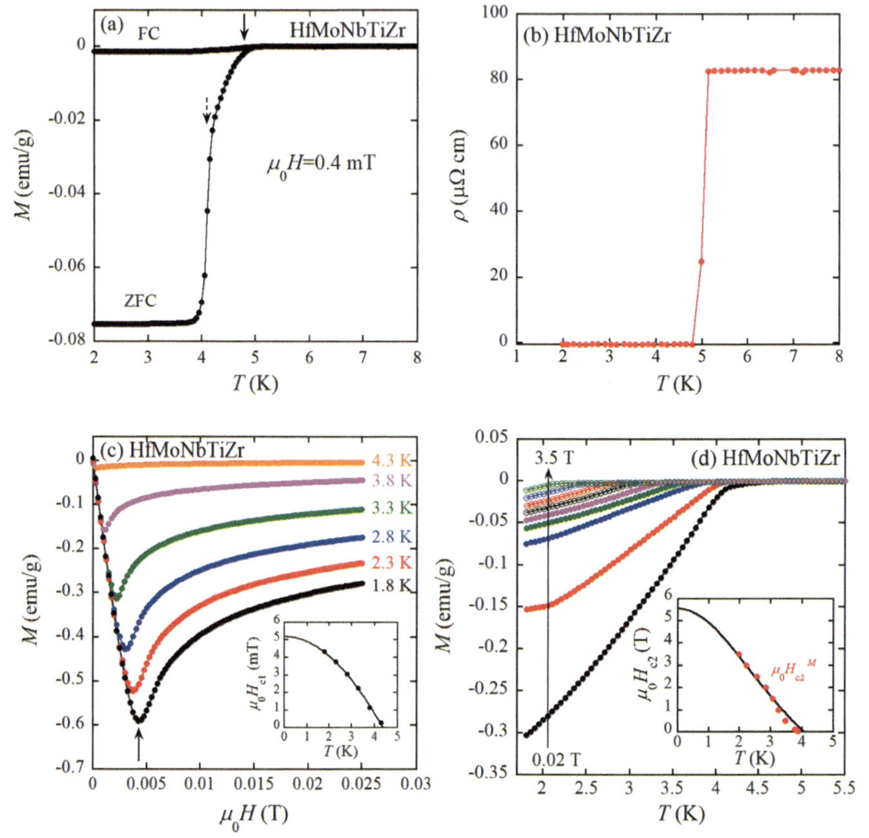

Fig. 3.5 **a** Temperature dependence of M for HfMoNbTiZr. **b** Temperature dependence of ρ for HfMoNbTiZr. **c** Isothermal magnetization curves of HfMoNbTiZr between 1.8 K and 4.3 K. The inset shows the temperature dependence of the lower critical field of HfMoNbTiZr. **d** Temperature dependence of ZFC magnetization under external fields of 0.02, 0.1, 0.5, 1, 1.5, 2, 2.5, 3, and 3.5 T. The inset shows the temperature dependence of the upper critical field determined by magnetization ($\mu_0 H_{c2}^M$). Reproduced with permission from [24]

3.4 HfMoNbTiZr

3.4.1 Fundamental Superconducting Properties

In [24], HfMoNbTiZr was fabricated using an arc furnace, without heat treatment. Figure 3.5a, b show the temperature dependence of the magnetization, $M(T)$, measured under zero-field cooled (ZFC) and field-cooled (FC) conditions and of ρ (T), respectively. The diamagnetic signal in the ZFC results and the zero resistivity suggest superconductivity. Based on M (T), $T_c = 4.8$ K is determined as the point where the difference between the ZFC and FC data becomes noticeable, and it corresponds

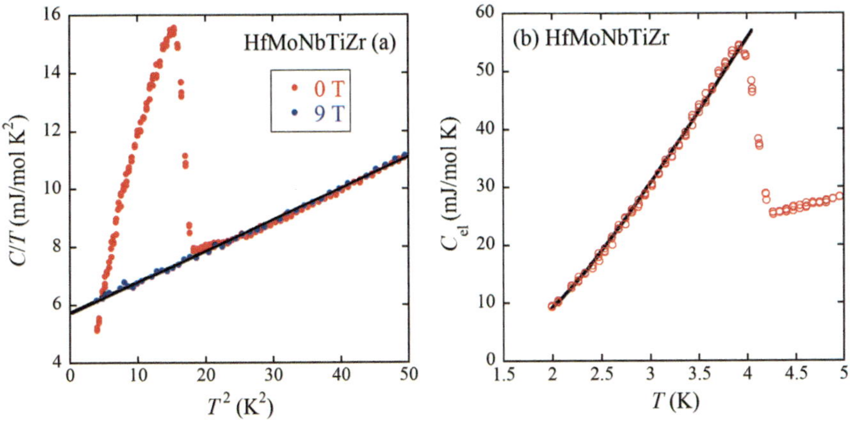

Fig. 3.6 **a** C/T versus T^2 plots of HfMoNbTiZr at 0 T and 9 T. The solid line represents the fitting result using $C(T) = \gamma T + \beta T^3$. **b** Low-temperature electronic specific heat of HfMoNbTiZr. C_{el} is obtained by subtracting the βT^3 term from $C(T)$. Reproduced with permission from [24]

to T_c^{onset} of 5 K based on ρ (T). The ZFC M curve shows an inflection at approximately 4.1 K (see the dotted arrow in Fig. 3.5a), which is a signature of sample inhomogeneity. The lower critical field, H_{c1}, is obtained from the isothermal M–H curves at several temperatures, as shown in Fig. 3.5c. In each curve, H_{c1} is defined as the field showing a negative peak. The temperature dependence of H_{c1}, as shown in the inset of Fig. 3.5c, can be reproduced using the Ginzburg–Landau equation as follows:

$$H_{c1}(T) = H_{c1}(0)\left(1 - \left(\frac{T}{T_c}\right)^2\right), \tag{3.2}$$

and $\mu_0 H_{c1}(0)$ was calculated as 5.2 mT. Figure 3.5d shows the ZFC $M(T)$ under various external fields used to obtain the upper critical field, H_{c2}. The inflection point is defined as T_c, and a $\mu_0 H_{c2}$ versus T plot was constructed, which is shown in the inset of Fig. 3.5d. The temperature dependence can be fitted using the following formula:

$$H_{c2}(T) = H_{c2}(0)\left(\frac{1 - (T/T_c)^2}{1 + (T/T_c)^2}\right), \tag{3.3}$$

and $\mu_0 H_{c2}(0)$ was estimated as 5.55 T. Thus, the Ginzburg–Landau coherence length, $\xi_{GL}(0)$, and the magnetic penetration depth, $\lambda_{GL}(0)$, were calculated as 7.7 nm and 348 nm, respectively. Therefore, the Ginzburg–Landau parameter, $\kappa_{GL}=\lambda_{GL}(0)/\xi_{GL}(0)$, was calculated to be 45, which is much greater than $1/\sqrt{2}$, suggesting that HfMoNbTiZr is a type II superconductor.

The temperature dependence of the specific heat, $C(T)$, shows bulk superconductivity at 4.1 K (Fig. 3.6a), which suggests that the $T_c = 4.8$ K superconducting phase detected by $M(T)$ is parasitic. Some HEAs exhibit phase segregation with slightly different chemical compositions during solidification after arc melting [26, 36]. In such a case, an HEA superconductor may possess a parasitic superconducting phase with a slightly different T_c. After depression of the superconducting transition under $\mu_0 H = 9$ T, $C(T)$ can be expressed as

$$C(T) = \gamma T + \beta T^3, \tag{3.4}$$

where γ (= 5.76 mJ/mol·K^2) and β (= 0.1069 mJ/mol·K^4) are the Sommerfeld coefficient and the lattice contribution, respectively. A Debye temperature θ_D of 263 K was derived from the β value. The electron–phonon coupling, λ_{e-p}, was evaluated using the McMillan formula [40] as follows:

$$T_c = \frac{\theta_D}{1.45} \exp\left[-\frac{1.04(1 + \lambda_{e-p})}{\lambda_{e-p} - \mu^*(1 + 0.62\lambda_{e-p})}\right]. \tag{3.5}$$

Employing the Coulomb pseudopotential $\mu^* = 0.13$ widely used for HEA superconductors, $\theta_D = 263$ K, and $T_c = 4.1$ K, λ_{e-p} was estimated to be 0.63. As shown in Fig. 3.6b, the temperature dependence of the electronic specific heat, $C_{el}(T)$, is fitted by a solid curve with $a\exp\left(-\frac{\Delta(0)}{k_B T}\right)$, where a is a proportionality constant, k_B is the Boltzmann constant, and $\Delta(0) = 0.62$ meV is the superconducting gap. The value of $\frac{2\Delta(0)}{k_B T_c} = 3.50$ agrees with 3.52 expected for an s-wave BCS superconductor. In addition, the normalized specific heat jump, $\frac{\Delta C_{el}}{\gamma T_c}$, is 1.58, which is close to 1.43 predicted using the BCS theory at the weak coupling limit.

3.4.2 *Electronic Structure Calculation*

Figure 3.7a–f show the electronic density of states (DOS) of HfMoNbTiZr obtained using the Akai-KKR software package [41]. The total DOS indicates a metallic nature. The partial DOSs reveal dominant d-electron components at the Fermi level, E_F. The d-electron DOSs of Mo, Nb, and Ti contribute relatively to the total DOS at E_F compared with those of Hf and Zr.

The Akai-KKR program can produce an electronic band structure, as shown in Fig. 3.8. A significant feature is the band broadening caused by atomic disorders. Band broadening at E_F reduces $N(E_F)$. Therefore, attention must be given to band broadening when designing HEA superconductors. Figure 3.8 essentially shows that the uncertainty principle yields band broadening. The uncertainty principle can be expressed as $\Delta E \Delta t \geq \hbar/2$, where ΔE is the energy uncertainty associated with band broadening, Δt is the lifetime, and $\hbar$ is the reduced Planck's constant. In a disordered atomic state, the inelastic scattering probability of an electron increases, and its lifetime remaining at an energy decreases. According to the uncertainty principle, a shorter electron lifetime leads to energy-band broadening. The uncertainty relationship also affects the phonon lifetime. Körmann et al. performed a pioneering study

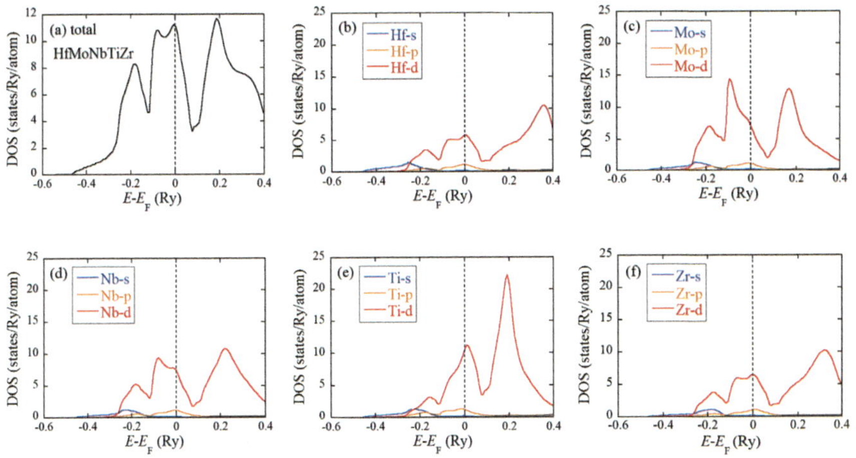

Fig. 3.7 **a** Total electronic DOS of HfMoNbTiZr. **b**–**f** Partial DOSs of Hf, Mo, Nb, Ti, and Zr, respectively. The Fermi energy is set as 0 Ry. Reproduced with permission from [24]

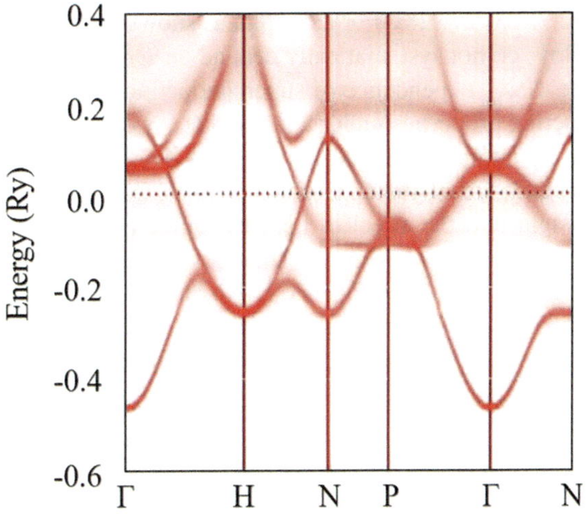

Fig. 3.8 Electronic band structure of HfMoNbTiZr. The Fermi energy is set as 0 Ry. Reproduced with permission from [24]

on phonon broadening in HEAs by ab initio calculations [42]. Figure 3.9 shows the phonon spectra of different alloys with increasing number of elements. From MoTa binary to MoTaNbWV quinary alloys, the phonon broadening at the H-point is gradually enhanced. This implies a shorter phonon lifetime with an increasing number of elements. As mentioned below, the phonon lifetime, which is related to the uncertainty principle, is a vital factor in determining the superconducting properties of HEAs.

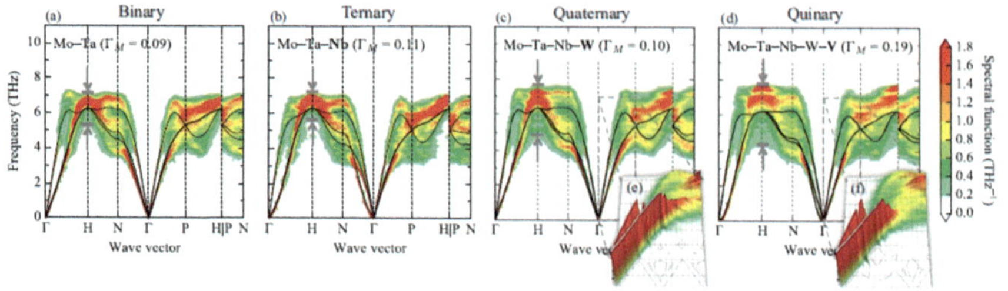

Fig. 3.9 Phonon spectra of **a** MoTa, **b** MoTaNb, **c** MoTaNbW, and **d** MoTaNbWV alloys. **e** and **f** show long-wavelength phonon spectra of MoTaNbW and MoTaNbWV. Reproduced from [42] under the Creative Common Attribution License

3.4.3 Comparison of the Superconducting Properties of Quinary Equiatomic HEAs

The effect of high-entropy on superconducting properties has been investigated in bcc HEA superconductors; however, these studies did not provide a clear conclusion [6, 7, 23, 27, 28]. For example, as shown in Fig. 3.10, the T_cs of Al–Nb–Ti–V–Zr alloys do not systematically depend on ΔS_{mix} [23]. Previous studies have examined the ΔS_{mix} dependence of superconducting properties [6, 7, 23, 27, 28]. Superconducting properties under a fixed ΔS_{mix} were not compared. Among quinary alloys, an equiatomic HEA has a maximum ΔS_{mix} value of $1.61R$. Therefore, the material dependence of the superconductivity of equiatomic quinary alloys might provide insight into the effect of high-entropy on their superconducting properties. In this subsection, we compare the superconducting properties of quinary equiatomic HEAs (HfNbTaTiZr, HfNbReTiZr, HfNbTaTiV, and HfMoNbTiZr).

The superconducting parameters of the above alloys are listed in Table 3.2. Figure 3.11a shows the VEC dependence of T_c, including those of several typi-

Table 3.2 Superconducting parameters of equiatomic quinary bcc HEA superconductors. The data are from [26] for HfNbTaTiZr, [31] for HfNbReTiZr, and [32] for HfNbTaTiV. Each T_c is determined by a specific heat measurement. Reproduced from [24] with permission

Superconducting Parameter	HfNbTaTiZr	HfNbReTiZr	HfNbTaTiV	HfMoNbTiZr
VEC	4.4	4.8	4.6	4.6
T_c (K)	6.0	5.3	4.37	4.1
γ (mJ/mol·K^2)	7.92	5.7	8.35	5.76
θ_D (K)	212	255	225	263
λ_{e-p}	0.77	0.69	0.67	0.63
$\mu_0 H_{c1}$ (mT)	–	33	19.5	5.2
$\mu_0 H_{c2}$ (T)	10.45	8.88	6.63	5.55
ξ_{GL} (nm)	5.6	6.1	7.1	7.7

Fig. 3.10 T_c versus $\Delta S_{mix}/R$ plots of $Al_5Nb_xTi_{35}V_5Zr_{55-x}$ and $Al_5Nb_xTi_{59-x}V_5Zr_{31}$. Reproduced with permission from [23]

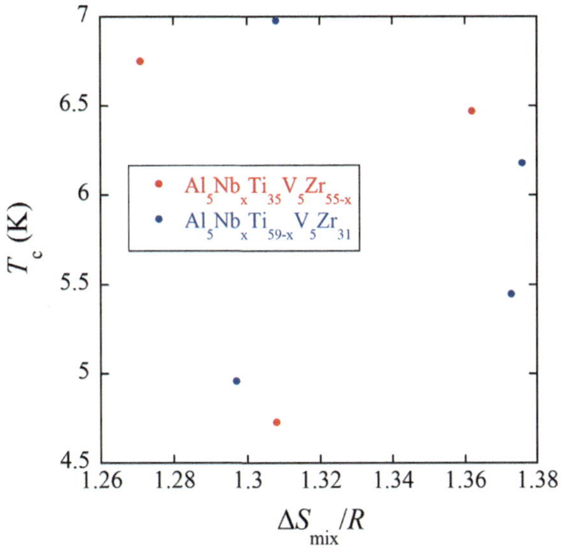

cal nonequiatomic superconductors [22, 23, 27]. As indicated by the dotted line, the T_c versus VEC plots of the typical nonequiatomic quinary HEAs exhibit universal behavior, which mimics the Matthias rule of conventional binary or ternary transition metal alloys (see the solid curve). This result suggests the vital role of $N(E_F)$ in determining T_c because the VEC reflects the information of $N(E_F)$. However, the T_c of the equiatomic HEA superconductors decreases even when the VEC increases above 4.4, and the equiatomic HEA superconductors do not seem to follow the Matthias rule, which suggests that factors other than $N(E_F)$ dominate the determination of T_c. This inference is supported by the T_c versus γ plot, as shown in Fig. 3.11b, which does not exhibit systematic behavior. Another parameter affecting T_c is the electron–phonon coupling, which is represented by λ_{e-p}. Figure 3.11c shows the λ_{e-p} dependence of T_c for the equiatomic HEA superconductors. T_c increases systematically with increasing λ_{e-p}. Moreover, λ_{e-p} is negatively correlated with θ_D. Therefore, the T_c of the equiatomic quinary bcc HEA superconductors decreases with increasing θ_D, which is contrary to our expectation for BCS superconductors, i.e., a compound with a higher θ_D tends to possess a higher T_c.

The uncertainty principle can explain the negative correlation between θ_D and T_c of equiatomic quinary bcc HEA superconductors. As mentioned above, the uncertainty principle can be established in terms of the phonon energy uncertainty and the phonon lifetime. The phonon band structure shows phonon broadening caused by atomic disorder (see Fig. 3.9), which becomes more significant with increasing phonon frequency. A broader phonon band indicates a shorter phonon lifetime. Therefore, a higher θ_D results in a broader phonon band and a shorter phonon lifetime. The shorter lifetime caused by a higher θ_D reduces the electron–phonon coupling. Figure 3.11d shows a T_c contour map as a function of θ_D and λ_{e-p} calculated using

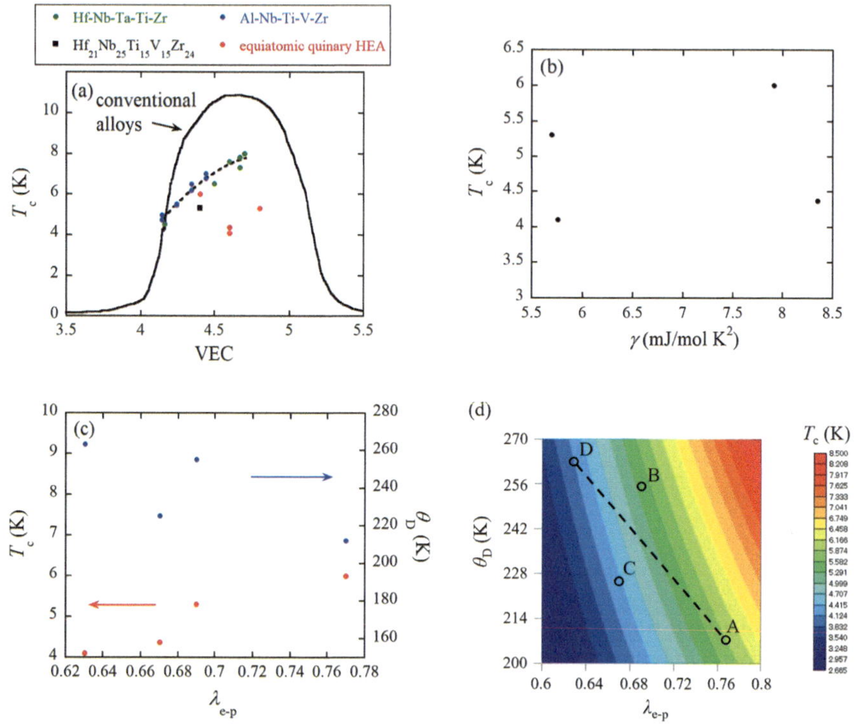

Fig. 3.11 **a** VEC dependence of T_c for typical quinary bcc HEA superconductors. The solid line represents the Matthias rule for conventional binary or ternary transition metal alloys. The dotted line is guide for eyes. Relationships between **b** T_c and γ and **c** T_c (or θ_D) and λ_{e-p} of equiatomic quinary bcc HEA superconductors. **d** T_c contour map as function of θ_D and λ_{e-p} (A: HfNbTaTiZr, B: HfNbReTiZr, C: HfNbTaTiV, and D: HfMoNbTiZr). The dotted line is guide for eyes. Reproduced with permission from [24]

the McMillan formula (Eq. (3.5)). In the figure, the results of four equiatomic HEAs (A: HfNbTaTiZr, B: HfNbReTiZr, C: HfNbTaTiV, and D: HfMoNbTiZr) are plotted. Conventional BCS superconductors exhibit enhanced T_c with increasing θ_D. However, in the HEAs from A to D along the dotted guideline, with increasing θ_D and decreasing λ_{e-p}, T_c is systematically decreased. Thus, the equiatomic quinary bcc HEAs exhibit a negative correlation between T_c and θ_D. Based on Eq. (3.5), T_c increases with increasing θ_D without the influence of the exponential term. The exponential term changes as a function of the electron–phonon coupling strength; thus, it decreases with decreasing λ_{e-p}. Therefore, if the reduction in T_c with decreasing λ_{e-p} dominates the enhancement of T_c with increasing θ_D, the resultant T_c can be decreased.

Table 3.3 Vickers microhardness, VEC, T_c, δ, and ΔS_{mix} of bcc HEA superconductors

Alloy	Hardness (HV)	VEC	T_c (K)	δ	ΔS_{mix} (/R)
HfMoNbTiZr	398(5)	4.6	4.1	6.09	1.609
$Hf_{21}Nb_{25}Ti_{15}V_{15}Zr_{24}$	389(5)	4.4	5.3	6.80	1.586
$(TaNb)_{0.7}(ZrHfTi)_{0.3}$	336(5)	4.7	8.0	4.34	1.426
$Al_5Nb_{24}Ti_{40}V_5Zr_{26}$	352(10)	4.24	5.5	5.28	1.359
$Al_5Nb_{14}Ti_{35}V_5Zr_{41}$	282(5)	4.14	4.73	5.72	1.308
$Al_5Nb_{24}Ti_{35}V_5Zr_{31}$	325(2)	4.24	5.45	5.53	1.373
$Al_5Nb_{34}Ti_{35}V_5Zr_{21}$	320(4)	4.34	6.47	5.06	1.362
$Al_5Nb_{44}Ti_{35}V_5Zr_{11}$	295(10)	4.44	6.75	4.18	1.271
$Al_5Nb_{14}Ti_{45}V_5Zr_{31}$	262(9)	4.14	4.96	5.40	1.297
$Al_5Nb_{34}Ti_{25}V_5Zr_{31}$	323(6)	4.34	6.18	5.65	1.376
$Al_5Nb_{44}Ti_{15}V_5Zr_{31}$	350(5)	4.44	6.98	5.77	1.308
$Nb_{76}Zr_{14}Mo_6Re_4$	331(5)	5.0	7.8	4.51	0.781
$V_{10}Nb_{66}Zr_{14}Mo_6Re_4$	388(7)	5.0	4.5	5.32	1.077
$Nb_{54}V_{10}Ti_{21}Mo_9Re_6$	382(3)	5.0	3.8	3.07	1.276
$Nb_{35}Ta_5Ti_{35}Mo_{15}Re_{10}$	428(6)	5.0	3.4	2.54	1.399

3.5 Hardness

Table 3.3 summarizes the Vickers microhardness of bcc HEA superconductors with VECs of 4.14~5. It also includes T_c, δ, and S_{mix}/R data. δ is a parameter representing the mismatch of the atomic radii of constituent elements and is expressed for an n-component alloy as follows:

$$\delta = 100 \times \sqrt{\sum_{i=1}^{n} c_i \left(1 - \frac{r_i}{\bar{r}}\right)^2}, \tag{3.6}$$

where r_i is the atomic radius of the i-th element and $\bar{r}$ denotes the composition-weighted average atomic radius. In the calculations, the values of r_i listed in Ref. [43] are employed.

Figure 3.12 shows the VEC dependence of the hardness of the bcc HEA superconductors listed in Table 3.3 with a guideline obtained for nonsuperconducting HEAs with bcc (VEC = 6.0~7.55) and face-centered cubic (fcc) (VEC:7.8~8.8) structures reported in [44]. The results for bcc refractory HEAs with VEC = 4.5~5.5 and fcc magnetic HEAs with VEC = 9.25 are also shown [45–49]. Figure 3.12 suggests a possible universal relationship between the hardness and VEC; the hardness shows a broad peak at approximately 6.8. This universal relation may reflect the fact that hard elements are prone to possess VECs = 5~7. Recent deep learning studies have also yielded results consistent with the above universal relation [50].

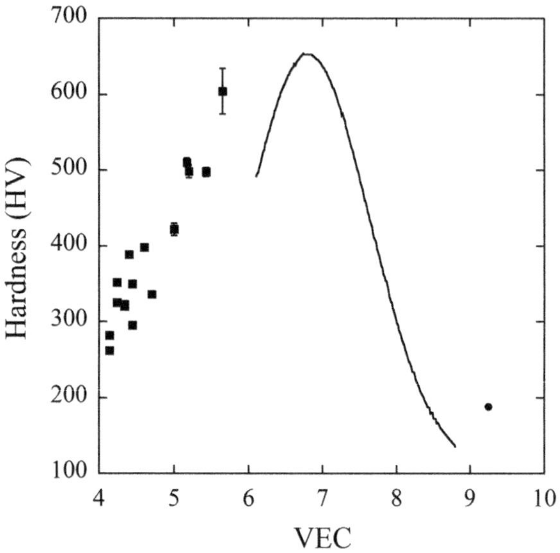

Fig. 3.12 VEC dependence of Vickers microhardness of HEAs. Refractory bcc HEAs reported in the literature [45–48] were included (VEC:4.5~5.5). Hardness of FeCoNiPd and FeCoNiPt with VEC = 9.25 are also plotted [49]. The solid curve represents the guideline obtained for HEAs with bcc (VEC: 6.0~7.55) and fcc (VEC: 7.8~8.8) structures taken from [44]. Reproduced with permission from [49]

Figure 3.13 shows a scatter matrix plot showing the correlation between five different features (hardness, VEC, δ, ΔS_{mix}, and T_c) of bcc HEA superconductors. As mentioned above, a positive correlation between the hardness and VEC ranging from 4 to 5 was confirmed. In addition, a remarkable feature is the negative correlation between the hardness and T_c, i.e., a higher hardness tends to depress T_c. Because θ_D is related to the hardness of a material, the mechanism behind the relation between θ_D and T_c is also appropriate for the negative correlation. Specifically, a higher hardness indicates a higher θ_D, and T_c is reduced based on the uncertainty principle. As shown in Fig. 3.12, the hardness of a bcc HEA increases with the VEC. Therefore, obtaining a high-T_c bcc HEA superconductor with high hardness may be challenging. Figure 3.13 also suggests a possible positive correlation between the hardness and δ (or ΔS_{mix}), which is caused by a severe lattice distortion effect.

The hardness of HEA superconductors was investigated for $(\mathrm{MoReRu})_{(1-2x)/3}(\mathrm{PdPt})_x\mathrm{C}_y$, which form hcp or fcc structures [51]. The C-free sample with $0.042 \leq x \leq 0.167$ had an hcp crystal structure, whereas C doping caused a structural transformation to a fcc structure. In $(\mathrm{MoReRu})_{(1-2x)/3}(\mathrm{PdPt})_x\mathrm{C}_y$, the covalent bonding between the metal and carbon atoms was identified to be responsible for the Vickers microhardness exceeding 1000 HV. The relationship between VEC(6.25~8.0) and hardness qualitatively agrees with the results shown in Fig. 3.12. For each crystal structure, T_c increases with decreasing VEC. Therefore, in $(\mathrm{MoReRu})_{(1-2x)/3}(\mathrm{PdPt})_x\mathrm{C}_y$, a higher hardness enhanced T_c, which is in contrast to the results obtained for bcc HEA superconductors. This discussion suggests that the effect of hardness on the T_c of bcc HEAs differs from that on hcp and fcc HEAs.

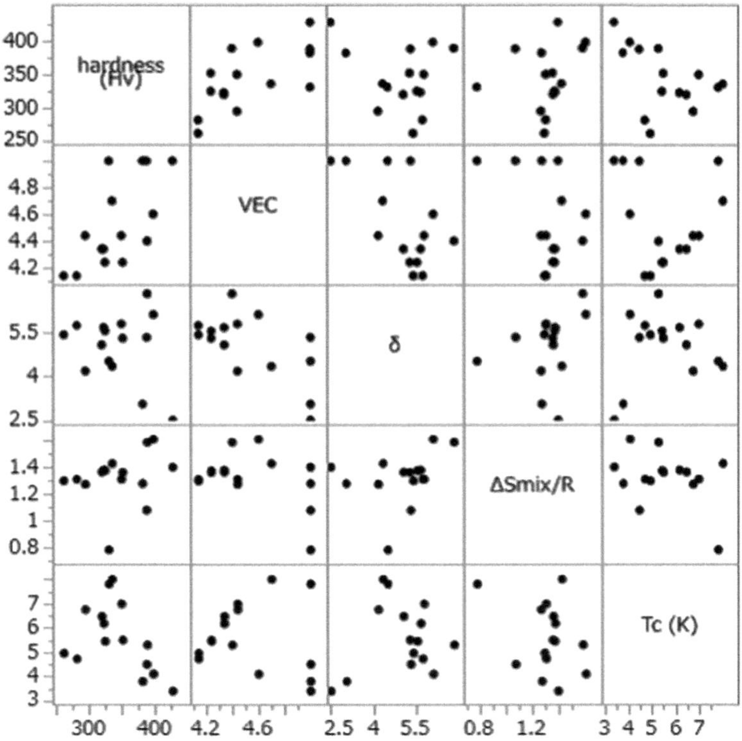

Fig. 3.13 Scatter matrix plot (5×5) showing correlation between five different features of bcc HEA superconductors

3.6 Simple Materials Design

The materials design of HEAs relies mainly on the mixing enthalpy, ΔS_{mix}, melting temperature, δ parameter, VEC, and electronegativity. For this purpose, machine learning approaches are promising. However, for experimental studies, a coarse and straightforward design with as few parameters as possible is preferable. Zhou et al. proposed a phase design of HEAs using a machine learning-guided approach [52]. They investigated the dependence of the design parameters on the phase selection of HEAs. The selected phases were amorphous, intermetallic, and solid solutions. The δ parameter was determined to be strongly correlated with the selection of the solid-solution phase [52]. Because herein, we are focusing on the superconductivity of a solid-solution phase, the VEC is a necessary design parameter. In the first part of this subsection, we present a simple materials design for bcc HEA superconductors based on VEC and δ, and their parameter dependences are compared with those of hcp HEA superconductors. The T_c, VEC, and δ data for hcp HEA superconductors are summarized in Table 3.4. The detailed materials design of hcp HEA superconductors is discussed in Chap. 7. Figure 3.14a, b show the δ and VEC dependence of T_c of for

Table 3.4 Superconducting critical temperature T_c, VEC, and δ parameter of hcp-related HEA superconductors reported in literature. Although simple hcp structure (Mg-type structure, $P6_3/mmc$) possesses only the $2c$ site, $Re_{0.56}Nb_{0.11}Ti_{0.11}Zr_{0.11}Hf_{0.11}$ and $Nb_{10+2x}Mo_{35-x}Ru_{35-x}Rh_{10}Pd_{10}$ ($0 \leq x \leq 5$) are multisite HEAs with the same space group. Therefore, $Re_{0.56}Nb_{0.11}Ti_{0.11}Zr_{0.11}Hf_{0.11}$ and $Nb_{10+2x}Mo_{35-x}Ru_{35-x}Rh_{10}Pd_{10}$ ($0 \leq x \leq 5$) are regarded as hcp-related HEA superconductors

HEA	T_c (K)	VEC	δ	Ref.
$Mo_{0.2375}Re_{0.2375}Ru_{0.2375}Rh_{0.2375}Ti_{0.05}$	3.6	7.33	1.99	[53]
$Mo_{0.225}Re_{0.225}Ru_{0.225}Rh_{0.225}Ti_{0.1}$	4.7	7.15	2.54	[53]
$Mo_{0.1}Re_{0.1}Ru_{0.55}Rh_{0.1}Ti_{0.15}$	2.1	7.2	3.14	[53]
$Mo_{0.105}Re_{0.105}Ru_{0.527}Rh_{0.105}Ti_{0.158}$	2.2	7.16	3.19	[53]
$Mo_{0.118}Re_{0.118}Ru_{0.47}Rh_{0.118}Ti_{0.176}$	2.5	7.06	3.29	[53]
$Re_{0.56}Nb_{0.11}Ti_{0.11}Zr_{0.11}Hf_{0.11}$	4.4	5.79	6.00	[54]
$Nb_{10}Mo_{35}Ru_{35}Rh_{10}Pd_{10}$	5.6	7.3	1.93	[55]
$Nb_{15}Mo_{32.5}Ru_{32.5}Rh_{10}Pd_{10}$	6.2	7.2	2.19	[55]
$Nb_{20}Mo_{30}Ru_{30}Rh_{10}Pd_{10}$	6.1	7.1	2.40	[55]
$Nb_5Mo_{35}Re_{15}Ru_{35}Rh_{10}$	7.5	7.1	1.57	[56]
$Nb_5Mo_{30}Re_{20}Ru_{35}Rh_{10}$	6.7	7.15	1.59	[56]
$Nb_5Mo_{25}Re_{25}Ru_{35}Rh_{10}$	6.5	7.2	1.61	[56]
$Nb_5Mo_{20}Re_{30}Ru_{35}Rh_{10}$	5.5	7.25	1.63	[56]
$(MoReRu)_{0.305}(PdPt)_{0.042}$	8.2	7.25	1.19	[51]
$(MoReRu)_{0.278}(PdPt)_{0.083}$	4.9	7.5	1.22	[51]
$(MoReRu)_{0.25}(PdPt)_{0.125}$	2.2	7.75	1.22	[51]
$(MoReRu)_{0.222}(PdPt)_{0.167}$	1.6	8.0	1.20	[51]

the bcc and hcp HEA superconductors, respectively. For bcc HEA superconductors, superconductivity appears at VECs = 4.1–5.1 and T_c tends to peak at approximately VEC = 4.67. However, although the number of samples of hcp-type HEAs is small, superconductivity emerges mainly when the VEC is in the range of 7–7.3. The superconductivity of the bcc-type material is robust against increasing δ, whereas the T_c of the hcp HEA superconductors decreases rapidly with increasing δ. These contrasting behaviors can be attributed to the atomic filling factor (68%) in the bcc structure being lower than that in the hcp structure (74%).

The elemental composition is an essential factor in the design of bcc HEA superconductors [28]. To demonstrate the importance of the elemental composition, the frequencies with which the elements are used in bcc HEA superconductors are shown on the periodic table in Fig. 3.15a. The primary sources are Hf, Zr, Ti, Ta, and Nb, each with a VEC = 4 or 5, because T_c forms a broad peak at VEC $\sim$ 4.67. A recent investigation of bcc HEA superconductors containing W explored VECs ranging from 4.8 to 5.11 [33, 35]. A bcc HEA superconductor with a VEC between 6 and 8 has still not been reported. Therefore, we must focus on the well-known relationships between the VEC and phase stability of bcc and fcc solid solutions of HEAs. Specifically, a single bcc phase has a VEC = 5.0$\sim$6.87, and a single fcc phase has

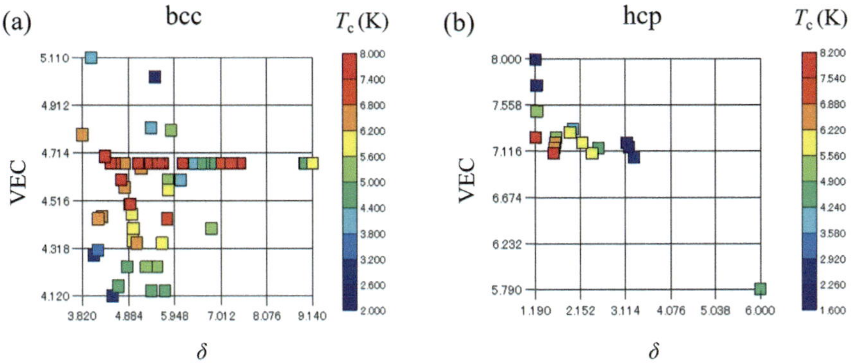

Fig. 3.14 δ and VEC dependence of T_c for **a** bcc HEA and (b) hcp-related HEA superconductors. Reproduced from [7] under the Creative Common Attribution License

a VEC > 8.0 [57]. Figure 3.15b shows the elements used for the hcp HEA superconductors. The constituent elements of the hcp HEA superconductors appear to be equally distributed over the VEC, in contrast with those of the bcc HEA superconductors. However, the frequencies of Mo, Ru, and Re are relatively large because many studies employ an hcp MoReRu superconductor with $T_c = 9.1$ K as the basic material. Moreover, the hcp elements Re and Ru are expected to play important roles in hcp stabilization. Al is also used in bcc HEA superconductors, and its addition might also be required to obtain additional hcp HEA superconductors.

According to the empirical Hume–Rothery electron concentration rule, the average number of itinerant electrons per atom, e/a, governs the structural stability of an alloy. This rule reflects the minimization of the kinetic energy of free electrons through the Fermi surface, constrained by the Brillouin zone. In contrast, the VECs considered in Figs. 3.14a, b provide information on the total DOS of the valence band integrated from the bottom up to a given energy. Thus, the VEC is a vital parameter for designing a superconductor. A possible usefulness of these two parameters is asssessed as a phase selection diagram based on the VEC and e/a, as shown in Fig. 3.16. The e/a values are calculated using the updated e/a value for each element listed in [58]. This figure shows the successful phase selection between bcc and hcp superconductors. The bcc and hcp HEA superconductors are located in the following regions: VEC = 4~4.8, e/a = 1.3~1.5 and VEC = 7~7.5, e/a = 1.1~1.25, respectively.

3.7 Summary

In this chapter, we introduced the materials research on bcc HEA superconductors. $Hf_{21}Nb_{25}Ti_{15}V_{15}Zr_{24}$ is a rare example maintaining a single bcc phase even after thermal annealing, whereas heat-treated HEAs frequently exhibit phase seg-

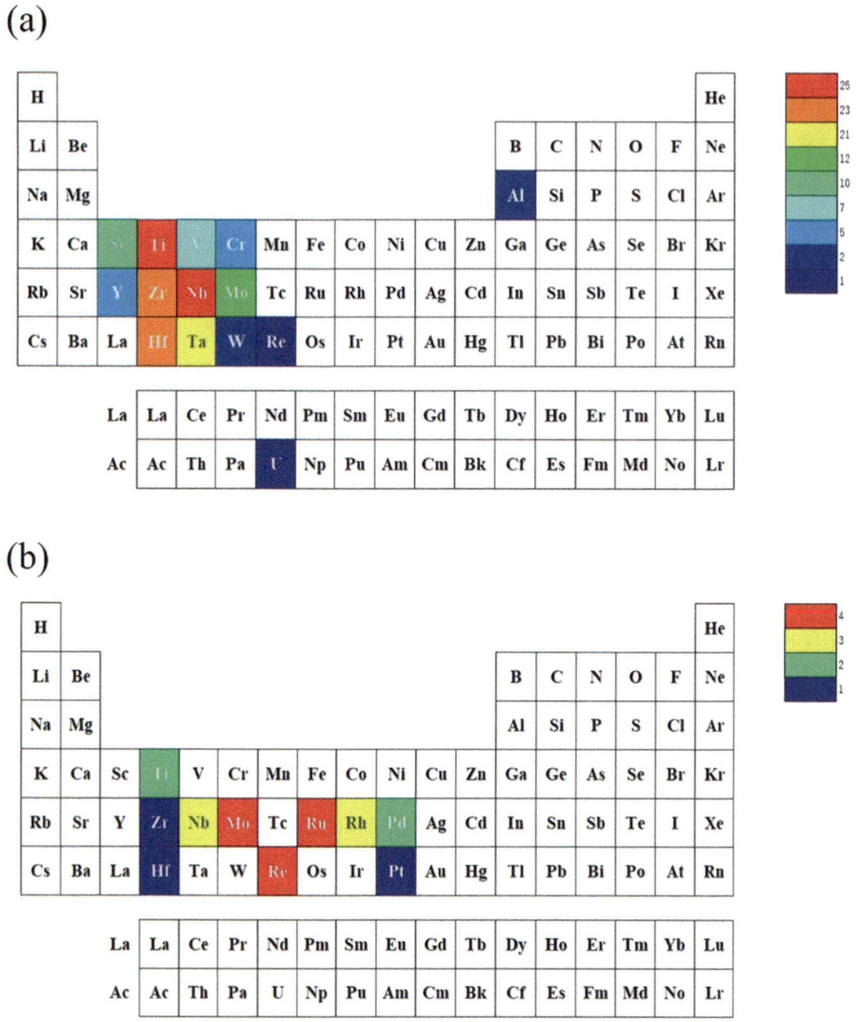

Fig. 3.15 Frequencies of use of elements in **a** bcc and **b** hcp HEA superconductors reported in literature thus far. The frequencies are counted for alloy systems with different combinations of elements listed in Tables 3.1 and 3.4. In **a**, the frequencies are 2 for Al, 10 for Sc, 25 for Ti, 7 for V, 5 for Cr, 5 for Y, 23 for Zr, 25 for Nb, 12 for Mo, 23 for Hf, 21 for Ta, 2 for W, 1 for Re, and 1 for U. In **b**, the frequencies are 2 for Ti, 1 for Zr, 3 for Nb, 4 for Mo, 4 for Ru, 3 for Rh, 2 for Pd, 1 for Hf, 4 for Re, and 1 for Pt. Reproduced from [7] under the Creative Common Attribution License

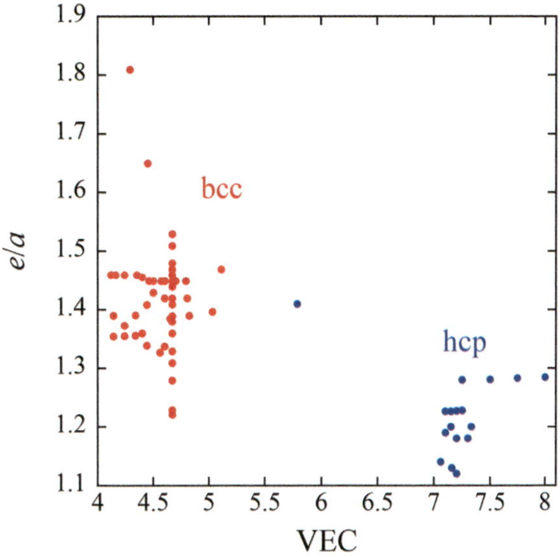

Fig. 3.16 Phase selection diagram of bcc and hcp HEA superconductors. Reproduced from [7] under the Creative Common Attribution License

regation with slightly different chemical compositions. An Al–Nb–Ti–V–Zr alloy system contains a gum metal with superconductivity. Gum-metal HEA superconductors possess excellent mechanical properties and good machinability, i.e., they are potential next-generation superconducting wire materials. $Al_5Nb_xTi_{35}V_5Zr_{55-x}$ and $Al_5Nb_xTi_{59-x}V_5Zr_{31}$ (x=14, 24, 34, 44) are superconductors with T_c ranging from 4.73 K to 6.98 K. A remarkable feature is the universal behavior of the VEC dependence of the T_c of Al–Nb–Ti–V–Zr and typical bcc HEA superconductors. Considering that some Al–Nb–Ti–V–Zr alloys are categorized as medium-entropy alloys, increasing the number of elements up to five can efficiently realize a high-entropy state. Equiatomic quinary bcc HEA superconductors are rare, and only three alloys have been reported. We discovered that HfMoNbTiZr is a new equiatomic bcc HEA superconductor; consequently, we could compare BCS superconducting properties under a fixed maximum ΔS_{mix}. The comparison revealed that the uncertainty principle concerning the energy and lifetime of phonons plays a vital role in determining T_c. A higher θ_D yields a broader phonon energy band and reduces the phonon lifetime based on the uncertainty principle. A shorter phonon lifetime leads to weakened electron–phonon coupling and a reduced T_c. The negative correlation between θ_D and T_c is contrary to our expectation that BCS superconductors show a positive correlation. The materials research has successfully revealed the enhanced influence of the uncertainty principle on a high-entropy state.

Hardness studies of HEA superconductors have been initiated and will be helpful for designing multifunctional superconductors and easily capturing phonon information. A negative correlation between the Vickers microhardness and T_c of bcc HEA superconductors was observed. A higher hardness implies a higher θ_D, which leads to a reduced T_c by applying the uncertainty principle to a larger-ΔS_{mix} alloy. This mechanism partially explains the negative correlation between the hardness and T_c. However, the fcc and hcp HEA superconductors exhibited inverse relationships, and the differences in their crystal structures probably need consideration.

Finally, in this chapter, a simple material design of HEA superconductors with bcc and hcp structures was discussed. The δ and VEC dependence of T_c of bcc HEAs showed the robustness of superconductivity against a significant mismatch between the atomic radii of their constituent elements, which probably originated from the lower atomic filling factor of a bcc structure than that of an hcp structure. The elemental compositions of the bcc and hcp structures also showed contrasting behavior; the elements in the bcc HEAs were mainly Hf, Zr, Ti, Ta, and Nb with VEC = 4 or 5 in bcc, whereas a relatively equal distribution of elements was observed for the hcp HEAs. We also focused on e/a, which is used in the Hume–Rothery rule to describe the structural stability based on the interaction between the Fermi surface and the Brillouin zone. A phase selection diagram based on the VEC and e/a for bcc and hcp HEA superconductors showed successful selection between them.

The effect of high entropy on superconductivity has not been sufficiently researched because HEAs possess a large compositional space and there are many unknown HEAs. Therefore, materials research is essential to further explore the remarkable properties. We expect that many material studies will be conducted and that novel phenomena will be discovered.

References

1. P. Sathiyamoorthi, H.S. Kim, High-entropy alloys with heterogeneous microstructure: Processing and mechanical properties. Prog. Mater. Sci. **123**, 100709 (2022)
2. X. Chang, M. Zeng, K. Liu, L. Fu, Phase engineering of high-entropy alloys. Adv. Mater. **32**, 1907226 (2020)
3. X. Wang, W. Guo, Y. Fu, High-entropy alloys: emerging materials for advanced functional applications. J. Mater. Chem. A **9**, 663–701 (2021)
4. F. Marques, M. Balcerzak, F. Winkelmann, G. Zepon, M. Felderhoff, Review and outlook on high-entropy alloys for hydrogen storage. Energy Environ. Sci. **14**, 5191–5227 (2021)
5. E.J. Pickering, A.W. Carruthers, P.J. Barron, S.C. Middleburgh, D.E.J. Armstrong, A.S. Gandy, High-entropy alloys for advanced nuclear applications. Entropy **23**, 98 (2021)
6. L. Sun, R.J. Cava, High-entropy alloy superconductors: Status, opportunities, and challenges. Phys. Rev. Mater. **3**, 090301 (2019)
7. J. Kitagawa, S. Hamamoto, N. Ishizu, Cutting edge of high-entropy alloy superconductors from the perspective of materials research. Metals **10**, 1078 (2020)
8. V. Chaudhary, R. Chaudhary, R. Banerjee, R.V. Ramanujan, Accelerated and conventional development of magnetic high entropy alloys. Mater. Today **49**, 231–252 (2021)
9. D. Castro, P. Jaeger, A.C. Baptista, J.P. Oliveira, An overview of high-entropy alloys as biomaterials. Metals **11**, 648 (2021)
10. S. Zherebtsov, N. Yurchenko, E. Panina, M. Tikhonovsky, N. Stepanov, Gum-like mechanical behavior of a partially ordered $Al_5Nb_{24}Ti_{40}V_5Zr_{26}$ high entropy alloy. Intermetallics **116**, 106652 (2020)
11. J. Kitagawa, M. Fukuda, S. Fukuda, K. Fujiki, Y. Nakamura, T. Nishizaki, Discovery of ferromagnetism in new multicomponent alloy Ti-Nb-Cr-Ru. APL Mater. **10**, 071101 (2022)

12. J. Guo, H. Wang, F. von Rohr, Z. Wang, S. Cai, Y. Zhou, K. Yang, A. Li, S. Jiang, Q. Wu, R.J. Cava, L. Sun, Robust zero resistance in a superconducting high-entropy alloy at pressures up to 190 GPa. Proc. Natl. Acad. Sci. **114**, 13144–13147 (2017)
13. M. Kasem, Y. Nakahira, H. Yamaoka, R. Matsumoto, A. Yamashita, H. Ishii, N. Hiraoka, Y. Takano, Y. Goto, Y. Mizuguchi, Robustness of superconductivity to external pressure in high-entropy-alloy-type metal telluride $AgInSnPbBiTe_5$. Sci. Rep. **12**, 7789 (2022)
14. D. Wu, K. Kusada, T. Yamamoto, T. Toriyama, S. Matsumura, S. Kawaguchi, Y. Kubota, H. Kitagawa, Platinum-group-metal high-entropy-alloy nanoparticles. J. Am. Chem. Soc. **142**, 13833–13838 (2020)
15. B. Jiang, Y. Yu, J. Cui, X. Liu, L. Xie, J. Liao, Q. Zhang, Y. Huang, S. Ning, B. Jia, B. Zhu, S. Bai, L. Chen, S.J. Pennycook, J. He, High-entropy-stabilized chalcogenides with high thermoelectric performance. Science **371**, 830–834 (2021)
16. S. Marik, D. Singh, B. Gonano, F. Veillon, D. Pelloquin, Y. Bréard, Enhanced magnetic frustration in a new high entropy diamond lattice spinel oxide. Scr. Mater. **186**, 366–369 (2020)
17. B. Yang, Y. Zhang, H. Pan, W. Si, Q. Zhang, Z. Shen, Y. Yu, S. Lan, F. Meng, Y. Liu, H. Huang, J. He, L. Gu, S. Zhang, L.-Q. Chen, J. Zhu, C.-W. Nan, Y.-H. Lin, High-entropy enhanced capacitive energy storage. Nat. Mater. **21**, 1074–1080 (2022)
18. R. Sogabe, Y. Goto, T. Abe, C. Moriyoshi, Y. Kuroiwa, A. Miura, K. Tadanaga, Y. Mizuguchi, Improvement of superconducting properties by high mixing entropy at blocking layers in BiS_2-based superconductor $REO_{0.5}F_{0.5}BiS_2$. Solid State Commun. **295**, 43-49 (2019)
19. B. Liu, J. Wu, Y. Cui, Q. Zhu, G. Xiao, S. Wu, G.-H. Cao, Z. Ren, Superconductivity and paramagnetism in Cr-containing tetragonal high-entropy alloys. J. Alloys Compd. **869**, 159293 (2021)
20. Y. Ge, S. Ma, K. Bao, Q. Tao, X. Zhao, X. Feng, L. Li, B. Liu, P. Zhu, T. Cui, Superconductivity with high hardness in Mo_3C_2. Inorg. Chem. Front. **6**, 1282–1288 (2019)
21. Y. Cui, J. Wu, B. Liu, Q. Zhu, G. Xiao, S. Wu, G. Cao, Z. Ren, Superconductivity and high hardness in metal-rich carbides $MoRe_2C$ and WRe_2C. J. Alloys Compd. **856**, 157314 (2021)
22. N. Ishizu, J. Kitagawa, New high-entropy alloy superconductor $Hf_{21}Nb_{25}Ti_{15}V_{15}Zr_{24}$. Res. Phys. **13**, 102275 (2019)
23. Y. Harayama, J. Kitagawa, Superconductivity in Al-Nb-Ti-V-Zr multicomponent alloy. J. Supercond. Nov. Magn. **34**, 2787–2794 (2021)
24. J. Kitagawa, K. Hoshi, Y. Kawasaki, R. Koga, Y. Mizuguchi, T. Nishizaki, Superconductivity and hardness of the equiatomic high-entropy alloy HfMoNbTiZr. J. Alloys Compd. **924**, 166473 (2022)
25. P. Koželj, S. Vrtnik, A. Jelen, S. Jazbec, Z. Jagličić, S. Maiti, M. Feuerbacher, W. Steurer, J. Dolinšek, Discovery of a superconducting high-entropy alloy. Phys. Rev. Lett. **113**, 107001 (2014)
26. S. Vrtnik, P. Koželj, A. Meden, S. Maiti, W. Steurer, M. Feuerbacher, J. Dolinšek, Superconductivity in thermally annealed Ta-Nb-Hf-Zr-Ti high-entropy alloys. J. Alloys Compd. **695**, 3530–3540 (2017)
27. F. von Rohr, M.J. Winiarski, J. Tao, T. Klimczuk, R.J. Cava, Effect of electron count and chemical complexity in the Ta-Nb-Hf-Zr-Ti high-entropy alloy superconductor. Proc. Natl. Acad. Sci. USA **113**, E7144–E7150 (2016)
28. F. von Rohr, R.J. Cava, Isoelectronic substitutions and aluminium alloying in the Ta-Nb-Hf-Zr-Ti high-entropy alloy superconductor. Phys. Rev. Mater. **2**, 034801 (2018)
29. G. Kim, M.-H. Lee, J.H. Yun, P. Rawat, S.-G. Jung, W. Choi, T.-S. You, S.J. Kim, J.-S. Rhyee, Strongly correlated and strongly coupled s-wave superconductivity of the high entropy alloy $Ta_{1/6}Nb_{2/6}Hf_{1/6}Zr_{1/6}Ti_{1/6}$ compound. Acta Mater. **186**, 250–256 (2020)
30. W.L. Nelson, A.T. Chemey, M. Hertz, E. Choi, D.E. Graf, S. Latturner, T.E. Albrecht-Schmitt, K. Wei, R.E. Baumbach, Superconductivity in a uranium containing high entropy alloy. Sci. Rep. **10**, 4717 (2020)
31. S. Marik, M. Varghese, K.P. Sajilesh, D. Singh, R.P. Singh, Superconductivity in equimolar Nb-Re-Hf-Zr-Ti high entropy alloy. J. Alloys Compd. **769**, 1059–1063 (2018)

32. N.K. Sarkar, C.L. Prajapat, P.S. Ghosh, N. Garg, P.D. Babu, S. Wajhal, P.S.R. Krishna, M.R. Gonal, R. Tewari, P.K. Mishra, Investigations on superconductivity in an equi-atomic disordered Hf-Nb-Ta-Ti-V high entropy alloy. Intermetallics **144**, 107503 (2022)
33. P. Sobota, R. Topolnicki, T. Ossowski, T. Pikula, A. Pikul, R. Idczak, Superconductivity in the high entropy alloy $(NbTa)_{0.67}(MoHfW)_{0.33}$. Phys. Rev. B **106**, 184512 (2022)
34. X. Zhang, N. Winter, C. Witteveen, T. Moehl, Y. Xiao, F. Krogh, A. Schilling, F. von Rohr, Preparation and characterization of high-entropy alloy $(TaNb)_{1-x}(ZrHfTi)_x$ superconducting films. Phys. Rev. Res. **2**, 013375 (2020)
35. R. Shu, X. Zhang, S. Rao, A. Febvrier, P. Eklund, Effects of alloying and deposition temperature on phase formation and superconducting properties of TiZrTaNb-based high entropy-alloy films. Appl. Phys. Lett. **120**, 151901 (2022)
36. V. Pacheco, G. Lindwall, D. Karlsson, J. Cedervall, S. Fritze, G. Ek, P. Berastegui, M. Sahlberg, U. Jansson, Thermal stability of the HfNbTiVZr high-entropy alloy. Inorg. Chem. **58**, 811–820 (2019)
37. T. Saito, T. Furuta, J.-H. Hwang, S. Kuramoto, K. Nishino, N. Suzuki, R. Chen, A. Yamada, K. Ito, Y. Seno, T. Nonaka, H. Ikehata, N. Nagasako, C. Iwamoto, Y. Ikuhara, T. Sakuma, Multifunctional alloys obtained via a dislocation free plastic deformation mechanism. Science **300**, 464–467 (2003)
38. F. Otto, Y. Yang, H. Bei, E.P. George, Relative effects of enthalpy and entropy on the phase stability of equiatomic high-entropy alloys. Acta Mater. **61**, 2628–2638 (2013)
39. B.W. Roberts, Survey of superconductive materials and critical evaluation of selected properties. J. Phys. Chem. Ref. Data **5**, 581–821 (1976)
40. P.H. Pan, D.K. Finnemore, A.J. Bevolo, H.R. Shanks, B.J. Beaudry, F.A. Schmidt, G.C. Danielson, Heat capacity of high-purity lanthanum. Phys. Rev. B **21**, 2809–2814 (1980)
41. H. Akai, Electronic structure Ni-Pd alloys calculated by the self-consistent KKR-CPA method. J. Phys. Soc. Jpn. **51**, 468–474 (1982)
42. F. Körmann, Y. Ikeda, B. Grabowski, M.H.F. Sluiter, Phonon broadening in high entropy alloys. Npj Comput. Mater. **3**, 36 (2017)
43. D.B. Miracle, O.N. Senkov, A critical review of high entropy alloys and related concepts. Acta Mater. **122**, 448–511 (2017)
44. F. Tian, L.K. Varga, N. Chen, J. Shen, L. Vitos, Empirical design of single phase high-entropy alloys with high hardness. Intermetallics **58**, 1–6 (2015)
45. Z.D. Han, N. Chen, S.F. Zhao, L.W. Fan, G.N. Yang, Y. Shao, K.F. Yao, Effect of Ti additions on mechanical properties of NbMoTaW and VNbMoTaW refractory high entropy alloys. Intermetallics **84**, 153–157 (2017)
46. Q. Li, H. Zhang, D. Li, Z. Chen, S. Huang, Z. Lu, H. Yan, W_xNbMoTa refractory high-entropy alloys fabricated by laser cladding deposition. Materials **12**, 533 (2019)
47. U. Bhandari, C. Zhang, C. Zeng, S. Guo, S. Yang, Computational and experimental investigation of refractory high entropy alloy $Mo_{15}Nb_{20}Re_{15}Ta_{30}W_{20}$. J. Mater. Res. Technol. **9**, 8929–8936 (2020)
48. S. Ge, H. Fu, L. Zhang, H. Mao, H. Li, A. Wang, W. Li, H. Zhang, Effects of Al addition on the microstructures and properties of MoNbTaTiV refractory high entropy alloy. Mater. Sci. Eng. A **784**, 139275 (2020)
49. J. Kitagawa, Magnetic properties, electrical resistivity, and hardness of high-entropy alloys FeCoNiPd and FeCoNiPt. J. Magn. Magn. Mater. **563**, 170024 (2022)
50. U. Bhandari, C. Zhang, C. Zeng, S. Guo, A. Adhikari, S. Yang, Deep learning-based hardness prediction of novel refractory high-entropy alloys with experimental validation. Crystals **11**, 46 (2021)
51. Q. Zhu, G. Xiao, Y. Cui, W. Yang, S. Song, G.-H. Cao, Z. Ren, Structural transformation and superconductivity in carbon-added hexagonal high-entropy alloys. J. Alloys Compd. **909**, 164700 (2022)
52. Z. Zhou, Y. Zhou, Q. He, Z. Ding, F. Li, Y. Yang, Machine learning guided appraisal and exploration of phase design for high entropy alloys. npj Comput. Mater. **5**, 128 (2019)

53. Y.-S. Lee, R.J. Cava, Superconductivity in high and medium entropy alloys based on MoReRu. Phys. C **566**, 1353520 (2019)
54. S. Marik, K. Motla, M. Varghese, K.P. Sajilesh, D. Singh, Y. Breard, P. Boullay, R.P. Singh, Superconductivity in a new hexagonal high-entropy alloy. Phys. Rev. Mater. **3**, 060602(R) (2019)
55. B. Liu, J. Wu, Y. Cui, Q. Zhu, G. Xiao, S. Wu, G. Cao, Z. Ren, Superconductivity in hexagonal Nb-Mo-Ru-Rh-Pd high-entropy alloys. Scr. Mater. **182**, 109–113 (2020)
56. B. Liu, J. Wu, Y. Cui, Q. Zhu, G. Xiao, S. Wu, G. Cao, Z. Ren, Structural evolution and superconductivity tuned by valence electron concentration in the Nb-Mo-Re-Ru-Rh high-entropy alloys. J. Mater. Sci. Technol. **85**, 11–17 (2021)
57. M.C. Gao, J.-W. Yeh, P.K. Liaw, Y. Zhang, *High-Entropy Alloys: Fundamentals and Applications* (Springer, Cham, Switzerland, 2015)
58. U. Mizutani, H. Sato, The physics of the Hume-Rothery electron concentration rule. Crystals **7**, 9 (2017)

Chapter 4
Superconductivity in the TiZrNbSn System

T. Nawaz, W. Abuzaid, and M. Egilmez

Abstract Modern materials science places a strong emphasis on exploring advanced materials such as high-entropy alloys due to their rich functionalities, such as high yield strength, cryogenic performance and shape memory properties. In particular, shape memory alloys (SMAs) have useful and distinct properties that make them appropriate for a wide range of applications in the aerospace, automobile, and biomedical sectors. The TiNbZrSn SMAs were designed primarily for biomedical applications due to their better biocompatibility over commonly available and extensively researched NiTi-based SMAs. While previous research has focused mostly on metallurgical and mechanical properties, this chapter focuses on the simultaneous occurrence of superelasticity and superconductivity within the same material at different temperatures within the TiZrNbSn system. We present an overview of the magnetic, electrical, and mechanical properties of single β-phase BCC-structured TiZrNbSn SMAs. Here, we showed that TiZrNbSn SMAs exhibit both superelasticity and superconductivity at different temperatures.

4.1 Introduction

High-entropy alloys (HEAs) and medium-entropy alloys (MEAs) have attracted great scientific and industrial interest during the last two decades due to their unique functionalities and rich compositional tuning potential [1, 2]. The term multiprincipal element alloys (MPEAs) is frequently used to describe this class of materials, which are typically developed to have 3–6 principal elements. The use of such a large number

T. Nawaz · M. Egilmez (✉)
Materials Science and Engineering Program, American University of Sharjah, 26666 Sharjah, United Arab Emirates
e-mail: megilmez@aus.edu

T. Nawaz · W. Abuzaid
Department of Mechanical Engineering, American University of Sharjah, 26666 Sharjah, United Arab Emirates

T. Nawaz · W. Abuzaid · M. Egilmez
Department of Physics, American University of Sharjah, 26666 Sharjah, United Arab Emirates

J. Kitagawa and Y. Mizuguchi (eds.), *High-Entropy Alloy Superconductors*, Springer Series in Solid-State Sciences 202, https://doi.org/10.1007/978-981-97-4129-8_4

of elements, in large percentages, leads to a high degree of disorder and hence a very large configurational entropy. This alloying methodology is in sharp contrast with the traditional alloying routine in which small amounts of secondary elements are added to a single principal element. However, it has a number of advantages, such as low- and high-temperature stability, high strength and hardness, superior corrosion resistance, shape memory properties, and tuneable magnetic and electrical functionalities, including superconductivity [3–5]. More specifically, such a new alloying technique may offer great potential for bulk superconductivity applications where mechanical functionalities such as shape memory effects, superelasticity, and high strength are desirable (e.g., high current-carrying wires used in bulk superconductors) [6].

Shape memory materials (SMMs) are a class of materials that can regain their initial shape after undergoing considerable deformation whenever a specific stimulus such as heat or magnetic field is applied [7]. In SMMs, properties such as superelasticity (in alloys) and viscoelasticity (in polymers) can be observed under certain conditions that are often optimized through complex thermomechanical treatments [8]. Shape memory alloys (SMAs) are of particular interest to many industries due to their high load-carrying capacity and functionality over a wide temperature range, including cryogenic temperatures [9]. The unique properties of SMAs are attributed to a reversible martensitic phase transition from high-symmetry austenite to a low-symmetry martensite phase under the influence of temperature or a stress field, resulting in a shape memory effect (SME) and superelasticity, respectively [10, 11]. Many applications across various industries have been developed capitalizing on the unique properties of SMAs.

Over the last decade, significant attention has been given to developing new superconducting materials using the multiprincipal element alloying technique. Superconducting transition temperatures (T_c) up to 11 K have been reported in these high-entropy alloy systems. Note that such a large transition temperature justifies further research on this group of materials for potential high strength materials requiring bulk applications. For example, Kim et al. [12] synthesized $Ta_{1/6}Nb_{2/6}Hf_{1/6}Zr_{1/6}Ti_{1/6}$ by spark plasma sintering to study the critical current density and vortex pinning. The important aspects of this work are further explained in Chap. 2 of this book. Similarly, the mechanical and transport properties of body-centered cubic HEAs based on AlNbTiVZr [13], HfMoNbTiZr [14], HfNbTiVZr [15], etc., have also been reported in the literature and are explained in Chap. 3 of this book. Moreover, the phonon-mediated superconductivity of TaNbHfZrTi and HEAs with hexagonal closed-packed crystal structures [16–20] are reported to exhibit type II superconductivity due to the weak electron–phonon coupling limit. These materials are reviewed in Chaps. 5, 6, and 7 of this book. Furthermore, chalcogenide-based HEAs with NaCl-type crystal structures have been studied for their superconducting properties, and the effects of multisite alloying on ΔS_{mix} and changing the composition of the alloyed materials on the carrier concentration of HEAs have been reported [21–24]. These HEAs are discussed in Chap. 8 of this book. Additionally, the literature reports special types of HEAs, such as zirconides [25–28] (further discussed in Chap. 10), unconventional superconductors, such as BiS_2, and rare earth metal-based superconductors [29, 30] (discussed in Chap. 11).

Table 4.1 presents a comprehensive review of the superconducting properties of high-entropy alloy systems discovered during the last 10 years.

The origins of superconductivity and important characteristic properties relevant to various applications have been widely investigated over the last few years, as summarized in Table 4.1. In particular, many of the related studies concentrated on the dependence of the valance electron concentration on the superconducting critical temperature. Note that Matthias was the first to reveal the relationship between the valence electron concentration and the superconducting transition temperature, and the highest T_c was revealed for valence electron concentrations between 5 and 7 [62]. However, as shown in Table 4.1, for high-entropy alloy superconductors, there is no clear relation between the valence electron concentration and the superconducting transition temperature. As we will discuss later, microscopic techniques such as muon spin resonance or scanning tunneling spectroscopy are essential for understanding the pairing mechanism in this new group of materials. Table 4.1 also summarizes the crystal structure upper critical field (H_{c2}), coherence length (ξ), and superconducting energy gap (Δ) values for this group of materials. The reported values for such parameters along with T_c place this group of materials as promising materials mostly for bulk applications of superconductivity. Although there are multiple concentrations of high-entropy alloys, in this chapter, we concentrate on the TiZrNbSn-based multi-principal element alloy system, which, in addition to being biocompatible, exhibits desirable superelastic properties.

4.2 Fabrication and Structural Analysis

The prescribed ratio of high purity (99.5 wt.%). Ti, Zr, Nb, and Sn are arc-melted in an inert atmosphere multiple times. Following casting, the ingot was homogenized at 1100 °C for 24 h in a vacuum furnace. Strips of TiZrNbSn were cut from the ingot and cold rolled to achieve a thickness reduction between 91 and 96.8%. Samples heat treated between 300 and 950 °C for 30 min followed by quenching in water were used in this work.

X-ray diffraction measurements obtained with CuK-α radiation are shown in Fig. 4.1a, where all the visible peaks are attributed to the body-centered cubic structure or β-phase (BCC) phase with a lattice parameter of 0.338 (0.001) nm. The XRD peaks, especially at higher angles, are very broad and are generally indicative of a distorted lattice. Note that the distortion originated from the different atomic radii of the constituent elements: the atomic radii mismatch between Zr and Ti is quite significant at approximately 9%. The BCC structure was further confirmed via transmission electron microscopy (TEM) analysis through the selective electron diffraction patterns shown in Fig. 4.1b. Electron dispersive X-ray (EDX) analysis was carried out to determine the elemental distribution at the surface of the specimen. Area scans of the elemental distribution revealed very uniform spreading of the elements over large surfaces. This is indicative of the uniformity of the entire specimen (Fig. 4.1c). The nominal atomic and weight compositions of the actual specimen

Table 4.1 Superconducting properties of MPEA systems discovered during the last 10 years. T_c: superconducting transition temperature, H_{c2}: upper critical field, VEC: valence electron concentration, ξ: coherence length, ΔS_{mix}: entropy of mixing

Sr. no.	High Entropy Alloy	Crystal Structure	T_c (K)	$H_{c2}(0)$ (kOe)	VEC	ξ (nm)	Band gap	ΔS_{mix}	References
1	$Ta_{34}Nb_{33}Hf_8Zr_{14}Ti_{11}$	BCC	7.2	82	4.67		2.2 meV	–	[16]
2	$Hf_{20}Nb_{20}Ta_{20}Ti_{20}V_{20}$	BCC	5	66.3	4.6		1.41 meV	1.609R	[31]
3	$Ta_{1/6}Nb_{2/6}Hf_{1/6}Zr_{1/6}Ti_{1/6}$	BCC	7.8	102	4.5	5.66	–	–	[12]
4	$Nb_{10+2x}Mo_{35-x}Ru_{35-x}Rh_{10}Pd_{10}$ $(0 \leq x \leq 5)$	Hexagonal			7.1–7.3				[17]
	x = 0		5.58	69	7.1–7.3	6.9	–	–	
	x = 2.5		6.19	81	7.1–7.3	6.4	–	–	
	x = 5		6.1	83	7.1–7.3	6.3	–	–	
5	$Cr_{5+x}Mo_{35-x}W_{12}Re_{35}Ru_{13}C_{20}$ $(0 \leq x \leq 9)$	Noncentrosymmetric cubic β-Mn type							[32]
	x = 0		5.49	97	–	5.8	$\Delta(0)/k_B T_C = 1.88$	1.40R	
	x = 3		4.73	84	–	6.3	$\Delta(0)/k_B T_C = 1.88$	1.45R	
	x = 6		3.83	73	–	6.7	–	1.49R	
	x = 9		3.35	53	–	7.9	–	1.51R	
6	$(TiZrNbTa)_{1-x}W_x$ and $(TiZrNbTa)_{1-x}V_x$	BCC			5.7–6.3				[33]
	$(TiZrNbTa)_{1-x}W_x$								
	x = 0		8	–	5.7–6.3	–	–	–	
	x = 0.1		6.2	–	5.7–6.3	–	–	–	
	x = 0.16		4.9	–	5.7–6.3	–	–	–	

(continued)

Table 4.1 (continued)

Sr. no.	High Entropy Alloy	Crystal Structure	T_c (K)	$H_{c2}(0)$ (kOe)	VEC	ξ (nm)	Band gap	ΔS_{mix}	References
	x = 0.21		4.1	–	5.7–6.3	–	–	–	
	x = 0.29		3	–	5.7–6.3	–	–	–	
	x = 0.35		2	–	5.7–6.3	–	–	–	
	$(TiZrNbTa)_{1-x}V_x$								
	x = 0.11		6	–	5.7–6.3	–	–	–	
	x = 0.21		5	–	5.7–6.3	–	–	–	
	x = 0.33		4.4	–	5.7–6.3	–	–	–	
7	$TrZr_2$	Tetragonal $CuAl_2$							[28]
	$CoZr_2$		5.5	–	–	–	–	0	
	$Co_{0.3}Ni_{0.4}Rh_{0.3}$		4.3	–	–	–	–	1.08R	
	$Fe_{0.2}Co_{0.3}Ni_{0.2}Rh_{0.3}$		4.9	–	–	–	–	1.36R	
	$Fe_{0.2}Co_{0.2}Ni_{0.2}Rh_{0.2}Ir_{0.2}$		4.7	–	–	–	–	1.6R	
	$Fe_{0.1}Co_{0.2}Ni_{0.2}Cu_{0.1}Rh_{0.2}Ir_{0.2}$		5	–	–	–	–	1.76R	
8	(Fe, Co, Ni, Cu, Ga)Zr_2	Tetragonal $CuAl_2$-type	2.9	15 @ $T_{c(zero)}$ 35 @ $T_{c(onset)}$	–	–	–	1.6R	[26]
9	$Mo_{0.11}W_{0.11}V_{0.11}Re_{0.34}B_{0.33}$	Tetragonal $CuAl_2$	4	19	5.24	13.1	$\Delta(0)/T_c =$ 1.81	–	[34]
10	$(TaNb)_{0.7}(HfZrTi)_{0.5}$	BCC	7.2	100.7	–	5.72	–	–	[35]
11	$Ta_5[Mo_{35-x}W_{5+x}]Re_{35}Ru_{20}$	Tetragonal			6.62–6.7		$\Delta(0)/T_c =$ 1.88	–	[36]

(continued)

Table 4.1 (continued)

Sr. no.	High Entropy Alloy	Crystal Structure	T_c (K)	$H_{c2}(0)$ (kOe)	VEC	ξ (nm)	Band gap	ΔS_{mix}	References
	x = y = 0		6.29	–	6.62–6.7	–	–	–	
	x = 5		6.2	–	6.62–6.7	–	–	–	
	x = 10		6.1	–	6.62–6.7	–	–	–	
	x = 15		5.69	–	6.62–6.7	–	–	–	
	x = 20		5.46	–	6.62–6.7	–	–	–	
	x = 25		5.45	–	6.62–6.7	–	–	–	
	x = 30		4.78	–	6.62–6.7	–	–	–	
	$[Ta_{5+y}Mo_{35-y}]W_5Re_{35}Ru_{20}$	Tetragonal			6.62–6.7		$\Delta(0)/T_c$ = 1.88		
	y = 2		6.13	–	6.62–6.7	–	–	–	
	y = 4		5.7	–	6.62–6.7	–	–	–	
	y = 6		5.32	–	6.62–6.7	–	–	–	
	y = 8		5.27	–	6.62–6.7	–	–	–	
12	$(Ag, In, Pb, Bi)Te_{1-x}Se_x$ (x = 0.0, 0.25, and 0.5)	NaCl-type							[21]
	x = 0		2.7	18	–	–	–	1.37R	
	x = 0.25		2.5	19	–	–	–	1.89R	
	x = 0.5		2	–	–	–	–	2R	
13	$V_{5+2x}Nb_{35-x}Mo_{35-x}Ir_{10}Pt_{15}$ ($0 \le x \le 10$)	A15-type							[37]
	x = 0		5.18	64	6.4–6.5	7.2	–	–	
	x = 5		4.49	57	6.4–6.5	7.6	–	–	

(continued)

Table 4.1 (continued)

Sr. no.	High Entropy Alloy	Crystal Structure	T_c (K)	$H_{c2}(0)$ (kOe)	VEC	ξ (nm)	Band gap	ΔS_{mix}	References
	x = 10		3.61	44	6.4–6.5	8.7	–	–	
14	$AgInSnPbBiTe_5$	NaCl-type	2.6	19	–	–	–	–	[22]
15	$(HfNb)_{0.10}(MoReRu)_{0.90}$	α-Mn structure	5.2	94	–	5.92	$\Delta(0)/k_B T_c = 1.85$	–	[38]
	$(ZrNb)_{0.10}(MoReRu)_{0.90}$		5.5	115	–	5.35	$\Delta(0)/k_B T_c = 1.82$	–	
16	$Nb_{60}Re_{10}Zr_{10}Hf_{10}Ti_{10}$	BCC	5.6	63.1	–	7.2	$\Delta(0)/k_B T_C = 2.57$	–	[39]
17	$Ta_{1/6}Nb_{2/6}Hf_{1/6}Zr_{1/6}Ti_{1/6}$	BCC	7.9	48.5	–	–	1.37 meV $2\Delta(0)/k_B T_c = 4.03$	1.56R	[40]
18	$(NbTa)_{0.67}(MoHfW)_{0.33}$	BCC	4.3	14.5	–	15.55	$2\Delta 0/k_B T_c = 2.4$	55 mJ K^{-1} mol^{-1}	[41]
19	$(TaNb)_{1-x}(ZrHfTi)_x$	Pseudo-BCC							[42]
	x = 0.43		4.61	104.4	4.24	–	–	1.55R	
	x = 0.54		5.6	110.5	4.35	–	–	1.602R	
	x = 0.65		6.14	99.3	4.46	–	–	1.604R	
	x = 0.76		6.76	87.7	4.57	–	–	1.55R	
20	$(Tr)Zr_2$	$CuAl_2$-type tetragonal							[27]
	$(Fe_{0.1}Co_{0.2}Ni_{0.1}Rh_{0.3}Ir_{0.3})$ Zr_2		7.8	–	–	–	–	1.50R	
	$(Fe_{0.1}Co_{0.3}Ni_{0.2}Rh_{0.1}Ir_{0.3})$ Zr_2		6.7	–	–	–	–	1.48R	

(continued)

Table 4.1 (continued)

Sr. no.	High Entropy Alloy	Crystal Structure	T_c (K)	$H_{c2}(0)$ (kOe)	VEC	ξ (nm)	Band gap	ΔS_{mix}	References
	$(Fe_{0.2}Co_{0.2}Ni_{0.2}Rh_{0.2}Ir_{0.2})\ Zr_2$		5.4	–	–	–	–	1.61R	
	$(Fe_{0.3}Co_{0.2}Ni_{0.3}Rh_{0.1}Ir_{0.1})\ Zr_2$		4.8	–	–	–	–	1.52R	
21	$Al_5Nb_{24}Ti_{40}V_5Zr_{26}$	BCC	5.5	–	4.24	–	–	1.35R	[13]
	$Al_5Nb_{14}Ti_{35}V_5Zr_{41}$		4.73	–	4.14	–	–	1.30R	
	$Al_5Nb_{24}Ti_{35}V_5Zr_{31}$		5.45	–	4.24	–	–	1.37R	
	$Al_5Nb_{34}Ti_{35}V_5Zr_{21}$		6.47	–	4.34	–	–	1.36R	
	$Al_5Nb_{34}Ti_{35}V_5Zr_{21}$		6.75	–	4.44	–	–	1.27R	
	$Al_5Nb_{14}Ti_{45}V_5Zr_{31}$		4.96	–	4.14	–	–	1.29R	
	$Al_5Nb_{14}Ti_{45}V_5Zr_{31}$		6.18	–	4.34	–	–	1.37R	
	$Al_5Nb_{14}Ti_{45}V_5Zr_{31}$		6.98	–	4.44	–	–	1.30R	
22	$(V_{0.5}Nb_{0.5})_{3-x}Mo_xAl_{0.5}Ga_{0.5}$	A15-type BCCs			4.55–4.85				[43]
	x = 0.2		10.2	201	4.55–4.85	4.0	$\Delta(0)/T_c =$ 2.07	–	
	x = 0.4		9.2	177	4.55–4.85	4.3	–	–	
	x = 0.6		8.9	170	4.55–4.85	4.4	–	–	
	x = 0.8		7.9	142	4.55–4.85	4.8	–	–	
	x = 1		6.1	99	4.55–4.85	5.8	–	–	
	x = 1.2		4.8	76	4.55–4.85	6.6	–	–	
	x = 1.4		3.2	48	4.55–4.85	8.3	–	–	
23	$(MoReRu)_{(1-2x)/3}(PdPt)_x$								[18]

(continued)

Table 4.1 (continued)

Sr. no.	High Entropy Alloy	Crystal Structure	T_c (K)	$H_{c2}(0)$ (kOe)	VEC	ξ (nm)	Band gap	ΔS_{mix}	References
	$x = 0.042$	HCP	8.17		7.25		–	1.2R–1.6R	
	$x = 0.083$		4.9		7.5		–	–	
	$x = 0.125$		2.2		7.75		–	–	
	$x = 0.167$		1.64		8		–	–	
	$x = 0.333$	FCC	–		9		–	1.5R–1.8R	
	$(MoReRu)_{(1-2x)/3}(PdPt)_xC_y$	FCC						1.5R–1.8R	
	$x = 0.042, y = 0.3$		2.46	–	6.5		–	–	
	$x = 0.042, y = 0.44$		2.72	–	6.26		–	–	
	$x = 0.083, y = 0.27$		2.28	–	6.75		–	–	
	$x = 0.083, y = 0.4$		2.3	–	6.5		–	–	
	$x = 0.125, y = 0.22$		1.86	–	7.07		–	–	
	$x = 0.125, y = 0.31$		1.91	–	6.86		–	–	
	$x = 0.167, y = 0.25$		1.87	–	7.2		–	–	
	$x = 0.167, y = 0.3$		1.75	–	7.08		–	–	
24	$(TaNb)_{1-x}(ZrHfTi)_x$ $(0.22 \leq x \leq 0.75)$	Amorphous and BCC nanocrystaline phase	5 @ x = 0.69	109 @ x = 0.69	4.36–4.54	–	–	–	[44]
25	HfMoNbTiZr ($Hf_{21.0}Mo_{19.8}Nb_{20.0}Ti_{19.1}Zr_{20.1}$)	BCC	4.1	55.5	4.6	7.7	$\Delta(0) =$ 0.62 meV	–	[14]
26	$(AgSnPbBi)_{(1-x)/4}In_xTe$	NaCl-type							[23]
	$x = 0.1$		2	–	–	–	–	1.57R	
	$x = 0.2$		2.4	–	–	–	–	1.61R	

(continued)

Table 4.1 (continued)

Sr. no.	High Entropy Alloy	Crystal Structure	T_c (K)	$H_{c2}(0)$ (kOe)	VEC	ξ (nm)	Band gap	ΔS_{mix}	References
	x = 0.3		2.7	–	–	–	–	1.57R	
	x = 0.4		2.7	–	–	–	–	1.51R	
27	Sc-Hf-Nb-Ta-Ti-Zr system	BCC and HCP							[45]
	ScHfNbTi		6.2	131	–	–	–	–	
	ScHfNbZr		6	111	–	–	–	–	
	ScHfTaTi		6	147	–	–	–	–	
	ScNbTiZr		7.3	137	–	–	–	–	
	ScTaTiZr		6.8	135	–	–	–	–	
	ScHfNbTaTi		6.1	131	–	–	–	–	
	ScHfNbTaZr		7.2	124	–	–	–	–	
	ScNbTaTiZr		7.2	173	–	–	–	–	
	ScHfNbTaTiZr		6.6	141	–	–	–	–	
28	$(TaNb)_{1-x}(ZrHfTi)_x$	BCC							[46]
	x = 0.35		6.7	81.8	–	6.3	–	−1.53R	
	x = 0.71		4.3	71.9	–	6.8	–	−1.56R	
29	$Co_{0.2}Ni_{0.1}Cu_{0.1}Rh_{0.3}Ir_{0.3}Zr_2$	Tetragonal $CuAl_2$-type	8	120	–	–	–	1.47R	[25]
30	$[TaNb]_{0.31}(TiUHf)_{0.69}$	BCC	3.2	64	4.31	–	$\Delta(0) \approx$ 0.36 meV	1.59R	[47]
31	$(TaNb)_{1-x}(HfZrTi)_x$ ($0.2 \leq x \leq 0.84$)	BCC							[48]

(continued)

Table 4.1 (continued)

Sr. no.	High Entropy Alloy	Crystal Structure	T_c (K)	$H_{c2}(0)$ (kOe)	VEC	ξ (nm)	Band gap	ΔS_{mix}	References
	x = 0.3		8	66.7	4.7	–	–	1.426R	
	x = 0.33		7.8	77.5	4.67	–	$2\Delta(0)/k_BT_c = 3.7$	1.461R	
	x = 0.4		7.6	84.3	4.6	–	–	1.528R	
	x = 0.5		6.5	116.7	4.5	–	–	1.589R	
	x = 0.84		4.5	90.2	4.16	–	2Δ(0)/kBTc = 3.6	1.473R	
	$[TaNbV]_{0.67}(ZrHfTi)_{0.33}$	BCC	4.3	–	4.67	–	$2\Delta(0)/k_BT_c = 3.2$	1.733R	
32	$(TaNb)_{0.67}(HfZrTi)_{0.33}$ substituted with {Sc-Cr}	BCC							[49]
	$(Sc_{0.33}Cr_{0.67}Nb)_{0.67}(HfZrTi)_{0.33}$		5.6	–	4.6	–	–	1.67R	
	$(TaSc_{0.33}Cr_{0.67})_{0.67}(HfZrTi)_{0.33}$		4.4	–	4.6	–	–	1.67R	
	$(TaNb)_{0.67}(Sc_{0.67}Cr_{0.33}ZrTi)_{0.33}$		7.5	–	4.6	–	–	1.53R	
	$(TaNb)_{0.67}(HfSc_{0.67}Cr_{0.33}Ti)_{0.33}$		7.4	–	4.6	–	–	1.53R	
	$(TaNb)_{0.67}(HfZrSc_{0.67}Cr_{0.33})_{0.33}$		7.6	–	4.6	–	–	1.53R	
	$(TaNb)_{0.67}(HfZrTi)_{0.33}Al_x$								
	x = 0.15		6.6	–	4.45	–	–	1.65R	
	x = 0.3		3	–	4.29	–	–	1.66R	
33	$Ta_{1/6}Nb_{2/6}Hf_{1/6}Zr_{1/6}Ti_{1/6}$	BCC	7.85	120.5	4.5	5.59	$2\Delta(0)/k_BT_c = 4.72$	–	[50]
34	$Nb_{22.1}Ta_{26.3}Ti_{16.6}Zr_{15.5}Hf_{19.5}$	BCC	7.12	20	4.4	20.6		1.59R	[51]

(continued)

Table 4.1 (continued)

Sr. no.	High Entropy Alloy	Crystal Structure	T_c (K)	$H_{c2}(0)$ (kOe)	VEC	ξ (nm)	Band gap	ΔS_{mix}	References
	$Nb_{21.5}Ta_{18.1}Ti_{15.9}Zr_{14.4}Hf_{16.6}V_{13.5}$		5.09	20	4.5			1.59R	
	NbTaTiZrFe		6.87	–	5.2			1.609R	
	NbTaTiZrGe		8.46	–	4.4			1.609R	
	NbTaTiZrGeV		8.4	–	4.5			1.79R	
	NbTaTiZrSiV		4.29	–	4.5			1.79R	
	NbTaTiZrSiGe		7.4	–	4.33			1.79R	
35	$Nb_{20}Re_{20}Zr_{20}Hf_{20}Ti_{20}$	BCC	5.3	88.8	4.8	6.1	$\Delta(0)/k_BT_c = 1.96$	1.609R	[52]
36	$Hf_{21}Nb_{25}Ti_{15}V_{15}Zr_{24}$	BCC	5.3	–	4.4	–	–	1.58R	[15]
37	Mo-Re-Ru-Rh-Ti	HCP							[19]
	$Mo_{0.2375}Re_{0.2375}Ru_{0.2375}Rh_{0.2375}Ti_{0.05}$		3.6	–	7.32	–	–	−1.52R	
	$Mo_{0.225}Re_{0.225}Ru_{0.225}Rh_{0.225}Ti_{0.1}$		4.7	–	7.15	–	–	−1.57R	
	$Mo_{0.1}Re_{0.1}Ru_{0.55}Rh_{0.1}Ti_{0.15}$		2.1	–	7.2	–	–	−1.30R	
	$Mo_{0.105}Re_{0.105}Ru_{0.527}Rh_{0.105}Ti_{0.158}$		2.2	–	7.15	–	–	−1.34R	
	$Mo_{0.118}Re_{0.118}Ru_{0.47}Rh_{0.118}Ti_{0.176}$		2.5	–	7.06	–	–	−1.42R	
38	$Re_{0.56}Nb_{0.11}Ti_{0.11}Zr_{0.11}Hf_{0.11}$	HCP	4.4	36	5.79	11	–	–	[53]
39	$(ScZrNb)_{1-x}[RhPd]_x$ and $(ScZrNbTa)_{1-x}[RhPd]_x$ systems	CsCl-type							[54]
	$(ScZrNb)_{1-x}[RhPd]_x$								
	$x = 0.35$		9.3	107	5.93	–	–	–	
	$x = 0.37$		8.1	96	6.04	–	–	–	

(continued)

Table 4.1 (continued)

Sr. no.	High Entropy Alloy	Crystal Structure	T_c (K)	$H_{c2}(0)$ (kOe)	VEC	ξ (nm)	Band gap	ΔS_{mix}	References
	$x = 0.39$		6.6	89	6.09	–	–	–	
	$x = 0.4$		4.2	21	6.2	–	–	–	
	$(ScZrNbTa)_{1-x}[RhPd]_x$					–	–	–	
	$x = 0.315$		6.2	88	5.9	–	–	–	
	$x = 0.33$		3.2	21	5.98	–	–	–	
40	$(ZrNb)_{1-x}(MoReRu)_x$ $(0.8 \leq x \leq 0.9)$	α-Mn type							[55]
	$x = 0.8$		4.2	–	6.5	–	–	–	
	$x = 0.9$		5.3	79	6.75	–	–	–	
	$(HfTaWIr)_{1-x}Re_x$ $(0.6 \leq x \leq 0.75)$	α-Mn type							
	$x = 0.6$		4	47	6.6	–	–	–	
	$x = 0.7$		4.5	–	6.7	–	–	–	
	$x = 0.75$		5.6	–	6.75	–	–	–	
	$(HfTaWPt)_{1-x}Re_x$ $(0.6 \leq x \leq 0.7)$	α-Mn type							
	$x = 0.6$		4.4	59	6.7	–	–	–	
	$x = 0.7$		5.7	–	6.78	–	–	–	
41	M-Te (M: Ag, In, Cd, Sn, Sb, Pb, Bi) system	NaCl-type							[24]
	$Ag_{0.20}Cd_{0.20}Sn_{0.20}Sb_{0.15}Pb_{0.20}Te_{1.05}$		1.2	–	6.63	–	–	–	
	$Ag_{0.24}In_{0.22}Sn_{0.18}Sb_{0.14}Pb_{0.19}Te_{1.03}$		1.4	–	5.83	–	–	–	

(continued)

Table 4.1 (continued)

Sr. no.	High Entropy Alloy	Crystal Structure	T_c (K)	$H_{c2}(0)$ (kOe)	VEC	ξ (nm)	Band gap	ΔS_{mix}	References
	$Ag_{0.22}Cd_{0.22}In_{0.23}Sn_{0.17}Sb_{0.14}Te_{1.02}$		0.7	–	6.63	–	–	–	
	$Ag_{0.19}Cd_{0.19}Sn_{0.20}Pb_{0.18}Bi_{0.21}Te_{1.03}$		1	–	6.56	–	–	–	
	$Ag_{0.21}Cd_{0.19}In_{0.25}Pb_{0.16}Bi_{0.18}Te_{1.00}$		1	–	6.44	–	–	–	
	$Ag_{0.21}Cd_{0.21}In_{0.24}Sn_{0.19}Bi_{0.19}Te_{0.97}$		1	–	6.54	–	–	–	
42	BiS_2-based superconductor $REO_{0.5}F_{0.5}BiS_2$								[29]
	RE = Pr		3.7	–	5.3	–	–	0	
	RE = $Ce_{0.5}Nd_{0.5}$		3.76	–	5.3	–	–	5.76(J/Kmol)	
	RE = $Ce_{1/3}Pr_{1/3}Nd_{1/3}$		3.76	–	5.3	–	–	9.13(J/Kmol)	
	RE = $La_{0.05}Ce_{0.25}Pr_{0.35}Nd_{0.35}$		3.69	–	5.3	–	–	9.93(J/Kmol)	
	RE = $La_{0.2}Ce_{0.2}Pr_{0.2}Nd_{0.2}Sm_{0.2}$		3.97	–	5.3	–	–	13.37(J/Kmol)	
43	$Y_{0.28}Nd_{0.16}Sm_{0.18}Eu_{0.18}Gd_{0.20}Ba_2Cu_3O_{7-\delta}$	NaCl-type	93	–	6.31	–	–	–	[30]
	$Y_{0.18}La_{0.24}Nd_{0.14}Sm_{0.14}Eu_{0.15}Gd_{0.15}Ba_2Cu_3O_{7-\delta}$		93	–	6.31	–	–	–	
44	$(Nb_{0.1}Mo_{0.3}W_{0.3}Re_{0.2}Ru_{0.1})_5Si_3$	W_5Si_3-type	3.3	50	5.43	8.1	–	–	[56]
	$(Nb_{0.2}Mo_{0.3}W_{0.3}Re_{0.1}Ru_{0.1})_5Si_3$		3.2	51	5.31	8	–	–	
45	Nb_3(Al, Sn, Ge, Ga, Si)	Cubic Cr_3Si-type							[57]
	$Nb_3(Al_{0.2}Sn_{0.2}Ge_{0.2}Ga_{0.2}Si_{0.2})$		9	104	–	–	–	1.59R	
	$Nb_3(Al_{0.3}Sn_{0.3}Ge_{0.2}Ga_{0.1}Si_{0.1})$		11	133	–	–	–	1.50R	
46	$Ta_{10}Mo_{35-x}Cr_xRe_{35}Ru_{20}$	Tetragonal							[58]
	x = 5		4.79	61	6.62–6.7	7.3	Δ_0/T_c = 1.83	–	

(continued)

Table 4.1 (continued)

Sr. no.	High Entropy Alloy	Crystal Structure	T_c (K)	$H_{c2}(0)$ (kOe)	VEC	ξ (nm)	Band gap	ΔS_{mix}	References
	x = 10		4.41	48	6.62–6.7	7.5	Δ_0/T_c = 1.71	–	
	x = 13		3.98	59	6.62–6.7	8.2	Δ_0/T_c = 1.55	–	
47	$Ti_{67}Zr_{19}Nb_{11.5}Sn_{2.5}$	BCC	4.65	92	–	6	Δ_0 = $1.764k_BT_{c0}$	–	[59]
48	$Ta_{10}Mo_5W_{30}Re_{35}Ru_{20}C_x$								[60]
	x = 0	Centrosymmetric tetragonal σ type structure for 0 ≤ x ≤ 2	4.87	67	6.60–6.65	7	–	1.43R	
	x = 20	Noncentrosymmetric cubic β-Mn type Structure for 16 ≤ x ≤ 20	5.34	93	6.21–6.28	6	–	1.43R	
49	$Nb_{67}Hf_{11}Ti_{11}Zr_{11}$	BCC							[61]
	600 nm film thickness		6.63	74	–	6.7	$2\Delta/k_BT_c$ = 3.54	–	
	100 nm film thickness		2.73	31	–	10.3	$2\Delta/k_BT_c$ = 3.54	–	
	30 nm film thickness		2.69	29	–	10.7	$2\Delta/k_BT_c$ = 3.54	–	
	$Nb_{35}Ta_{35}Hf_{10}Ti_{10}Zr_{10}$	BCC							

(continued)

Table 4.1 (continued)

Sr. no.	High Entropy Alloy	Crystal Structure	T_c (K)	$H_{c2}(0)$ (kOe)	VEC	ξ (nm)	Band gap	ΔS_{mix}	References
	600 nm film thickness		5.31	58	–	7.5	$2\Delta/k_BT_c = 4.21$	–	
	100 nm film thickness		2.69	32	–	10.1	$2\Delta/k_BT_c = 4.21$	–	
50	Nb-Mo-Re-Ru-Rh System								[20]
	$Nb_{25}Mo_{5+x}Re_{35}Ru_{25-x}Rh_{10}$	α-Mn-type							
	x = 0		4.66	75	6.9	6.6	$\Delta_0/T_c = 1.5$	–	
	x = 5		5.1	83	6.8	6.3	$\Delta_0/T_c = 1.5$	–	
	x = 10		5.1	79	6.7	6.4	$\Delta_0/T_c = 1.764$	–	
	$Nb_5Mo_{35-y}Re_{15+y}Ru_{35}Rh_{10}$	HCP-type							
	y = 0		7.54	89	7.1	6.1	$\Delta_0/T_c = 2$	–	
	y = 5		6.69	75	7.15	6.6	$\Delta_0/T_c = 1.88$	–	
	y = 10		6.51	75	7.2	6.6	$\Delta_0/T_c = 1.764$	–	
	y = 15		5.46	61	7.25	7.3	$\Delta_0/T_c = 1.764$	–	

are given in Fig. 4.1d. The actual weight distribution of the specimen was found to be slightly different than the starting composition of the powders. The grain orientation map of the polished specimen obtained from electron backscatter diffraction (EBSD) measurements is given in Fig. 4.1e. The presented EBSD result, which was obtained following homogenization treatment, revealed a relatively coarse grain structure. A significant grain refinement that is conducive to superelasticity was introduced into the microstructure through cold rolling. Consistent with the XRD measurements, the specimen has a single β-phase. Figure 4.1f shows the inverse pole figures (IPFs) along the X, Y, and Z directions. In general, the grain orientation map in Fig. 4.1e and the IPF map in Fig. 4.1f point to a coarse grain structure with a random grain orientation.

4.3 Magnetic Properties

The magnetic properties of the TiZrNbSn system were studied using vibrating sample magnetometry. In Fig. 4.2a, we plot the *T*-dependence of the zero-field cooling (ZFC) and field cooling (FC) magnetization (*M*) of the TiZrNbSn strip in a magnetic field of 0.5 mT. Magnetization measurements reveal a T_{c0} of 4.65 K. M(H). Loops in the superconducting state for a relatively small magnetic field range are shown in Fig. 4.2b, in which the lower critical field (H_{c1}) is determined from the point of deviation from linearity in the low magnetic field regime. We have plotted the determined H_{c1} values in Fig. 4.2c. To determine the ground-state lower critical field, we fitted the experimentally determined H_{c1} values to $H_{c1}(T) = H_{c1}(0)[1 - (T/T_{c0})^2]$, and the expression and solid line in Fig. 4.2c represent the fit. Extrapolation through the fit given in Fig. 4.2c resulted in a ground state lower critical field value of $H_{c1}(0) = 13.7(2)$ mT. The high magnetic field behavior of the magnetization revealed an anomaly at intermediate fields as seen in Fig. 4.2d. Such an anomaly is often called the peak effect in the superconductivity literature.

It should be noted here that this effect (i.e., the anomaly shown in Fig. 4.2d) is observed in type 2 superconductors. In the literature, various terminologies have been used to describe peak effects, such as the fishtail effect or the second peak in magnetization. The feature (M(H)) shown in Fig. 4.2b is the most common feature of the field dependence of the magnetization in type II superconductors. At magnetic fields larger than the lower critical field, vortex lines are most likely to be trapped at the pinning centers in spatially inhomogeneous matter; therefore, a net magnetic moment will be observed. A further increase in the magnetic field will increase the interaction of the vortex lines and the chance of trapping at pinning centers; therefore, the magnetization will increase. At large fields, the interaction forces between the vortex lines become stronger than the pinning forces, vortex lines creep out of the sample, and the magnetization decreases. Interestingly, for some type II superconductors, such as YBCO (Y123) [63], BISCCO [64], iron-based superconductors [65], and $NbSe_2$ [66], a second peak in magnetization occurs close to the upper critical field (H_{c2}). This behavior is often referred to as second peak magnetization or, more

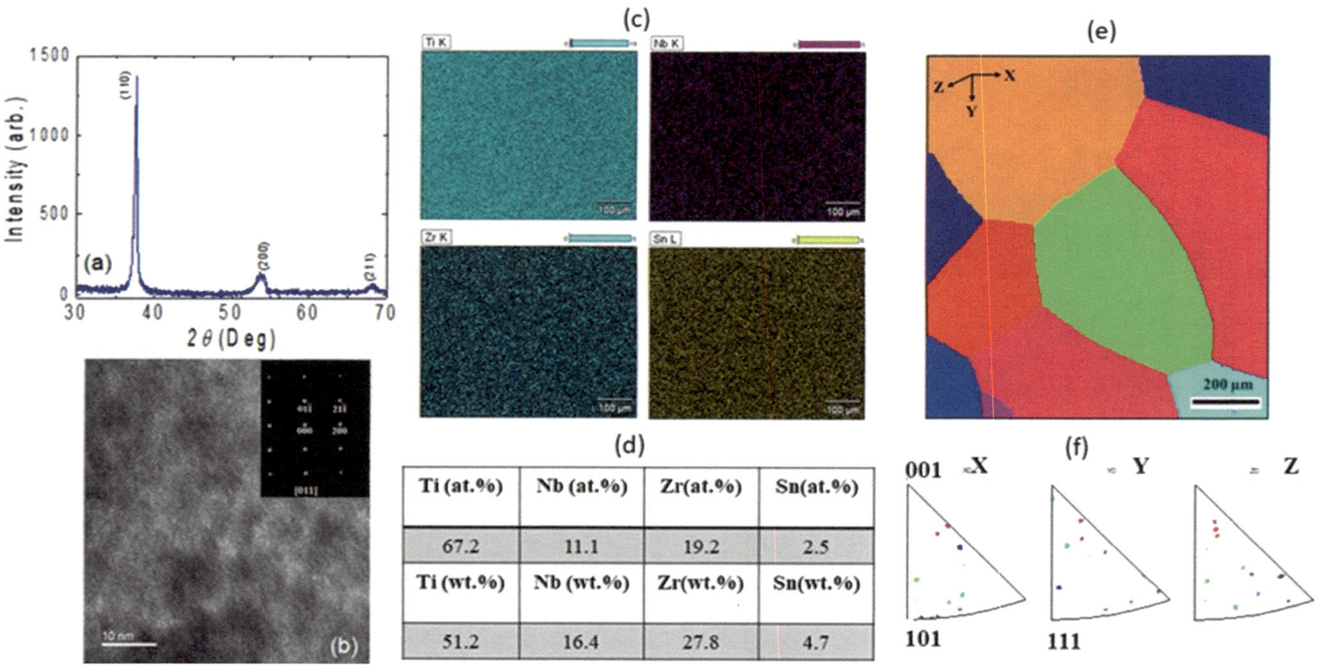

Ti (at.%)	Nb (at.%)	Zr(at.%)	Sn(at.%)
67.2	11.1	19.2	2.5
Ti (wt.%)	Nb (wt.%)	Zr(wt.%)	Sn(wt.%)
51.2	16.4	27.8	4.7

Fig. 4.1 **a** X-ray diffraction (XRD) and **b** high-resolution TEM image. The inset shows a selective electron diffraction pattern. **c** EDX area maps of the constituent elements. **d** Composition of the alloy obtained from the EDX spectrum. **e** EBSD image of the surface. **f** Inverse pole figures obtained from the surface of the TiZrNbSn alloy. Reproduced with permission from [58]

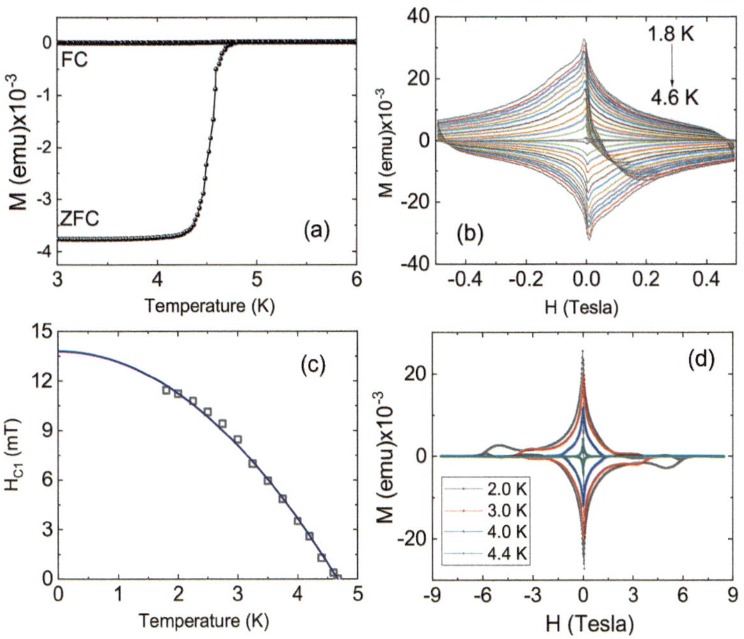

Fig. 4.2 **a** Temperature dependence of the magnetization measured in a magnetic field of 0.5 mtesla in zero-field and field-cooled configurations. **b** Magnetic field dependence of the magnetization measured between 1.8 and 4.6 K. **c** Lower critical field values derived from the field dependence of magnetizations. **d** Magnetic field dependence of the magnetization measured in magnetic fields up to 9 T. Reproduced with permission from [58]

commonly, the "fist tail effect". There are multiple interpretations of the fishtail effect in type II superconductors. For instance, vortex dynamical mechanisms, including crossover from elastic to plastic (E–P) vortex creep, vortex order–disorder phase transition, vortex lattice structural phase transition (VL), surface barrier, and vortex clustering, were suggested to explain the fishtail effect in the literature. TiZrNbSn SMA superconductors exhibit interesting vortex dynamics, as shown in Fig. 4.1. Figure 4.2d. Gao et al. recently reported the fishtail effect in $(TaNb)_{0.7}(HfZrTi)_{0.5}$ and interpreted the origin as the presence of small normal core pinning centers in the material. However, understanding the origin of the fishtail effect in high-entropy alloys requires further work, such as building a vortex phase diagram for many compositions and understanding the flux creep mechanism in this intermediate range.

4.4 Electrical Transport Properties

The temperature dependence of the resistivity for various magnetic fields is shown in Fig. 4.3a. The TiZrNbSn alloy exhibited the typical behavior of type 2 superconductors. The temperature at which the resistance of the specimen reaches zero resistance is defined as the zero resistance critical temperature (T_{c0}), and for each magnetic field, the critical temperature is determined. The magnetic field dependence of the critical temperature is shown in Fig. 4.3b. The experimental data (shown in Fig. 4.3b) were used to determine the upper critical field for the considered TiZrNbSn SMAs. An empirical expression, $H_{c2}(T) = H_{c2}(0)(1 - (T/T_c)^2)$, was used to fit the experimental data and predict the critical magnetic field through extrapolation, as shown in Fig. 4.3b, and the upper critical field of the alloy was found to be 9.20 T. Such a value of the upper critical field is comparable to the Pauli-paramagnetic limit for weak electron–phonon coupling for low-temperature superconductors, $H_{c2} = 1.86 \times T_{c0} \approx 8.65$ T [67, 68], when T_{c0} is approximately 4.65 K. It is well accepted that H_{c2} also provides important information on the nature of superconductivity in the studied system. In particular, H_{c2} provides information on coherence length ξ, the fundamental length scale for the superconducting state: $\xi(T) = \sqrt{\frac{\Phi_0}{2\pi H_{c2}(T)}}$ where Φ_0 is the flux quantum [69]. Using $H_{c2}(0) = 9.20$ T, one can calculate $\xi(0) = 6$ nm. This value is within the reported ranges of coherence lengths for Nb-family superconductors. The temperature dependence of the resistivity of the alloy for a larger temperature range is plotted in Fig. 4.3c. The temperature dependence of the resistivity revealed an interesting reversible temperature anomaly as seen in Fig. 4.3c in the temperature range of 150–225 K. Shape memory properties are generally linked to a structural phase transition from martensite to austenite phase [70]. The transition from the martensite to the austenite phase with temperature induces readjustment of the crystal lattice and therefore the electronic structure of the material [71]. Hence, a reversible change in the electrical resistance of the SMA is expected across the phase transition temperatures [71]. This response of the electrical resistance of the material to the structural phase transition is often attributed to the formation of super-zone boundary gaps, which alter the density of the electronic states near the Fermi surface [72–75]. Such behavior could be useful for low-temperature actuator applications. It is also noted here that the phase transformation temperatures observed from Fig. 4.3c point to an austenite finish temperature (A_f) of approximately 225 K. This finding is consistent with the observation of a full austenite structure at room temperature (EBSD and XRD data) and thus the SE response, as shown in the mechanical properties section below.

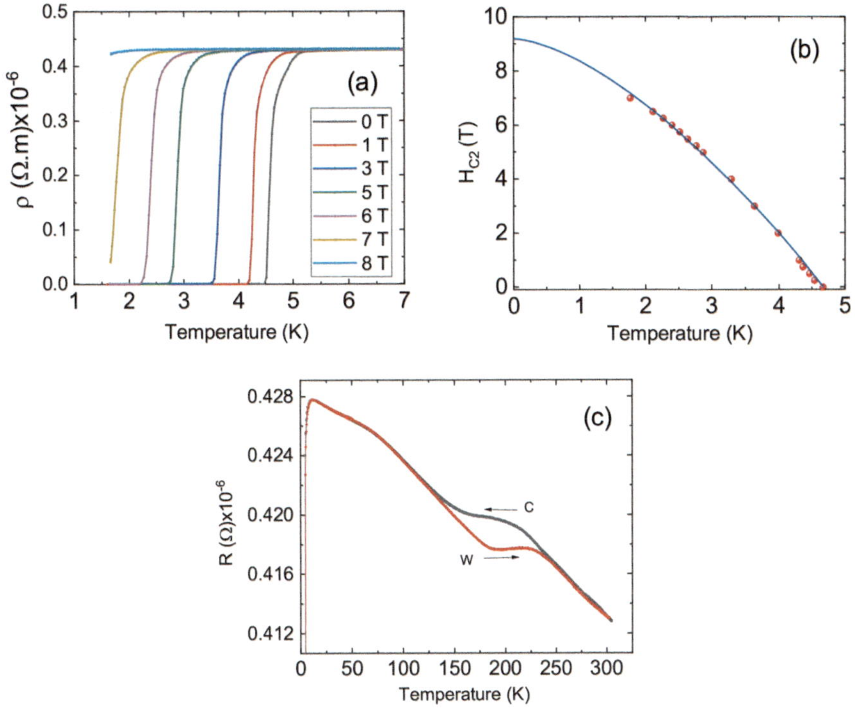

Fig. 4.3 **a** Temperature dependence of the resistivity measured in various magnetic fields up to 9 T at low temperatures. **b** Upper critical field (H_{c2}) estimated from the zero-resistance superconducting transition temperature. The solid line is fit. **c** Temperature dependence of the resistivity measured in the 2–300 K range. Reproduced with permission from [58]

4.5 Specific Heat and Local Magnetic Properties

The temperature dependencies of the specific heat measured in magnetic fields up to 9 T are shown in Fig. 4.4a. In a zero magnetic field, an abrupt jump occurs at $\Delta C =$ 49.05 mJ/mol. K has been observed at a superconducting transition temperature of 4.65 K. The response of the specific heat to the superconducting transition is a well-established subject; at the superconducting transition, the electronic contribution to the specific heat is more important than the lattice contribution. At low temperatures in the normal state, the specific heat of a metal can be written as $C/T = \gamma + \alpha T^2$, where γ and α are the electronic and lattice-specific heat coefficients, respectively [76]. As we have confirmed from our field dependencies of magnetization and resistivity at a magnetic field of 9 T, our specimen is in a normal state at temperatures higher than 1.6 K. Therefore, we can use 9-T specific heat data to estimate the electronic and lattice-specific heat coefficients. The coefficients γ and α can be easily determined using a linear fit when the low-temperature specific heat is written in the $C/T = \gamma + \alpha T^2$ form. Therefore, a linear fit to C/T versus T^2 data in the normal state (9 T) would reveal the γ and α values. In Fig. 4.4b, we show the specific heats in zero fields and 9 T in a plot of C/T versus T^2, and the linear fit reveals an intercept of $\gamma = 7.36$ mJ/mol K^2 and a slope of $\alpha = 0.51$ mJ/mol K^4. In Fig. 4.4b, we also plot the zero magnetic field data; note that the intercept for zero magnetic field data yields zero, which means that the entire specimen becomes superconductive below T_c. The field dependence of the specific heat curves for the selected temperatures is given in Fig. 4.4c–f. The magnetic fields in which the samples turned into a normal state were consistent with the resistivity and magnetization measurements. Such temperature and field dependences of the specific heat suggest typical Bardeen–Schaeffer copper (BCS) superconductivity. The implications of BCS-type superconductivity will be discussed below.

The Muon spin rotation (μSR) technique is especially suited for the study of superconducting materials since it allows one to investigate the local magnetic field distribution on a microscopic scale and to directly access the volume fraction of the

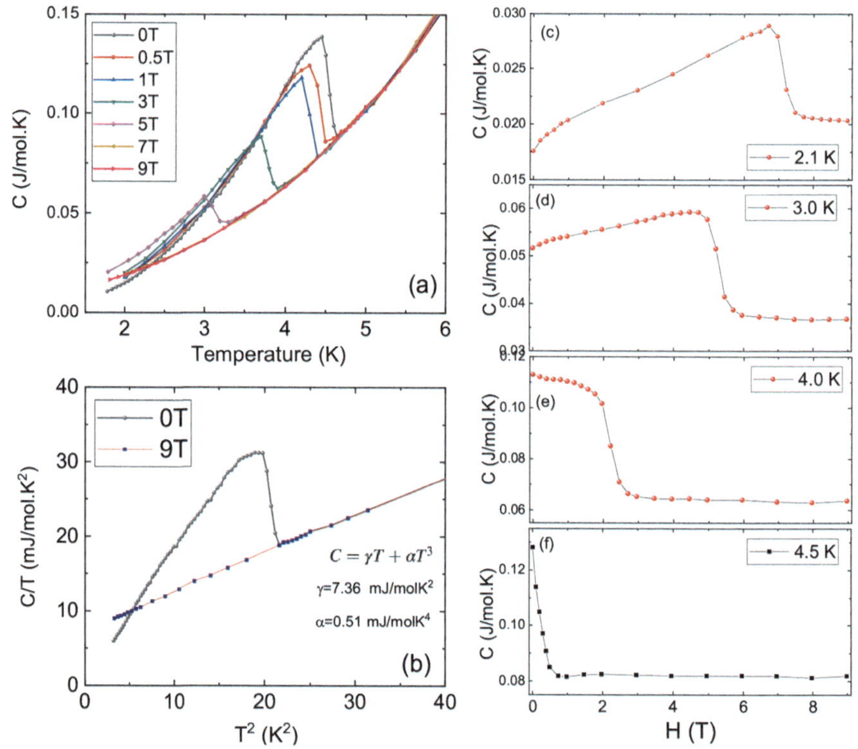

Fig. 4.4 **a** Temperature dependence of the specific heat measured in various magnetic fields up to 9 T at low temperatures. **b** Scaled specific heat data for a zero magnetic field and 9 T. **c**–**f** Magnetic field dependence of the specific heat at the indicated temperatures. Reproduced with permission from [58]

superconducting phase. The μSR measurements were performed on a GPS spectrometer at the Paul Scherer Institute, Switzerland. In these measurements, fully spin-polarized positive muons (μ^+) are implanted into the alloy. Muons decay into a positron with a lifetime of $\tau_\mu \sim 2.2\ \mu s$, which is emitted preferentially in the direction of the muon's spin at the time of decay. The collective average of the polarizations of the implanted muons is proportional to the asymmetry, A(t), which is monitored via asymmetric decay using appropriately positioned detectors (see Fig. 4.5a, b). A(t) can be used to extract information regarding the internal magnetic fields at the muon site. Such measurements, especially in a transverse magnetic field (500 G) configuration (with respect to initial muon polarization axes), are essential for superconducting properties since below T_c, a vortex lattice is formed that results in an inhomogeneous broadening of the field distribution sensed by the muons. Egilmez et al. performed such measurements, and a summary of these measurements is given in Fig. 4.5 [59]. The measured order parameter fits well with the predictions of BCS theory, as revealed in Fig. 4.5c. In their work, they reported an experimentally measured isotropic gap value of 0.98 meV, while the BCS theory prediction $\Delta_0 = 1.764\ k_B T_c$ yields a value of 0.70 meV. Moreover, the msr measurements confirmed bulk superconductivity. Note that such observations are in line with our specific heat data. As mentioned above, our specific heat data revealed an abrupt increase of $\Delta C = 49.05$ mJ/mol K at a superconducting transition temperature of 4.65 K. In BCS theory, the transition from the normal state to the superconducting state can be regarded as a disorder-order transition, and the entropy lost in this process can be calculated as $\Delta C/\gamma T_{c0} = 1.43$ if the superconductivity is a bulk effect. Such a value is predicted for weakly coupled superconducting materials and is found to be the case for the vast majority of low-temperature type II superconductors [76]. Considering our $\Delta C = 49.05$ mJ/mol K and $\gamma = 7.36$ mJ/mol K^2, one can obtain $\Delta C/\gamma T_{c0} = 1.4315$ for our sample. This value is in line with the predictions of BCS theory.

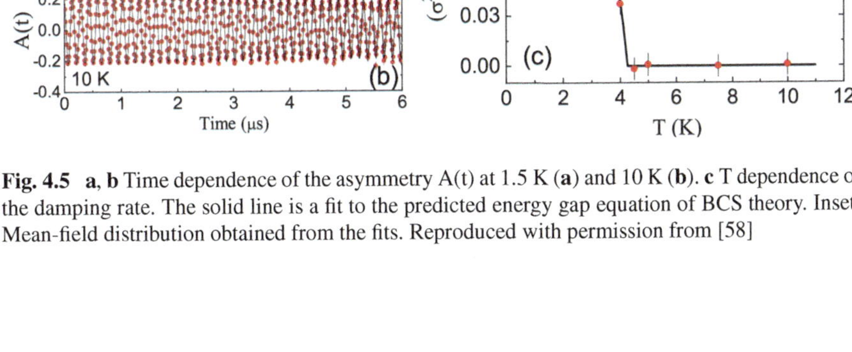

Fig. 4.5 **a**, **b** Time dependence of the asymmetry A(t) at 1.5 K (**a**) and 10 K (**b**). **c** T dependence of the damping rate. The solid line is a fit to the predicted energy gap equation of BCS theory. Inset: Mean-field distribution obtained from the fits. Reproduced with permission from [58]

4.6 Mechanical Properties

Mechanical testing was performed on the cold-rolled TiZrNbSn alloy specimens. Dogbone-shaped tensile samples with a 0.3×3 mm^2 gauge section and 8 mm gauge length were machined from the rolled sheets. All the samples were encapsulated in quartz tubes, evacuated to a pressure <50 mTorr, and backfilled with Ar before sealing. The encapsulated specimens were heat-treated at 900 °C for 30 min followed by quenching with water. To avoid oxidation, the tubes were broken inside a water bath. Following heat treatment, the dogbone specimens were polished using SiC, and a fine black paint speckle pattern was applied for full-field strain measurements via digital image correlation (DIC). Tensile loading was conducted using a screw-driven universal tensile machine (UTM), which was operated at a crosshead speed of 1 mm/min ($\approx 1^{-3}$ s^{-1}). Optical images were captured every 1 s for DIC strain measurements.

Figure 4.6a shows a representative stress–strain curve of a specimen deformed to approximately 2% strain. A clear superelastic (SE) response was observed with a high level of recovery ($\varepsilon_{recoverable} = 1.85\%$) and an insignificant accumulation of residual strains ($\varepsilon_{residual} = 0.07\%$). Upon further loading to higher strain levels, the SE strain magnitudes increased; however, this also triggered a finite accumulation of residual strains, as shown in Fig. 4.6b. The evolution of the recoverable ($\varepsilon_{recoverable}$) and residual ($\varepsilon_{residual}$) strains as a function of the applied strain is shown in Fig. 4.6c. Recoverable strains reaching up to 3% were measured for this particular rolling level and heat treatment.

It should be noted here that the cold rolling process had a significant impact on the development of SE properties by introducing a desirable grain refinement, which increased the slip resistance. The stress–strain curves shown in Fig. 4.7 highlight the development of favorable SE properties with increasing rolling levels (i.e., a reduction in the magnitude of irrecoverable strains). As noted above, this difference is attributable to the decrease in grain size from approximately 400–500 μm following homogenization to an average grain diameter of approximately 50 μm for the 96.6% cold-rolled samples.

4.7 Conclusions

The exploration of new materials, such as high-entropy alloys, for their functional properties has driven modern materials science. Among the various functional groups, superelasticity is very important. In particular, shape memory alloys 'memorize' their previous form following thermomechanical stimulus, making them highly important materials for a vast number of industrial applications in the aerospace, automotive, biomedical, and robotic sectors. Research on SMAs has largely focused on metallurgical, mechanical, structural, or phase transformation properties, and less is known about their electrical and magnetic properties. We have shown that superelasticity and superconductivity exist within the same sample at different temperatures

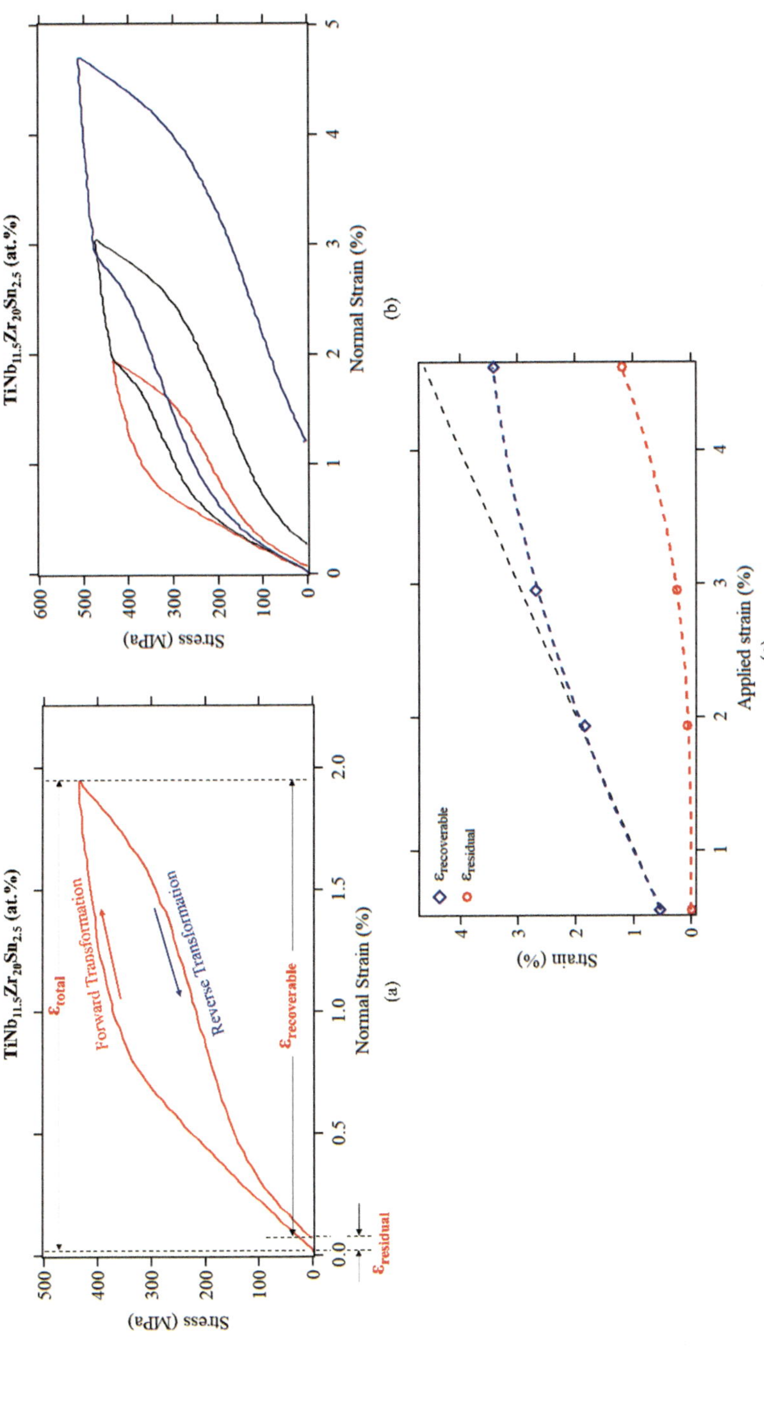

Fig. 4.6 a–c Strain–strain response showing the superelastic response of the TiZrNbSn SMAs. Reproduced with permission from [58]

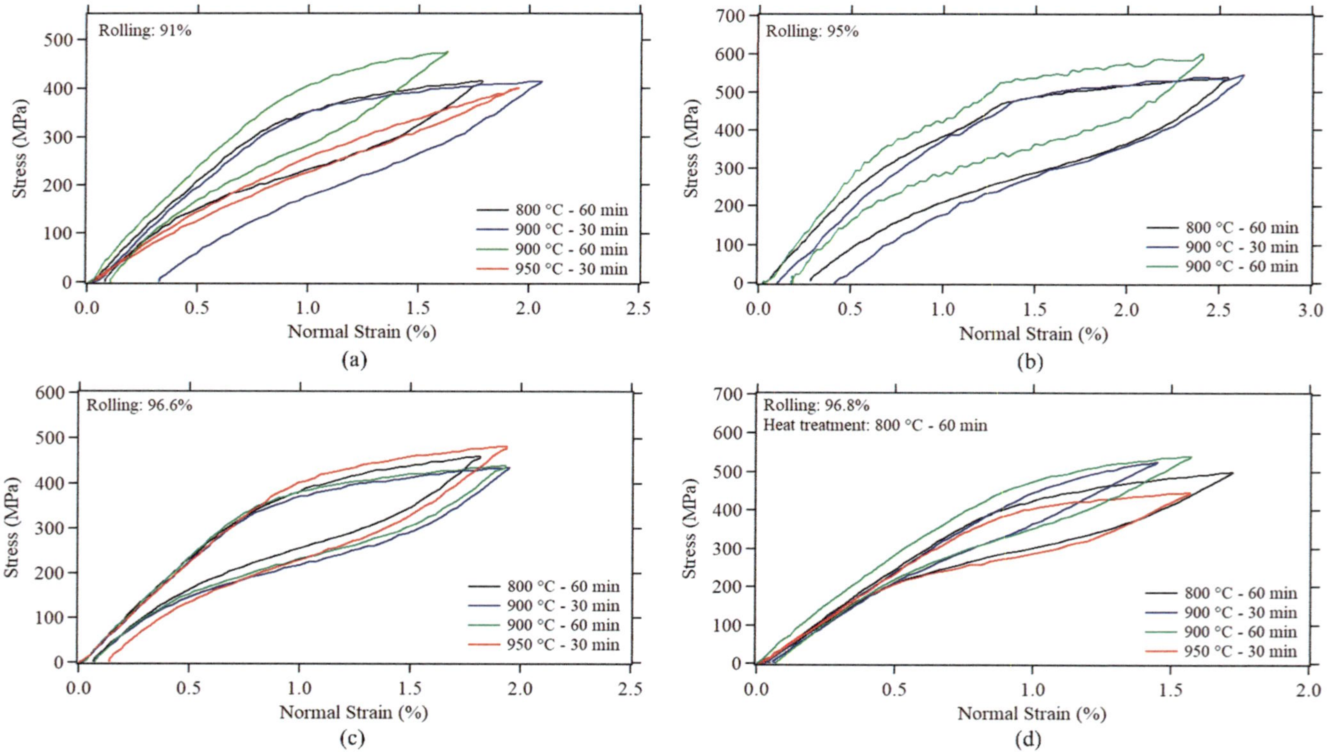

Fig. 4.7 Strain–strain response showing the superelastic response of the TiZrNbSn SMAs following different rolling levels and solution heat treatments. Reproduced with permission from [58]

in the TiZrNbSn system. The compositional study of the superconducting properties along with its superelastic functionalities in TiZrNbSn may lead to superconducting transition temperatures comparable to those of the commercial binary alloy TiNb. More generally, research on high-entropy alloy superconductors has revealed their great potential for bulk superconductivity applications.

References

1. E.P. George, D. Raabe, R.O. Ritchie, High-entropy alloys. Nat. Rev. Mater. **4**, 515–534 (2019). https://doi.org/10.1038/s41578-019-0121-4
2. M. Egilmez, W. Abuzaid, Magnetic, electrical and mechanical properties of entropy alloy. Sci. Rep., 1–12 (2021). https://doi.org/10.1038/s41598-021-87527-x
3. Y.F. Ye, Q. Wang, J. Lu, C.T. Liu, Y. Yang, High-entropy alloy: challenges and prospects. Mater. Today **19**, 349–362 (2016). https://doi.org/10.1016/j.mattod.2015.11.026
4. J. Kitagawa, S. Hamamoto, N. Ishizu, Cutting edge of high-entropy alloy superconductors from the perspective of materials research. Metals (Basel) **10**, 1–22 (2020). https://doi.org/10.3390/met10081078
5. Y. Harayama, J. Kitagawa, Superconductivity in Al-Nb-Ti-V-Zr multicomponent alloy. J. Supercond. Nov. Magn. **34**, 2787–2794 (2021). https://doi.org/10.1007/s10948-021-05966-z
6. J.H. Durrell, M.D. Ainslie, D. Zhou, P. Vanderbemden, T. Bradshaw, S. Speller, et al., Bulk superconductors: a roadmap to applications. Supercond. Sci. Technol. **31**, 2018. https://doi.org/10.1088/1361-6668/aad7ce.
7. W.M. Huang, Z. Ding, C.C. Wang, J. Wei, Y. Zhao, H. Purnawali, Shape memory materials. Mater. Today **13**, 54–61 (2010). https://doi.org/10.1016/S1369-7021(10)70128-0
8. Owen, T., Shape memory alloys edited by Hiroyasu Funakubo, translated from the Japanese by J.B. Kennedy Gordon and Breach Science Publishers, New York, 1986 (Price $135.00, circa £75.00). Robotica **6**, 259–259 (1988). https://doi.org/10.1017/S0263574700004458.
9. C. Naresh, P.S.C. Bose, C.S.P. Rao, Shape memory alloys: a state of art review, in *IOP Conference Series. Materials Science and Engineering*, vol. 149, p. 012054 (2016). https://doi.org/10.1088/1757-899X/149/1/012054
10. D. Xue, R. Yuan, Y. Zhou, D. Xue, T. Lookman, G. Zhang et al., Design of high temperature Ti-Pd-Cr shape memory alloys with small thermal hysteresis. Sci. Rep. **6**, 28244 (2016). https://doi.org/10.1038/srep28244
11. K. Otsuka, T. Kakeshita, Science and technology of shape-memory alloys: new developments. MRS Bull. **27**(2), 91–100 (2002). https://doi.org/10.1557/MRS2002.43
12. J.H. Kim, R. Hidayati, S.G. Jung, Y.A. Salawu, H.J. Kim, J.H. Yun et al., Enhancement of critical current density and strong vortex pinning in high entropy alloy superconductor $Ta_{1/6}Nb_{2/6}Hf_{1/6}Zr_{1/6}Ti_{1/6}$ synthesized by spark plasma sintering. Acta Mater. **232**, 117971 (2022). https://doi.org/10.1016/J.ACTAMAT.2022.117971
13. Y. Harayama, J. Kitagawa, Superconductivity in Al-Nb-Ti-V-Zr multicomponent alloy. J. Supercond. Nov. Magn. **34**, 2787–2794 (2021). https://doi.org/10.1007/S10948-021-05966-Z/TABLES/1
14. J. Kitagawa, K. Hoshi, Y. Kawasaki, R. Koga, Y. Mizuguchi, T. Nishizaki, Superconductivity and hardness of the equiatomic high-entropy alloy HfMoNbTiZr. J. Alloys Compd. **924**, 166473 (2022). https://doi.org/10.1016/J.JALLCOM.2022.166473
15. N. Ishizu, J. Kitagawa, New high-entropy alloy superconductor $Hf_{21}Nb_{25}Ti_{15}V_{15}Zr_{24}$. Results Phys. **13**, 102275 (2019). https://doi.org/10.1016/J.RINP.2019.102275
16. P. Koželj, S. Vrtnik, A. Jelen, S. Jazbec, Z. Jagličić, S. Maiti et al., Discovery of a superconducting high-entropy alloy. Phys. Rev. Lett. **113**, 107001 (2014). https://doi.org/10.1103/PHYSREVLETT.113.107001/FIGURES/4/MEDIUM

17. B. Liu, J. Wu, Y. Cui, Q. Zhu, G. Xiao, S. Wu et al., Superconductivity in hexagonal Nb-Mo-Ru-Rh-Pd high-entropy alloys. Scr. Mater. **182**, 109–113 (2020). https://doi.org/10.1016/J.SCRIPTAMAT.2020.03.004
18. Q. Zhu, G. Xiao, Y. Cui, W. Yang, S. Song, G.H. Cao et al., Structural transformation and superconductivity in carbon-added hexagonal high-entropy alloys. J. Alloys Compd. **909**, 164700 (2022). https://doi.org/10.1016/J.JALLCOM.2022.164700
19. Y.S. Lee, R.J. Cava, Superconductivity in high and medium entropy alloys based on MoReRu. Phys. C: Supercond. Its Appl. **566**, 1353520 (2019). https://doi.org/10.1016/J.PHYSC.2019.1353520
20. B. Liu, J. Wu, Y. Cui, Q. Zhu, G. Xiao, S. Wu et al., Structural evolution and superconductivity tuned by valence electron concentration in the Nb-Mo-Re-Ru-Rh high-entropy alloys. J. Mater. Sci. Technol. **85**, 11–17 (2021). https://doi.org/10.1016/J.JMST.2021.02.002
21. A. Yamashita, R. Jha, Y. Goto, T.D. Matsuda, Y. Aoki, Y. Mizuguchi, An efficient way of increasing the total entropy of mixing in high-entropy-alloy compounds: a case of NaCl-type (Ag, In, Pb, Bi)$Te_{1-x}Se_x$ (x = 0.0, 0.25, 0.5) superconductors. Dalton Trans. **49**, 9118–9122 (2020). https://doi.org/10.1039/D0DT01880E
22. Y. Mizuguchi, Superconductivity in High-Entropy-Alloy Telluride $AgInSnPbBiTe_5$, 88 (2019). https://doi.org/10.7566/JPSJ.88.124708
23. M.R. Kasem, R. Ishii, T. Katase, O. Miura, Y. Mizuguchi, Tuning of carrier concentration and superconductivity in high-entropy-alloy-type metal telluride $(AgSnPbBi)_{(1-x)/4}In_xTe$. J. Alloys Compd. **920**, 166013 (2022). https://doi.org/10.1016/J.JALLCOM.2022.166013
24. M.R. Kasem, K. Hoshi, R. Jha, M. Katsuno, A. Yamashita, Y. Goto, et al., Superconducting properties of high-entropy-alloy tellurides M-Te (M: Ag, In, Cd, Sn, Sb, Pb, Bi) with a NaCl-type structure. Appl. Phys. Express **13**, 033001 (2020). https://doi.org/10.35848/1882-0786/AB7482
25. Y. Mizuguchi, M.R. Kasem, T.D. Matsuda, Superconductivity in $CuAl_2$-type $Co_{0.2}Ni_{0.1}Cu_{0.1}Rh_{0.3}Ir_{0.3}Zr_2$ with a high-entropy-alloy transition metal site **9**, 141–147 (2020). http://McManuscriptcentral.com/Tmrl. https://doi.org/10.1080/21663831.2020.1860147
26. M.R. Kasem, H. Arima, Y. Ikeda, A. Yamashita, Y. Mizuguchi, Superconductivity of high-entropy-alloy-type transition-metal zirconide (Fe Co, Ni, Cu, Ga)Zr_2. J. Phys.: Mater. **5**, 045001 (2022). https://doi.org/10.1088/2515-7639/AC8E34
27. M.R. Kasem, A. Yamashita, Y. Goto, T.D. Matsuda, Y. Mizuguchi, Synthesis of high-entropy-alloy-type superconductors (Fe Co, Ni, Rh, Ir)Zr_2 with tunable transition temperature. J. Mater. Sci. **56**, 9499–9505 (2021). https://doi.org/10.1007/S10853-021-05921-2/FIGURES/5
28. M.R. Kasem, A. Yamashita, T. Hatano, K. Sakurai, N. Oono-Hori, Y. Goto et al., Anomalous broadening of specific heat jump at T c in high-entropy-alloy-type superconductor $TrZr_2$. Supercond. Sci. Technol. **34**, 125001 (2021). https://doi.org/10.1088/1361-6668/AC2554
29. R. Sogabe, Y. Goto, T. Abe, C. Moriyoshi, Y. Kuroiwa, A. Miura, et al., Improvement of superconducting properties by high mixing entropy at blocking layers in BiS_2-based superconductor $REO_{0.5}F_{0.5}BiS_2$. Solid State Commun. **295**, 43–49 (2019). https://doi.org/10.1016/J.SSC.2019.04.001
30. Y. Shukunami, A. Yamashita, Y. Goto, Y. Mizuguchi, Synthesis of RE123 high-Tc superconductors with a high-entropy-alloy-type RE site. Phys. C: Supercond. Its Appl. **572**, 1353623 (2020). https://doi.org/10.1016/J.PHYSC.2020.1353623
31. N.K. Sarkar, C.L. Prajapat, P.S. Ghosh, N. Garg, P.D. Babu, S. Wajhal et al., Investigations on superconductivity in an equi-atomic disordered Hf-Nb-Ta-Ti-V high entropy alloy. Intermetallics (Barking) **144**, 107503 (2022). https://doi.org/10.1016/J.INTERMET.2022.107503
32. G. Xiao, W. Yang, Q. Zhu, S. Song, G.H. Cao, Z. Ren, Superconductivity with large upper critical field in noncentrosymmetric Cr-bearing high-entropy alloys. Scr. Mater. **223**, 115099 (2023). https://doi.org/10.1016/J.SCRIPTAMAT.2022.115099
33. R. Shu, X. Zhang, S.G. Rao, A. Le Febvrier, P. Eklund, Effects of alloying and deposition temperature on phase formation and superconducting properties of TiZrTaNb-based high entropy-alloy films. Appl. Phys. Lett. **120**, 151901 (2022). https://doi.org/10.1063/5.0091777

34. K. Motla, V. Soni, P.K. Meena, R.P. Singh, Boron based new high entropy alloy superconductor $Mo_{0.11}W_{0.11}V_{0.11}Re_{0.34}B_{0.33}$. Supercond. Sci. Technol. **35**, 074002 (2022). https://doi.org/10.1088/1361-6668/AC5BEA
35. L. Gao, T. Ying, Y. Zhao, W. Cao, C. Li, L. Xiong et al., Fishtail effect and the vortex phase diagram of high-entropy alloy superconductor. Appl. Phys. Lett. **120**, 092602 (2022). https://doi.org/10.1063/5.0083365
36. B. Liu, J. Wu, Y. Cui, Q. Zhu, G. Xiao, H. Wang et al., Formation and superconductivity of single-phase high-entropy alloys with a tetragonal structure. ACS Appl. Electron. Mater. **2**, 1130–1137 (2020). https://doi.org/10.1021/ACSAELM.0C00108/ASSET/IMAGES/LARGE/EL0C00108_0004.JPEG
37. B. Liu, J. Wu, Y. Cui, Q. Zhu, G. Xiao, S. Wu et al., Superconductivity in cubic A15-type V-Nb–Mo–Ir–Pt high-entropy alloys. Front. Phys. **9**, 73 (2021). https://doi.org/10.3389/FPHY.2021.651808/BIBTEX
38. K. Motla, Arushi, P.K. Meena, D. Singh, P.K. Biswas, A.D. Hillier, et al., Probing the superconducting ground state of noncentrosymmetric high-entropy alloys using muon-spin rotation and relaxation. Phys. Rev. B **104**, 094515 (2021). https://doi.org/10.1103/PHYSREVB.104.094515/FIGURES/8/MEDIUM
39. K. Motla, P.K. Meena, Arushi, D. Singh, P.K. Biswas, A.D. Hillier, et al., Superconducting and normal-state properties of the high-entropy alloy Nb-Re-Hf-Zr-Ti investigated by muon spin relaxation and rotation. Phys. Rev. B **105**, 144501 (2022). https://doi.org/10.1103/PHYSREVB.105.144501/FIGURES/7/MEDIUM
40. V.T.A. Hong, H. Jang, S.G. Jung, Y. Han, J.H. Kim, R. Hidayati et al., Probing superconducting gap of the high-entropy alloy $Ta_{1/6}Nb_{2/6}Hf_{1/6}Zr_{1/6}Ti_{1/6}$ via Andreev reflection spectroscopy. Phys. Rev. B **106**, 024504 (2022). https://doi.org/10.1103/PHYSREVB.106.024504/FIGURES/5/MEDIUM
41. P. Sobota, R. Topolnicki, T. Ossowski, T. Pikula, A. Pikul, R. Idczak, Superconductivity in the high-entropy alloy $(NbTa)_{0.67}(MoHfW)_{0.33}$. Phys. Rev. B **106**, 184512 (2022). https://doi.org/10.1103/PHYSREVB.106.184512/FIGURES/9/MEDIUM.
42. X. Zhang, N. Winter, C. Witteveen, T. Moehl, Y. Xiao, F. Krogh et al., Preparation and characterization of high-entropy alloy $(TaNb)_{1-x}(ZrHfTi)_x$ superconducting films. Phys. Rev. Res. **2**, 013375 (2020). https://doi.org/10.1103/PHYSREVRESEARCH.2.013375/FIGURES/4/MEDIUM
43. J. Wu, B. Liu, Y. Cui, Q. Zhu, G. Xiao, H. Wang et al., Polymorphism and superconductivity in the V-Nb-Mo-Al-Ga high-entropy alloys. Sci. China Mater. **63**, 823–831 (2020). https://doi.org/10.1007/S40843-019-1237-5/METRICS
44. X. Zhang, R. Shu, H. Liu, A. Elsukova, P.O.Å. Persson, A. Schilling, et al., Suppression of the transition to superconductivity in crystal/glass high-entropy alloy nanocomposites. Commun. Phys. **5**(1), 1–7 (2022). https://doi.org/10.1038/s42005-022-01059-y
45. M. Krnel, A. Jelen, S. Vrtnik, J. Luzar, D. Gačnik, P. Koželj et al., The effect of scandium on the structure, microstructure and superconductivity of equimolar Sc-Hf-Nb-Ta-Ti-Zr refractory high-entropy alloys. Materials **15**, 1122 (2022). https://doi.org/10.3390/MA15031122/S1
46. C.K.W. Leung, X. Zhang, von F. Rohr, R. Lortz, B. Jäck, Evidence for isotropic s-wave superconductivity in high-entropy alloys. Sci. Rep. **12**(1), 1–7 (2022). https://doi.org/10.1038/s41598-022-16355-4
47. W.L. Nelson, A.T. Chemey, M. Hertz, E. Choi, D.E. Graf, S. Latturner, et al., Superconductivity in a uranium containing high entropy alloy. Sci. Rep. **10**(1), 1–8 (2020). https://doi.org/10.1038/s41598-020-61666-z.
48. Von F. Rohr, M.J. Winiarski, J. Tao, T. Klimczuk, R.J. Cava, Effect of electron count and chemical complexity in the Ta-Nb-Hf-Zr-Ti high-entropy alloy superconductor. Proc. Natl. Acad. Sci. USA **113**, E7144–E7150 (2016)https://doi.org/10.1073/PNAS.1615926113/ASSET/CD431D96-5CF8-4A0D-B2F5-47607FCCF188/ASSETS/GRAPHIC/PNAS.1615926113FIG07.JPEG
49. F.O. Von Rohr, R.J. Cava, Isoelectronic substitutions and aluminum alloying in the Ta-Nb-Hf-Zr-Ti high-entropy alloy superconductor. Phys. Rev. Mater. **2**, 034801 (2018). https://doi.org/10.1103/PHYSREVMATERIALS.2.034801/FIGURES/4/MEDIUM

50. G. Kim, M.H. Lee, J.H. Yun, P. Rawat, S.G. Jung, W. Choi et al., Strongly correlated and strongly coupled s-wave superconductivity of the high entropy alloy $Ta_{1/6}Nb_{2/6}Hf_{1/6}Zr_{1/6}Ti_{1/6}$ compound. Acta Mater. **186**, 250–256 (2020). https://doi.org/10.1016/J.ACTAMAT.2020.01.007
51. K.-Y. Wu, S.-K. Chen, J.-M. Wu, K.-Y. Wu, S.-K. Chen, J.-M. Wu, Superconducting in equal molar NbTaTiZr-based high-entropy alloys. Nat. Sci. (Irvine) **10**, 110–124 (2018). https://doi.org/10.4236/NS.2018.103012
52. S. Marik, M. Varghese, K.P. Sajilesh, D. Singh, R.P. Singh, Superconductivity in equimolar Nb-Re-Hf-Zr-Ti high entropy alloy. J. Alloys Compd. **769**, 1059–1063 (2018). https://doi.org/10.1016/J.JALLCOM.2018.08.039
53. S. Marik, K. Motla, M. Varghese, K.P. Sajilesh, D. Singh, Y. Breard et al., Superconductivity in a new hexagonal high-entropy alloy. Phys. Rev. Mater. **3**, 060602 (2019). https://doi.org/10.1103/PHYSREVMATERIALS.3.060602/FIGURES/9/MEDIUM
54. K. Stolze, F.A. Cevallos, T. Kong, R.J. Cava, High-entropy alloy superconductors on an α-Mn lattice. J. Mater. Chem. C Mater. **6**, 10441–10449 (2018). https://doi.org/10.1039/C8TC03337D
55. B. Liu, W. Yang, G. Xiao, Q. Zhu, S. Song, G.H. Cao et al., High-entropy silicide superconductors with W_5Si_3-type structure. Phys. Rev. Mater. **7**, 014805 (2023). https://doi.org/10.1103/PHYSREVMATERIALS.7.014805/FIGURES/5/MEDIUM
56. A. Yamashita, T.D. Matsuda, Y. Mizuguchi, Synthesis of new high-entropy alloy-type Nb_3 (Al, Sn, Ge, Ga, Si) superconductors. J. Alloys Compd. **868**, 159233 (2021). https://doi.org/10.1016/J.JALLCOM.2021.159233
57. B. Liu, J. Wu, Y. Cui, Q. Zhu, G. Xiao, S. Wu et al., Superconductivity and paramagnetism in Cr-containing tetragonal high-entropy alloys. J. Alloys Compd. **869**, 159293 (2021). https://doi.org/10.1016/J.JALLCOM.2021.159293
58. M. Egilmez, O. Batal, W. Abuzaid, Z. Salman, M. Alkhader, H. Akamine, et al., Superconductivity in $Ti_{67}Zr_{19}Nb_{11.5}Sn_{2.5}$ shape memory alloy. Phys. Rev. Mater. **5**, 074802 (2021). https://doi.org/10.1103/PHYSREVMATERIALS.5.074802/FIGURES/6/MEDIUM
59. G. Xiao, Q. Zhu, W. Yang, Y. Cui, S. Song, G.H. Cao et al., Centrosymmetric to noncentrosymmetric structural transformation in a superconducting high-entropy alloy due to carbon addition. Sci. China Mater. **66**, 257–263 (2023). https://doi.org/10.1007/S40843-022-2144-X/METRICS
60. G. Pristáš, J. Bačkai, M. Orendáč, S. Gabáni, F. Košuth, M. Kuzmiak et al., Superconductivity in medium- and high-entropy alloy thin films: impact of thickness and external pressure. Phys. Rev. B **107**, 024505 (2023). https://doi.org/10.1103/PHYSREVB.107.024505/FIGURES/9/MEDIUM
61. B.T. Matthias, Empirical relation between superconductivity and the number of valence electrons per atom. Phys. Rev. **97**, 74–76 (1955). https://doi.org/10.1103/PhysRev.97.74
62. H. Küpfer, T. Wolf, C. Lessing, A. Zhukov, X. Lançon, R. Meier-Hirmer et al., Peak effect and its evolution from oxygen deficiency in single crystals. Phys. Rev. B Condens. Matter. Mater. Phys. **58**, 2886–2894 (1998). https://doi.org/10.1103/PhysRevB.58.2886
63. L. Gozzelino, D. Botta, A. Chiodoni, R. Gerbaldo, G. Ghigo, F. Laviano et al., Magneto-optical analysis of second-peak effect in BSCCO single crystals. Europhys. Lett. **60**, 910–916 (2002). https://doi.org/10.1209/epl/i2002-00304-5
64. T. Terashima, N. Kikugawa, A. Kiswandhi, E.S. Choi, K. Kihou, S. Ishida, et al., Anomalous peak effect in iron-based superconductors $Ba_{1-x}K_xFe_2As_2$ ($x \approx 0.69$ and 0.76) for magnetic-field directions close to the ab plane and its possible relation to the spin paramagnetic effect. Phys. Rev. B **99**, 1–7 (2019). https://doi.org/10.1103/PhysRevB.99.094508
65. N. Kokubo, K. Kadowaki, K. Takita, Peak effect and dynamic melting of vortex matter in $NbSe_2$ crystals. Phys. Rev. Lett. **95**, 2–5 (2005). https://doi.org/10.1103/PhysRevLett.95.177005
66. B.S. Chandrasekhar, A note on the maximum critical field of highfield superconductors **7** (1962). https://doi.org/10.1063/1.1777362
67. P. Ko, S. Vrtnik, A. Jelen, S. Jazbec, Z. Jagli, S. Maiti, et al., Discovery of a superconducting high-entropy alloy **107001**, 1–5 (2014)https://doi.org/10.1103/PhysRevLett.113.107001

68. G. Grissonnanche, O. Cyr-Choinière, F. Laliberté, S. René De Cotret, A. Juneau-Fecteau, S. Dufour-Beauséjour, et al., Direct measurement of the upper critical field in cuprate superconductors. Nat. Commun. **5**, 1–8 (2014). https://doi.org/10.1038/ncomms4280
69. R. Sewak, C.C. Dey, Martensitic phase transformation in TiNi. Sci. Rep. **9**, 1–8 (2019). https://doi.org/10.1038/s41598-019-49605-z
70. B.D. Ingale, W.C. Wei, P.C. Chang, Y.K. Kuo, S.K. Wu, Anomalous transport and thermal properties of NiTi and with Cu and Fe-doped shape memory alloys near the martensitic transition. J. Appl. Phys. **110** (2011). https://doi.org/10.1063/1.3666029.
71. S.M. Shapiro, B.X. Yang, G. Shirane, Y. Noda, L.E. Tanner, Neutron scattering study of the martensitic transformation in a Ni-Al β-phase alloy. Phys. Rev. Lett. **62**, 1298–1301 (1989). https://doi.org/10.1103/PhysRevLett.62.1298
72. G.L. Zhao, T.C. Leung, B.N. Harmon, M. Keil, M. Müllner, W. Weber, Electronic origin of the intermediate phase of NiTi. Phys. Rev. B **40**, 7999–8001 (1989). https://doi.org/10.1103/PhysRevB.40.7999
73. X. Huang, I.I. Naumov, K.M. Rabe, Phonon anomalies and elastic constants of cubic NiAl from first principles. Phys. Rev. B Condens. Matter. Mater. Phys. **70**, 1–7 (2004). https://doi.org/10.1103/PhysRevB.70.064301
74. I.I. Naumov, O.I. Velikokhatniy, Simultaneous manifestations of the 2D van Hove singularity and the Fermi surface nesting in the acoustic soft phonon mode of β-NiAl alloys. J. Phys. Condens. Matter **9**, 10339–10351 (1997). https://doi.org/10.1088/0953-8984/9/47/004
75. M. Tinkham, Introduction to Superconductivity, 2nd ed. (Dover Publications, RR Donnelley, Mineola, N.Y., 2015)
76. K. Stolze, J. Tao, F.O. Von Rohr, T. Kong, R.J. Cava, Sc-Zr-Nb-Rh-Pd and Sc-Zr-Nb-Ta-Rh-Pd high-entropy alloy superconductors on a CsCl-type lattice. Chem. Mater. **30**, 906–914 (2018). https://doi.org/10.1021/ACS.CHEMMATER.7B04578/SUPPL_FILE/CM7B04578_SI_001.PDF

Chapter 5
Electronic Structure and Electron-Phonon Coupling Calculations for *bcc* HEA Superconductors Ta-Nb-Hf-Zr-Ti

Kinga Jasiewicz, Sylwia Gutowska, Janusz Tobola, and Bartlomiej Wiendlocha

Abstract Results of the electronic structure and electron-phonon coupling calculations for the first superconducting HEA, *bcc*-type $Ta_{0.34}Nb_{0.33}Hf_{0.08}Zr_{0.14}Ti_{0.11}$ and its variant $(TaNb)_{0.67}(HfZrTi)_{0.33}$ are reviewed. The Korringa-Kohn-Rostoker method with the coherent potential approximation is used to calculate the electronic structure in the presence of the chemical disorder. The electron-phonon coupling constant λ is calculated using the rigid muffin tin approximation. Next, the electronic structure, phonons and electron-phonon coupling are calculated using the virtual crystal approximation, which allows us to discuss the effects of disorder on the properties of the HEA. The analysis suggested that disorder influences the phonon structure more strongly than the electronic structure. The effect of external pressure on electron-phonon coupling and superconductivity is studied, and the results correctly reproduce the observed increase in T_c. Finally, we studied the local distortions of the crystal structure and found that they tended to decrease λ.

5.1 Introduction and Methods

In this chapter, we present the results of the electronic structure and electron-phonon coupling calculations for the first superconducting HEA, $Ta_{0.34}Nb_{0.33}Hf_{0.08}Zr_{0.14}Ti_{0.11}$ [1] ($T_c = 7.3$ K, called here TNHZT-1) and its later-reported variant with slightly higher $T_c = 7.7$ K [2] $(TaNb)_{0.67}(HfZrTi)_{0.33}$,[1] which we will label here as TNHZT-2. Some of the results presented here were originally reported in [3, 4].[2]

[1] By this notation the full chemical formula is expanded to $Ta_{0.335}Nb_{0.335}Hf_{0.11}Zr_{0.11}Ti_{0.11}$.

[2] Density of states (DOS) calculations were repeated for the purpose of this article to keep the numerical details the same for both TNHZT variants; thus, some of the computed values ($N(E_F)$

K. Jasiewicz · S. Gutowska · J. Tobola · B. Wiendlocha (✉)
Faculty of Physics and Applied Computer Science, AGH University of Krakow, Aleja Mickiewicza 30, 30-059 Krakow, Poland
e-mail: wiendlocha@fis.agh.edu.pl

J. Kitagawa and Y. Mizuguchi (eds.), *High-Entropy Alloy Superconductors*, Springer Series in Solid-State Sciences 202, https://doi.org/10.1007/978-981-97-4129-8_5

Calculations were performed using the Korringa-Kohn-Rostoker method with the coherent potential approximation (KKR-CPA) [5–9], which allows one to compute the electronic structure of alloys with substitutional (chemical) disorder. KKR-CPA computations were performed in a primitive cell of the material, and the presence of different atoms randomly occupying the given crystal site was simulated using multiple-scattering theory. As a result, configurationally averaged single-particle quantities, such as the density of states, are obtained. Electron scattering due to the presence of disorder is taken into account and its magnitude may be quantified through calculations of electronic dispersion relations. Here, two possible ways of studying E(**k**) are possible. One is the complex energy band technique [10] and the other is the Bloch spectral function method [9, 11, 12]. Both methods provide information on how strongly the band structure is smeared due to the presence of disorder.

In a perfect crystalline material, because of the Bloch theorem, electrons are not scattered by ordered atomic potentials; thus, the lifetime of the electronic state is infinite. As a consequence, the electronic dispersion relations may be defined with infinitely sharp electronic bands, and the crystalline system has zero resistivity at T = 0 K. In the case of a chemically disordered alloy, electrons are scattered on a random lattice of atomic potentials and the Bloch theorem fails, leading to the appearance of residual resistivity, the finite lifetime of the electronic states τ, and the smearing of dispersion relations. In the complex energy band technique [10, 13], the energy eigenvalue has a real part (the "virtual" band center) and an imaginary part, corresponding to the lifetime of electronic states:

$$\tau = \frac{\hbar}{2\mathrm{Im}(E)}. \tag{5.1}$$

Alternatively, one can use the Bloch spectral function $A^B(\mathbf{k}, E)$ (BSF) to describe electronic dispersions in alloys. For a perfect crystalline material, BSF at a selected wavevector **k** is a Dirac delta function, wchich is nonzero only at the (**k**, E) points where the electron in the band ν has an energy eigenvalue $E_{\nu,\mathbf{k}}$, thus here BSF defines the usual dispersion relation:

$$A^B(\mathbf{k}, E) = \sum_{\nu} \delta(E - E_{\nu,\mathbf{k}}). \tag{5.2}$$

BSF is also a **k**-resolved density of electronic states. For a disordered system, BSF becomes smeared, and in most alloys it takes the form of the Lorentz function, with the full width at half maximum (FWHM) value Γ corresponding to the lifetime of the given electronic state [14]:

$$\tau = \frac{\hbar}{\Gamma}. \tag{5.3}$$

---

and McMillan-Hopfield parameters) are slightly different from those in the original papers, but the differences are minor, leaving the qualitative conclusions unaffected.

In this case, the center of the "virtual" energy band is at the energy point, where the BSF Lorentzian has the maximum, and its bandwidth corresponds to Γ. Comparing Eqs. 5.3 with 5.1 we see that $\Gamma = 2\mathrm{Im}(E)$. The stronger the electron scattering by impurities and chemical disorder are, the wider the BSF is, leading to strong smearing of energy bands and decreasing the lifetime of electrons. For the case of very strong scattering (e.g., by so-called resonant impurities [12, 15, 16]), the BSF becomes very broad and can lose its Lorentzian shape. Classic examples of such systems are transition-metal noble-metal alloys, e.g., Cu-Ni [14].

Here, two implementations of the KKR-CPA method were used: the RCPA code [6, 8, 17] and the SPR- KKR package [9, 18]. Most of the results were obtained using the RCPA code, where semirelativistic and spherical potential approximations are employed. The local density approximation was used in the parameterizations of Perdew and Wang [19] as well as von Barth and Hedin [20] and no significant differences were observed between the two formulas of the exchange-correlation potentials. The muffin-tin sphere radius R_{MT} was set to $\sqrt{3}a_0/4$, where a_0 is the lattice constant. To study the magnitude of the electron-phonon interaction, the McMillan-Hopfield parameters were calculated using rigid muffin-tin approximation (RMTA) [21] (see below), and electronic dispersion relations were investigated using the complex energy band technique. To verify the precision of the spherical potential approximation and to analyze the impact of spin-orbit coupling, full-potential SPR-KKR calculations were performed. This approach also allowed us to inspect the complex energy band KKR-CPA results from the RCPA package with the Bloch spectral functions, which were computed using the SPR- KKR package. In both cases, an angular momentum cutoff $l_{max} = 3$ was used and the Fermi level was computed using the Lloyd formula [6, 9].

To facilitate the discussion of electron-phonon coupling and explain the relationship between the so-called Eliashberg function and the McMillan-Hopfield parameters, we introduce here some of the basic equations from this formalism. The electron-phonon interaction matrix element is calculated as [22, 23]:

$$g_{\mathbf{q}\nu}(\mathbf{k}, i, j) = \sum_{J} \left(\frac{\hbar}{2M_J\omega_{\mathbf{q}\nu}}\right)^{\frac{1}{2}} \left\langle \psi_{j,\mathbf{k}+\mathbf{q}} \left| \epsilon_{\mathbf{q}J\nu} \cdot \nabla_u V \right| \psi_{i,\mathbf{k}} \right\rangle \tag{5.4}$$

where V_{BZ} is the volume of the Brillouin zone, $\mathbf{q}$, ν are the wave vector and the phonon mode index, $\epsilon_{\mathbf{q}J\nu}$ is the phonon polarization vector of the J-th atom, $\nabla_u V$ is the potential gradient with respect to the displacement of this atom, $\mathbf{k}, i$ are the wave vector and the band index of the electronic state with one electron energy $E_{\mathbf{k},i}$, E_F represents the Fermi energy, and $N(E_F)$ is the total density of states. Once $g_{\mathbf{q}\nu}(\mathbf{k}, i, j)$ is known, the phonon linewidths $\gamma_{\mathbf{q}\nu}$ [24] may be calculated:

$$\gamma_{\mathbf{q}\nu} = 2\pi\omega_{\mathbf{q}\nu} \sum_{ij} \int \frac{d^3k}{V_{BZ}} |g_{\mathbf{q}\nu}(\mathbf{k}, i, j)|^2 \delta(E_{\mathbf{k},i} - E_F)\delta(E_{\mathbf{k}+\mathbf{q},j} - E_F). \tag{5.5}$$

Based on $g_{\mathbf{q}\nu}$, the Eliashberg function is defined as:

$$\alpha^2 F(\omega) = \frac{1}{\hbar N(E_F)} \sum_{\nu \mathbf{q} i j} \int \frac{d^3k}{V_{bz}} |g_{\mathbf{q}\nu}(\mathbf{k}, i, j)|^2 \times \\ \times \delta(\hbar\omega - \hbar\omega_{\mathbf{q}\nu}) \delta(E_{\mathbf{k},i} - E_F) \delta(E_{\mathbf{k}+\mathbf{q},j} - E_F). \tag{5.6}$$

The electron-phonon interaction parameter λ is next defined based on the integration of the Eliashberg function as

$$\lambda = 2 \int_0^{\omega_{\max}} \frac{\alpha^2 F(\omega)}{\omega} d\omega. \tag{5.7}$$

To fully characterize the electron-phonon interaction in superconductors, the electronic structure, phonon spectrum, phonon linewidths, Eliashberg function, and λ must be calculated. As this was not directly possible in the past and is still very demanding for alloys, simplified methods have been developed, mainly with the aim of decoupling calculations of λ into independent electronic structures and phonon contributions. The most successful method of reformulation of λ involves the use of the so-called McMillan-Hopfield parameters η [25, 26], which were defined for the monoatomic system as:

$$\eta = 2M \int_0^{\omega_{\max}} \omega \cdot \alpha^2 F(\omega) \mathrm{d}\omega \equiv 2MI. \tag{5.8}$$

The advantage of multiplying the Eliashberg function by ω in integral is that η (and I) plays the role of the electronic contribution to the electron-phonon interaction, as phonon frequencies cancel out:

$$I = \int_0^{\omega_{\max}} \omega \cdot \alpha^2 F(\omega) \mathrm{d}\omega = \frac{1}{N(E_F)} \int_0^{\omega_{\max}} d\omega \sum_{\mathbf{q}\nu} \delta(\omega - \omega_{\mathbf{q}\nu}) \times \\ \sum_{ij} \int \frac{d^3k}{V_{BZ}} \left| \sum_J \frac{1}{\sqrt{2M_J}} \langle \psi_{j,\mathbf{k}+\mathbf{q}} \left| \epsilon_{\mathbf{q}J\nu} \cdot \nabla_u V \right| \psi_{i,\mathbf{k}} \rangle \right|^2 \delta(E_{\mathbf{k},j} - E_F) \delta(E_{\mathbf{k}+\mathbf{q},i} - E_F). \tag{5.9}$$

Now, the second moment of the Eliashberg function is defined as:

$$\langle \omega^2 \rangle = \int_0^{\omega_{\max}} \omega \cdot \alpha^2 F(\omega) d\omega \Big/ \int_0^{\omega_{\max}} \alpha^2 F(\omega) \frac{d\omega}{\omega}, \tag{5.10}$$

and combining with the definition (5.7) of λ we obtain the following identity:

$$\lambda = \frac{\eta}{M \langle \omega^2 \rangle}. \tag{5.11}$$

Note that $\langle\omega^2\rangle$, although frequently called the average square frequency, is defined as the second moment of the $\frac{\alpha^2 F(\omega)}{\omega}$ function. The benefits of writing the equation for λ in the form (5.11) are twofold.

First, with additional assumptions of the spherical crystal potential and the "rigid" behavior of the crystal potential under the displacement of the atoms (the potential moves rigidly with the atomic vibrations), the McMillan-Hopfield parameter η may be computed directly from the electronic band structure of the system. In the RMTA approach (see Refs. [17, 21, 27–32] for more details), η is calculated as:

$$\eta = \sum_l \frac{2(2l+2)\, n_l(E_F)\, n_{l+1}(E_F)}{(2l+1)(2l+3)N(E_F)} \left| \int_0^{R_{\rm MT}} r^2 R_l \frac{dV}{dr} R_{l+1} \right|^2. \tag{5.12}$$

$V(r)$ is the self-consistent potential, $R_{\rm MT}$ is the radius of the MT sphere, $R_l(r)$ is a regular solution of the radial Schrödinger equation (normalized to unity inside the MT sphere), $n_l(E_F)$ is the l–th partial DOS per spin at the Fermi level E_F, and $N(E_F)$ is the total DOS per primitive cell.

Second, because $\langle\omega^2\rangle$ is the ratio of two integrals, in the material where the electron-phonon interaction weakly depends on phonon frequencies (e.g., in niobium), it may be estimated using the "bare" phonon density of states $F(\omega)$ function. In such a case, we may write $\alpha^2 F(\omega) = \alpha^2 \times F(\omega)$ ($\alpha^2 \simeq const$), and

$$\langle\omega^2\rangle = \int_0^{\omega_{\rm max}} \omega\alpha^2 F(\omega) d\omega \Big/ \int_0^{\omega_{\rm max}} \alpha^2 F(\omega) \frac{d\omega}{\omega} \simeq \int_0^{\omega_{\rm max}} \omega F(\omega) d\omega \Big/ \int_0^{\omega_{\rm max}} F(\omega) \frac{d\omega}{\omega}, \tag{5.13}$$

as the α^2 factor cancels out.

It turns out that RMTA can effectively decouple calculations of λ into electronic and phonon parts. Moreover, if $F(\omega)$ is unknown and difficult to compute directly, as in the case of high entropy alloys, further simplifications can be made. If we assume that the phonon DOS follows the Debye model, then $F(\omega) = \frac{9\omega^2}{\omega_D^3}$, and one has

$$\langle\omega^2\rangle = \int_0^{\omega_D} \omega \frac{9\omega^2}{\omega_D^3} d\omega \Big/ \int_0^{\omega_D} \frac{9\omega^2}{\omega_D^3} \frac{d\omega}{\omega} = \frac{\omega_D^2}{2}. \tag{5.14}$$

This allows us to use the experimental Debye temperature θ_D ($k_B\theta_D = \hbar\omega_D$) to estimate $\langle\omega^2\rangle$ in the event that other methods are not accessible.

5.2 Results and Discussion

5.2.1 Electronic Structure

Figure 5.1 compares the total and partial electronic density of states of the two compositions of TNHZT. There are practically no differences in the $N(E)$ shape. In

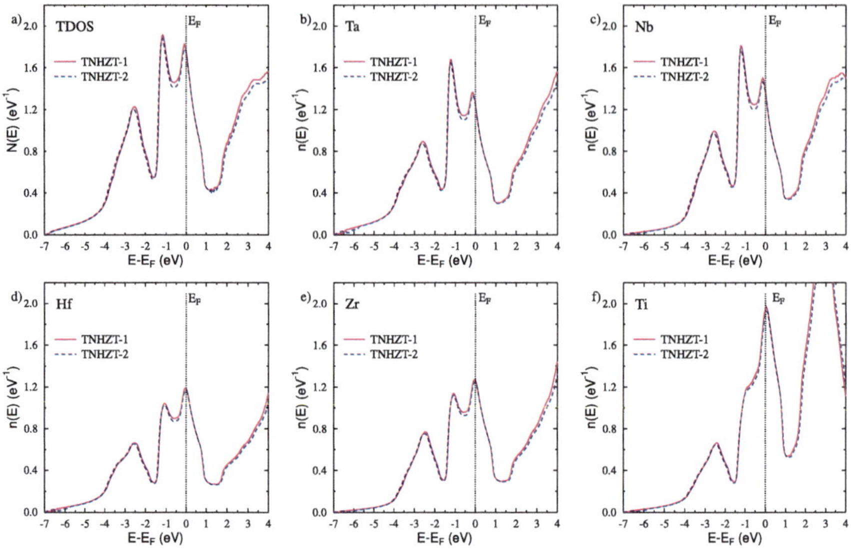

Fig. 5.1 Electronic DOS of $Ta_{0.34}Nb_{0.33}Hf_{0.08}Zr_{0.14}Ti_{0.11}$ (TNHZT-1) and $Ta_{0.335}Nb_{0.335}Hf_{0.11}Zr_{0.11}Ti_{0.11}$ (TNHZT-2). The total and partial atomic DOS are compared between the two compositions, and only minor differences are observed

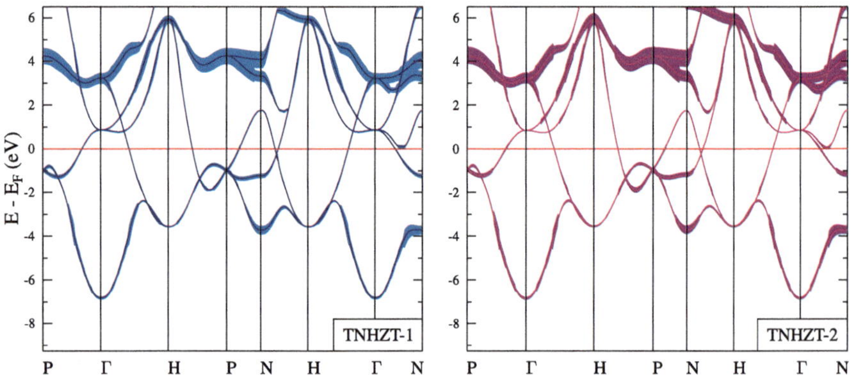

Fig. 5.2 Comparison of the electronic dispersion relations of TNHZT-1 and TNHZT-2. Shading represents the imaginary part of the complex energy E(**k**) and describes the band smearing effect

both cases, the Fermi level is located at the peak of the DOS, and the total DOS at the Fermi level $N(E_F)$, is slightly lower in TNHZT-2 (see Table 5.1). When inspecting the partial DOS, it appears that the dominant contribution arises from the *d* orbitals for all atoms: 3*d* for Ti, 4*d* for Zr and Nb, and 5*d* for Ta and Hf (see [3, 4]).

The complex energy bands are shown in Fig. 5.2. Again, no significant differences were detected between TNHZT-1 and TNHZT-2. The most important and striking result was that despite the large chemical disorder (fully random distribution of

Table 5.1 Densities of states $N(E_F)$ (in eV^{-1}), weighted average $\langle\eta\rangle$ and atomic McMillan-Hopfield parameters η_i (in mRy/a_B^2) for the two variants of TNHZT

alloy	$N(E_F)$	$\langle\eta\rangle$	η_i^{Ta}	η_i^{Nb}	η_i^{Hf}	η_i^{Zr}	η_i^{Ti}
TNHZT-1	1.78	158.28	155.00	163.91	161.88	178.92	128.27
TNHZT-2	1.75	164.08	160.84	169.47	168.77	184.53	132.38

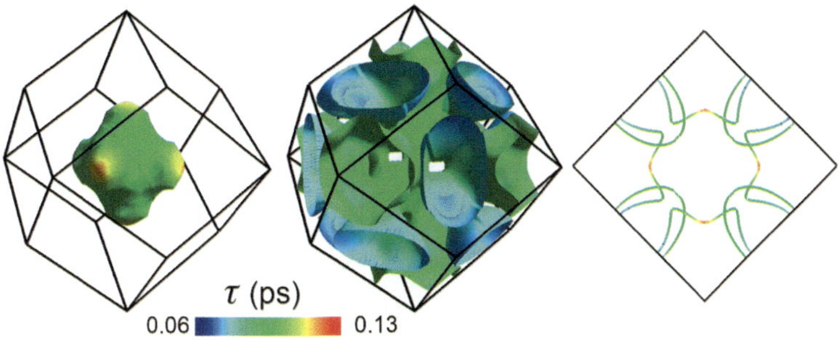

Fig. 5.3 Fermi surface of TNHZT-1 at ambient pressure and its cross-section at $k_z = 0$, colored according to the lifetime of the electronic states

atoms), the electronic bands are very sharp with a small smearing effect. This is especially well seen for the electronic states near the Fermi level. The relatively small band-smearing effect may be qualitatively understood as the consequence of alloying of the isoelectronic elements (Ti, Zr, and Hf) with their nearest neighbors (Nb and Ta) in the periodic table. Therefore, the effective crystal potentials experienced by the valence electrons are not very different, and the influence of chemical disorder on the electronic dispersion relations is not strong. This differs TNHZT from the CsCl-type $(ScZrNb)_{1-x}(RhPd)_x$ superconductors, for which a strongly smeared bandstructure was calculated [33].

The Fermi surfaces (FS) of the studied alloys are shown in Figs. 5.3 and 5.4, and are very similar for the two compositions. FS consists of two sheets, a single pocket centered on Γ and a set of pockets centered on N. To quantify the strength of electron scattering, we calculated the electronic lifetime τ and the mean free path ℓ of electrons. The average values on the Fermi surface are: $\ell = 459$ Å and $\tau = 89$ fs in TNHZT-1 and $\ell = 282$ Å and $\tau = 57$ fs in TNHZT-2. In regard to the fine details, there are some differences in the lifetimes for these two compositions that appear due to slightly different atomic concentrations and Fermi level positions (the differences in band smearing are not visible at the scale of Fig. 5.2). For both alloys, the mean free path is on the order of 100 interatomic distances, proving that the electron scattering induced by the random atomic occupation of lattice sites is not stronger than that in a typical alloy. Consequently, the condition $\hbar/\tau \gg \hbar\omega_D$, which would place the TNHZT alloys in the localized regime (refer to [34]), is not fulfilled. Here, ω_D represents the Debye frequency, and for TNHZT-1 and TNHZT-2, $\hbar/\tau$ equals 7.5 meV and 11.5 meV, respectively, while $\hbar\omega_D$ is 21 meV and 19 meV, respectively.

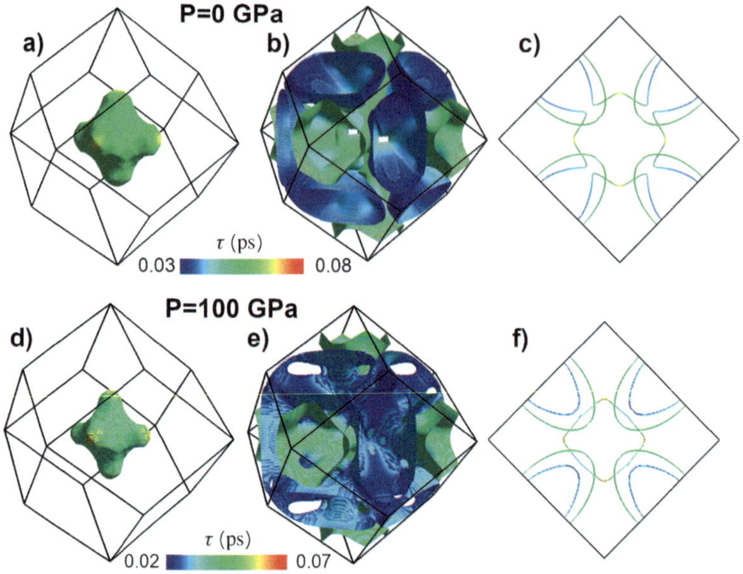

Fig. 5.4 Fermi surface of TNHZT-2 at ambient pressure and 100 GPa, together with its cross section at $k_z = 0$, colored according to the lifetime of the electronic states

Fig. 5.5 Comparison of the electronic DOS for TNHZT-1 obtained from full potential calculations with spin orbit coupling (SOC) and full potential calculations (both using the SPR- KKR code); and spherical potential calculations (using the RCPA code)

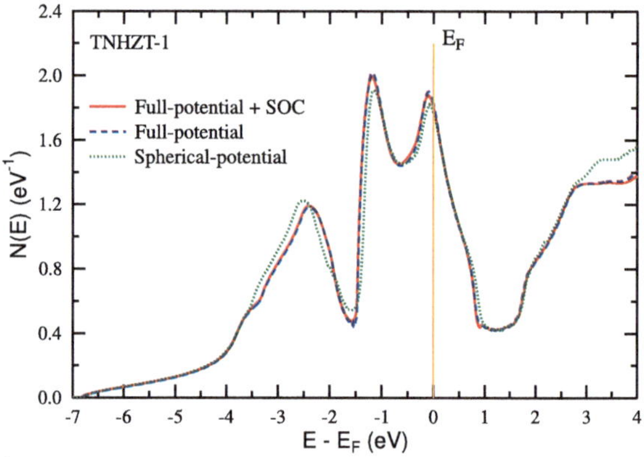

To ensure that the semirelativistic and spherical potential approximations, used in the DOS and electronic band computations in Figs. 5.1 and 5.2 provide an accurate description of the electronic structure, we verified them using the full potential and full relativistic KKR-CPA method, implemented in the Munich SPR- KKR package [9, 18]. The DOS with and without spin-orbit coupling (SOC) are presented in Fig. 5.5, and we can conclude that the SOC does not affect the $N(E)$ shape. Furthermore, the $N(E_F) = 1.82\ eV^{-1}$ computed using the SPR- KKR code is also very close to the value obtained from the RCPA code ($N(E_F) = 1.78\ eV^{-1}$).

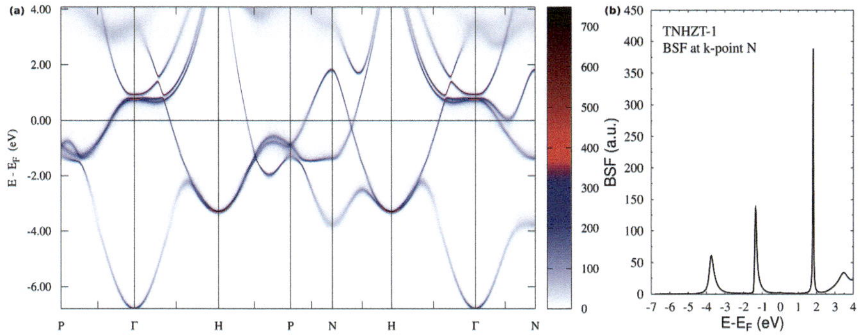

Fig. 5.6 Fully relativistic Bloch spectral functions of TNHZT-1: **a** plotted as dispersion relations with color marking of the value of BSF and **b** plotted at the k-point N as a function of energy. The BSFs for states near E_F are narrow

Next, we calculated the Bloch spectral functions along the same path as in Fig. 5.2 and they are plotted as a color map in Fig. 5.6a. Both the $E(\mathbf{k})$ dependence and the relatively small band smearing are confirmed. Spin-orbit coupling leads to anticrossing effects, for example, in the P-Γ direction, but does not affect the bands near the Fermi level. Although Fig. 5.6 may give an impression of slightly stronger band smearing, this is related rather to a different graphical representation of the results, as in the BSF plot a common color scale is used to show the spectral function values for bands with different smearing. To compare them more precisely, in Fig. 5.6b we plot the BSFs at point N. The FWHM values of the Lorentzian-like BSF are: 0.30 eV (BSF with the peak, i.e., the band center, at –3.71 eV), 0.11 eV (band at –1.34 eV), and 0.04 eV (band at 1.82 eV). FWHM values agree well with the 2Im(E) values of the complex bands, which are equal to 0.39 eV (band at –3.71 eV), 0.20 eV (band at –1.22 eV), and 0.035 eV (band at 1.76 eV).

5.2.2 *Electron-Phonon Coupling and Superconductivity*

Now, the question arises as to what effect a chemical disorder has on superconductivity in the investigated HEA system, as the disorder seems to only weakly affect the electronic structure. Firstly, alloying elements with different atomic numbers allow one to tune the Fermi level position to reach the peak in the $N(E)$ function, as shown in Fig. 5.1, but does it influence superconductivity further?

According to Anderson's theorem [35, 36], due to time-reversal symmetry, superconductivity in a conventional superconductor is usually robust with respect to the presence of nonmagnetic defects and weak disorder. Unconventional superconductors, such as Sr_2RuO_4 [37], where T_c rapidly decreases in the presence of nonmagnetic impurities, are more sensitive to defects. Suppression of superconductivity occurs even in samples with a large mean free path of electrons, $\ell > 500$ Å, that is, approximately hundreds of interatomic distances. However, in conventional superconductors, disorder may also lead to the suppression of superconductivity. This requires strong disorder [34, 36] and has been observed in A-15 superconductors [36], in granular and highly disordered metals [38], and in thin films [39]. Effectively, strong disorder enhances the depairing Coulomb pseudopotential μ^* [36] and not only T_c can be affected. For disordered two-band anisotropic superconductors, the specific heat jump at the superconducting transition $\Delta C/\gamma T_c$ (γ is the Sommerfeld coefficient) may be reduced below the weak-coupling BCS limit of 1.43 [40].

Before proceeding further, we should specify that the effects of disorder should be analyzed on two levels. The first level is the description of how disorder influences the electronic structure and phonons of the HEA system, leading to modifications of the electronic and phononic dispersion relations, densities of states, and, consequently, the strength of electron-phonon interaction. Considering the basic parameters used to quantify this interaction, the first step is to understand how disorder directly influences the value of the electron-phonon coupling constant λ.

More difficult is the second level, which we define as the description of the influence of disorder on the thermodynamic properties of the superconducting phase in HEAs. This includes the value of the critical temperature T_c or the specific heat jump, to mention the basic characteristic parameters. More particularly, using the McMillan formula (5.17) as a picture of passing from basic electron-phonon properties to superconductivity, the second level task includes understanding how the $T_c(\lambda, \theta_D)$ relation is affected by disorder in HEAs, e.g., whether the disorder effectively enhances μ^* to lower T_c below the value we would observe in the ordered system, having the same λ, $N(E_F)$ and θ_D? We will attempt to address these issues here, bearing in mind that the answer is still incomplete.

KKR-CPA calculations have already shown that the disorder effect on the electronic structure, described using band smearing and the electronic lifetime, is not strong in TNHZT. Based on these results, the McMillan-Hopfield parameters were calculated using Eq. (5.12) for each constituent atom in the alloy (Table 5.1). By generalizing equation (5.11) to the i-component alloys [17] we can write λ as

$$\lambda = \sum_i \frac{c_i \eta_i}{M_i \langle \omega_i^2 \rangle}, \tag{5.15}$$

where i refers to the type of atom and c_i is its concentration. Using the KKR-CPA method, the η_i parameter can be calculated for each atomic component of the alloy. However, the phonon part still remains a problem, since phonon calculations for alloys are much more complex and time consuming and have not yet been performed for TNHZT. To avoid this difficulty, the atom-dependent product $M_i \langle \omega_i^2 \rangle$ was replaced by the average term $\langle M \rangle \langle \omega^2 \rangle$ (see [3, 4] for more details), and the average phonon frequency was deduced in the Debye approximation from the experimental Debye temperature, $\langle \omega^2 \rangle = \frac{1}{2}\theta_D^2$. Defining $\langle \eta \rangle = \sum c_i \eta_i$ we may write λ as:

Table 5.2 Average atomic mass (in u), experimental Debye temperature (in K), and dimensionless electron-phonon coupling constant, calculated on the basis of McMillan-Hopfield parameters λ and electronic specific heat λ_γ, theoretical and experimental T_c (both in K) for two variants of TNHZT

alloy	$\langle M\rangle$	θ_D^{expt}	λ	λ_γ	$T_c(\mu^* = 0.13)$	T_c^{expt}
TNHZT-1	124.55	243	1.16	0.98	15.5	7.3
TNHZT-2	126.77	216	1.19	0.87	14.0	7.7

$$\lambda = \frac{\langle\eta\rangle}{\frac{1}{2}\langle M\rangle\theta_D^2}. \tag{5.16}$$

The experimental Debye temperature [1, 4] is 243 K (TNHZT-1) and 216 K (TNHZT-2), which together with the calculated McMillan-Hopfield parameters (see Table 5.1) gives $\lambda = 1.16 - 1.19$ for the two alloys (see Table 5.2). This can be directly compared with the value extracted from the specific heat measurements, as the Sommerfeld coefficient γ is renormalized from the calculated bandstructure value $\gamma_{\text{calc}} = \frac{\pi^2}{3}k_B^2 N(E_F)$ to $\gamma_{\text{expt}} = \gamma_{\text{calc}}(1+\lambda_\gamma)$. The experimental values are $\gamma_{\text{expt}} = 8.3$ and 7.7 mJ/(mol K^2) for TNHZT-1 [1] and TNHZT-2 [4], respectively. When taking into account the KKR-CPA $N(E_F)$ values, we arrive at $\lambda_\gamma \simeq 0.87 - 0.98$.[3] These results are slightly lower than the values computed from the McMillan-Hopfield parameters and the Debye temperature using Eq. (5.16), nevertheless they confirm that $\lambda \simeq 1.0$ in TNHZT.

Finally, the superconducting critical temperature can be calculated using the McMillan formula [25]:

$$T_c = \frac{\theta_D}{1.45}\exp\left[-\frac{1.04(1+\lambda)}{\lambda - \mu^*(1+0.62\lambda)}\right], \tag{5.17}$$

where μ^* is the Coulomb pseudopotential parameter, which effectively describes the depairing effects. When the most frequently used value of $\mu^* = 0.13$ is taken, $\lambda \approx 1.20$ leads to overestimated critical temperatures of $T_c > 14$ K (see Table 5.2). To reproduce the experimental critical temperatures with $\lambda = 1.16 - 1.19$ enhanced values of $\mu^* \simeq 0.25$ must be used. If lower values of λ_γ are used, $\mu^* \simeq 0.20$ and 0.14 are required for THNZT-1 and TNHZT-2, respectively. Thus, we see a need to use the increased value of μ^*, which could indicate the suppression of superconductivity due to disorder, decreasing T_c from the values expected for the ordered compound. However, it should be noted that in some ordered systems, including Nb or Nb_3Ge [41, 42], $\mu^* \sim 0.2$ was needed to reproduce the experimental critical temperatures, so a clear conclusion cannot be drawn based only on this observation.

[3] The subscript γ is used to denote the λ value obtained from the Sommerfeld parameter.

To extend the discussion of the effect of disorder on both the electron and phonon spectra, we performed calculations using the virtual crystal approximation, as described in the next section.

5.2.3 Virtual Crystal Approximation

In the next step, we employed the virtual crystal approximation (VCA) to calculate the electronic and phonon structures of TNHZT-1 [43]. Initially, we did not expect the VCA approach to adequately describe the electronic structure of this HEA; however, the KKR-CPA calculations showed the presence of a sharp Fermi surface and small band smearing; therefore, verification of the effectiveness of VCA was a truly intriguing issue. We used the QUANTUM ESPRESSO code [44, 45], which is based on the pseudopotential and plane-wave methods. Fully relativistic ultrasoft pseudopotentials ([46]) with Perdew-Burke-Ernzerhof (PBE) [47] exchange-correlation functionals were used. The alloying effect was simulated by VCA, where in the case of the pseudopotential technique, a single effective pseudopotential is created by adding the pseudopotentials of the constituent atoms with appropriate weights, according to their concentrations.[4] The unit cell volume was optimized, giving a lattice parameter of $a = 3.385$ Å, which is only 0.7% larger than the experimental. Spin-orbit coupling was included in the calculations, and the kinetic energy and charge density cutoffs of 80 Ry and 800 Ry were used. A mesh of 20^3 k-points was used in the self-consistent cycle, while the electronic densities of states were calculated on a mesh of 48^3 k-points. The phonon structure was calculated using density functional perturbation theory on a mesh of 10^3 $\mathbf{q}$ points.

The VCA electronic structure of TNHZT-1 is presented in Fig. 5.7, and, quite surprisingly, it is very similar to that obtained from the CPA calculations. Apart from the blurring effect, which is absent in the ordered VCA medium, the dispersion relations (Fig. 5.7a) are similar to those presented in Figs. 5.2 and 5.6, including band extreme positions and anticrossings, which appear due to spin-orbit coupling. In Fig. 5.7b the VCA density of states is compared to the DOS obtained with the fully relativistic KKR-CPA method. The shape of the DOS is only slightly different, as the CPA density is blurred due to the band smearing effect, which generally reduces the height of the $N(E)$ peaks. As a consequence, the VCA gives a 10% greater $N(E_F)$ (1.99 eV^{-1}) than KKR-CPA (1.82 eV^{-1}). The VCA Fermi surface is shown in panels (c) and (d) and is also very similar to that calculated by KKR-CPA in Fig. 5.3. This confirms that quite surprisingly, the VCA reproduces the electronic structure of the TNHZT alloy well. This behavior is likely related to the fact that the investigated alloy consists of three isoelectronic elements and two of their nearest neighbors in the periodic table; thus, the screened atomic potentials "seen" by the valence electrons are not very different, leading to an almost rigid-band-like evolution of the electronic structure.

[4] See the QUANTUM ESPRESSO documentation and the virtual.x program.

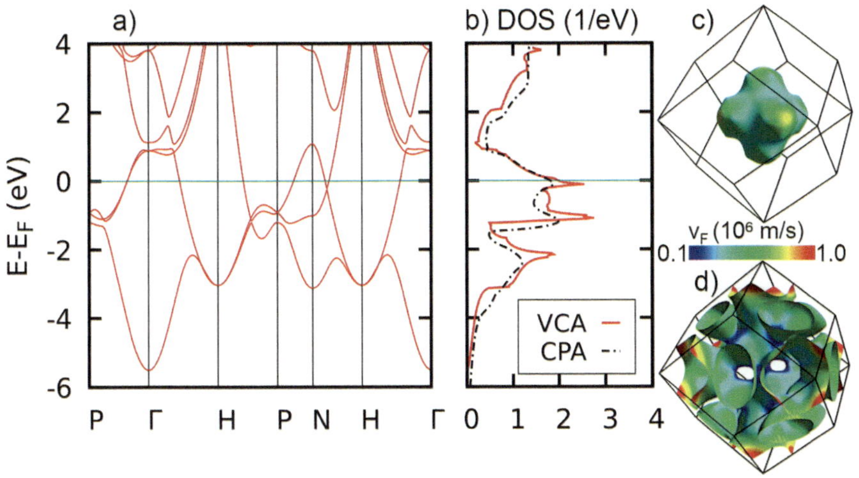

Fig. 5.7 Electronic structure of TNHZT-1 calculated via the VCA approach: **a** band structure plotted along the high-symmetry directions; **b** DOS (comparison of the VCA and CPA data both computed with spin-orbit coupling); **c–d** Fermi surface colored with the Fermi velocity

The next interesting task was to obtain the phonon spectrum and the electron-phonon interaction parameter as predicted by the VCA. The average atomic mass was used to calculate the phonon frequencies, with the force constants obtained for the uniform VCA medium. The mass- and force-constant fluctuation effects, neglected here, are expected to have a strong influence on phonons, which may be relatively stronger than the atomic potential disorder has on the electronic structure. The mass differences are large: from 47.9 u for Ti to 180.9 u for Ta. The effect of disorder on phonons should appear in real materials in the same way as for electronic dispersions, with a broadening of phonon dispersion relations due to the finite lifetime of the phonon [48]. The mass-fluctuation phonon scattering parameter may be used to predict whether phonons are strongly scattered in alloys [48, 49]:

$$\Gamma_M = \sum_i \frac{c_i (M_i - \langle M \rangle)^2}{\langle M \rangle^2}, \tag{5.18}$$

where c_i and M_i are the concentration and atomic mass of the i-th component of the alloy and $\langle M \rangle$ is the average mass. Γ_M is proportional to the scattering rate, that is, the inverse of the relaxation time of the phonon [49]. This parameter is 0.158 for TNHZT-1 and 0.155 for TNHZT-2, which are moderate values. In [48] phonon broadening for various alloys in the V-W-Nb-Ta-Ti system was theoretically investigated, from binaries to five-component HEAs, with Γ_M ranging from 0.09 to 0.32. In [50] phonons for $Fe_{0.75}Pd_{0.25}$ ($\Gamma_M = 0.10$) and $Fe_{0.75}Pt_{0.25}$ ($\Gamma_M = 0.44$) were studied. However, these materials also exhibit magnetic disorder.

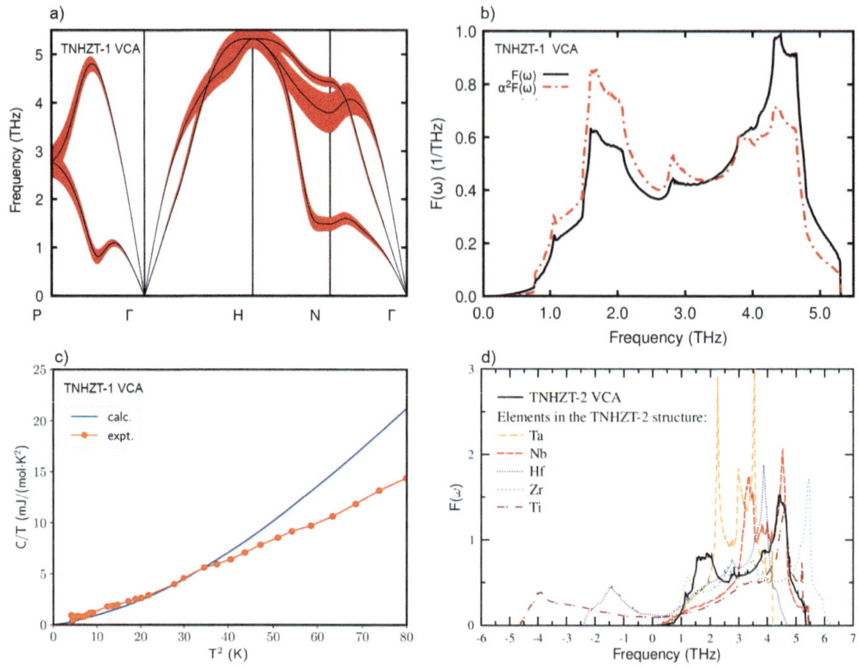

Fig. 5.8 Phonon structure of TNHZT-1 calculated within the VCA approach: **a** dispersion relations plotted along high symmetry directions, together with the phonon linewidths (multiplied by a factor of 8) shown as the bandwidth; **b** phonon density of states $F(\omega)$ and Eliashberg function $\alpha^2F(\omega)$ ($F(\omega)$ is renormalized to have the same area under the curve as the Eliashberg function); **c** calculated phonon contribution to the specific heat of TNHZT-1 together with experimental data (obtained from the measured specific heat [1] and subtracting the electronic contribution γT); **d** $F(\omega)$ of TNHZT-2 together with $F(\omega)$ of elements calculated in the crystal structure of this alloy

The above-mentioned works showed that although Γ_M cannot fully cover variations in phonon broadening with alloy composition (e.g., it does not take into account the fluctuations in force constants), we may expect that in systems with lower Γ_M the influence of disorder on phonons should be weaker, and the phonon spectrum should be closer to that predicted by the averaged mass and averaged force constant model, that is, a VCA-like model applied to phonons.

In Fig. 5.8a, b, we present the VCA phonon structure of TNHZT-1 in terms of the phonon dispersion relation $\omega(\mathbf{q})$ and DOS $F(\omega)$. The monoatomic unit cell of this compound results in three acoustic phonon branches, which span in the frequency range from 0 to nearly 5.5 THz. What seems unusual in this calculated phonon spectrum is the softened longitudinal mode with a dip along the $\Gamma - P$ direction and a flat dispersion near N. These features enhance the phonon DOS below and at approximately 1.5 THz, leading to a rather soft phonon spectrum. To compare the calculations with the experimental findings, the low-temperature phonon-specific heat is calculated on the basis of the theoretical phonon DOS and compared with the experimental data from [1] in Fig. 5.8c. The calculated lattice specific heat is considerably larger than the experimental one above 6 K, which means that the theoretical phonon DOS is too large in the low-frequency region. Disagreement with the experimental results proves that the computed VCA phonon spectrum is too soft.

In Fig. 5.8d we present the phonon DOS of TNHZT-2 calculated in the same way. It is very similar to the one of TNHZT-1. Furthermore, in this figure, we have added the DOS of hypothetical structures of constituent atoms calculated in the *bcc* unit cell with the TNHZT-2 alloy lattice parameter. Among these elements, only Nb and Ta naturally crystallize in the *bcc* structure (but with a smaller unit cell volume), while Hf, Zr, and Ti form a hexagonal close-packed structure, *hcp*. The phonon spectra of Nb, Ta, and Zr in the TNHZT-2 structure are stable, with average frequencies of 3.57 THz, 2.85 THz and 3.67 THz, respectively, while the Ti and Hf spectra are unstable (frequencies are imaginary, denoted here as negative). The mean phonon frequencies of Nb and Ta mentioned above are lower than those of their natural structures (4.23 THz and 3.24 THz, respectively, calculated using the same methods as for TNHZT with optimized lattice constants equal to 3.313 Å and 3.30 Å respectively). This is due to the smaller unit cell volume of the elements. The instability of the phonon spectrum in the *bcc* structure for Ti and Hf confirms that it is not the preferred structure for these elements at this volume per atom. Ti and Hf undergo a transition to *bcc* crystal structure at the pressure of 243 GPa [51] and 71 GPa [52], respectively, while the TNHZT-2 lattice parameter is equivalent to pressures of –10 GPa and 15.2 GPa, respectively. For Zr, the pressure of the *hcp*-*bcc* transition is much lower, equal to 30 GPa [52], which explains why the calculated spectrum is stable (the TNHZT-2 lattice parameter for *bcc* Zr generates a pressure of 18.1 GPa). As one can see, the VCA phonon spectrum of TNHZT is certainly not the average of the phonon spectra of the constituent elements. The instability of the Hf and Ti phonon spectra in the cubic structure may qualitatively explain the softening of the VCA phonon spectrum and may also suggest that in real samples the Hf and Ti atoms locally tend to avoid the cubic *bcc*-like geometry.

Let us now examine the VCA effect on the electron-phonon interaction in TNHZT. In Fig. 5.8a phonon dispersion relations have shading added, which corresponds to the phonon linewidth $\gamma_{\mathbf{q}\nu}$ (inversely proportional to the phonon lifetime, in the same way as the electronic lifetime in Eq. (5.1); however, here the electron-phonon interaction is the reason for phonon scattering). In Fig. 5.8b the Eliashberg function $\alpha^2F(\omega)$, which describes the frequency-dependent magnitude of the electron-phonon interaction, is presented. We note that the computed Eliashberg function does not follow the shape of the phonon DOS. Phonon linewidths are larger in the low-frequency range for the lowest longitudinal mode, resulting in an enhancement of the $\alpha^2F(\omega)$ function above the bare phonon DOS below 2.5 THz. This has a strong influence on the average electron-phonon coupling constant λ. As $\lambda \propto \frac{\gamma_{\mathbf{q}\nu}}{\omega^2_{\mathbf{q}\nu}}$, the soft phonon spectrum, with large $\alpha^2F(\omega)$ values in the low-frequency range, results in a very large value of $\lambda = 1.70$, significantly overestimating the experimental and theoretical findings ($\lambda \sim 1$).

Such a large disagreement in λ shows that the VCA applied to phonons fails to describe the electron-phonon coupling in TNHZT. This markedly contrasts with a fairly good VCA description of the electronic structure, which appeared to be consistent with the coherent potential calculations in which chemical disorder is taken into account.

To confirm the hypothesis that the excessively soft phonon structure causes the overestimation of λ in the Virtual Crystal Approximation (VCA) description of the TNHZT high-entropy alloy, we calculated the electronic and phonon contributions to λ as predicted by the VCA Eliashberg function. The VCA value of the McMillan-Hopfield parameter, based on the definition (5.8), is equal to $\eta = 132$ mRy/a_B^2. This is not far from the value $\langle\eta\rangle = 158$ mRy/a_B^2 computed using the rigid muffin-tin approximation in KKR-CPA, being only 16% lower (see Table 5.1). By definition in Eq. (5.10), the VCA value of $\langle\omega^2\rangle$ is 6.29 THz^2 for TNHZT-1 and 6.38 THz^2 for TNHZT-2. These values are much lower than those used to estimate $\langle\omega^2\rangle$ via Eq. (5.14) from the experimental Debye temperatures: 12.8 THz^2 and 10.1 THz^2 respectively. We cannot directly compare these numbers because neither the phonon Density of States (DOS) is exactly Debye-like, nor is the $\alpha^2 \simeq$ const approximation fulfilled. However, the much lower VCA $\langle\omega^2\rangle$ values confirm that the overestimation of λ computed from VCA $\alpha^2 F(\omega)$ is mainly due to the underestimation of the phonon frequencies. On the other hand, the close agreement of the two estimates of η by independent methods, VCA and Rigid Muffin-Tin Approximation (RMTA) + KKR-CPA, suggests that the electronic contribution to λ is calculated with good accuracy.

5.2.4 Effect of Pressure

An important discovery for superconducting high entropy alloys was made in [53], where the evolution of superconducting properties under external pressure was investigated. In that work, the superconducting properties of TNHZT-2 were measured up to extreme pressures of 190 GPa and T_c was found to increase to approximately 10 K at 50-60 GPa and then remained almost constant to approximately 100 GPa. At higher pressures, it slowly decreased to 9 K at 190 GPa. In our work [4] we investigated the effects of pressure on the electronic structure and superconductivity of TNHZT-2 up to 100 GPa, that is, in the pressure range in which the crystal structure was experimentally determined and confirmed as *bcc*, with no phase transitions or atomic ordering observed.

Calculations were performed using the spherical-potential implementation of KKR-CPA in the RCPA code [6–8], with similar technical details as previously used in studies under ambient conditions. The influence of pressure on the total density of states, $N(E)$, and the atomic contributions, $n(E)$, is presented in Fig. 5.9. With a decrease in the unit cell volume, hybridization becomes stronger and the DOS peaks decrease. The separation between the DOS peaks, the one at the Fermi level and the next one below, also increases. The density of states in the lowest energy region

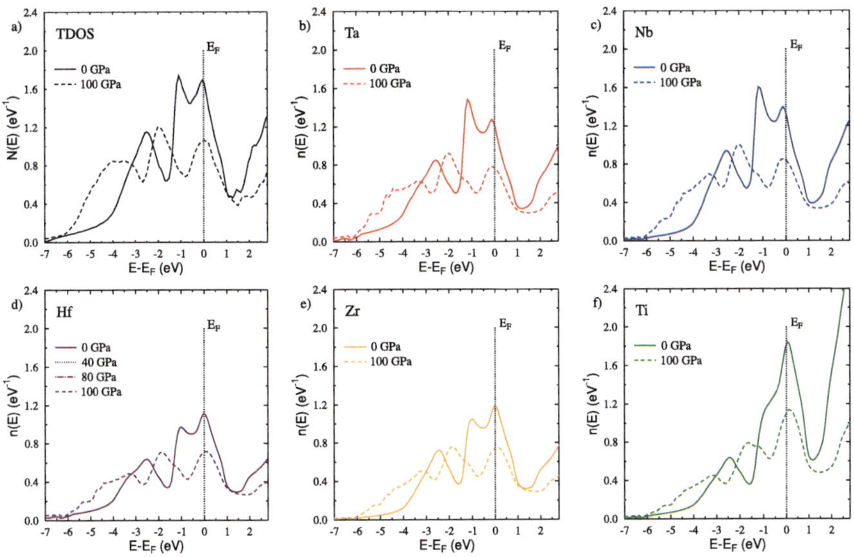

Fig. 5.9 Evolution of the electronic structure under pressure in TNHZT-2: total **a** and atomic **b–f** densities of states

(~ -5 eV) increases with pressure. The change in unit cell volume does not have an impact on the order of relative atomic contributions to the total density of states (the highest contributions for all pressures are those from Ta and Nb because of their highest atomic concentration).

The complex energy bands for TNHZT-2 at ambient pressure and 100 GPa are presented in Fig. 5.10 and the Fermi surface is presented in Fig. 5.4. In the dispersion relations, we may distinguish two opposite trends, which explain the evolution in the DOS. Some of the electronic bands move to lower energies, for example, at Γ, N or H, whereas the hole-like bands at P or N move up. The most interesting change occurs in the $\Gamma - N$ direction in the vicinity of the Fermi level, where the Lifshitz transition is observed (the topology of the Fermi surface changes as the additional band starts to contribute to the FS). This is illustrated in Fig. 5.11. The center of the band crosses E_F at approximately 70 GPa; however, due to band smearing the blurred band contributes to $N(E_F)$ even at lower pressures. Between 30 and 40 GPa there is a noticeable increase in the band smearing, while between 20 and 30 GPa the difference in the imaginary part of the energy is negligible. This transition is correlated with the pressure evolution of $N(E_F)$ (see [4]), which generally decreases with pressure, but between 40 GPa and 70 GPa this trend slows.

Notably, the Lifshitz transition was reported for one of the TNHZT constituent atoms, Nb. The transition pressure is 60 GPa; additional states at the Fermi level appear in $\Gamma - N$, as in the TNHZT case. For a more detailed discussion, see ref. [4] and the references therein.

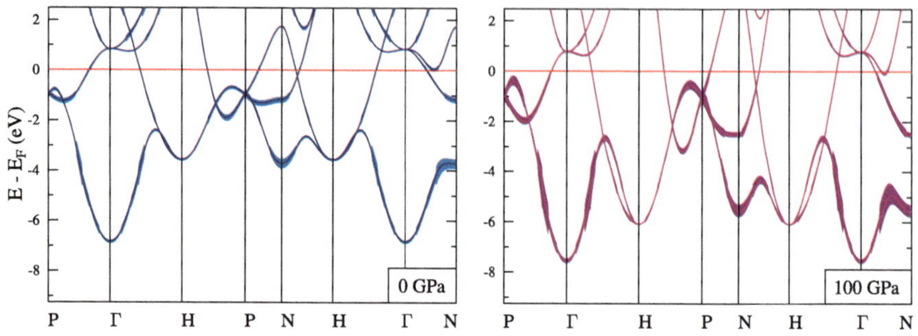

Fig. 5.10 Pressure evolution of electronic dispersion curves in TNHZT-2

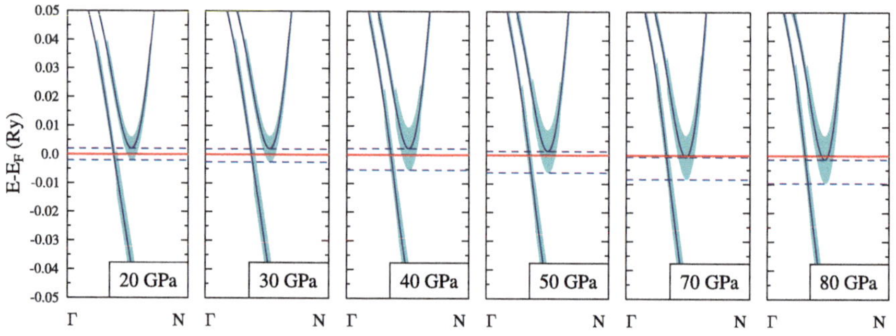

Fig. 5.11 Electronic dispersion relations in the $\Gamma - N$ high symmetry direction for selected hydrostatic pressures in TNHZT-2. Band smearing is marked with light blue. Dashed lines show the conduction band minimum (center and half-bandwidth)

Next, the pressure evolution of the McMillan-Hopfield parameters was determined, and a typical result was obtained, namely, that they increase with pressure [4]. Figure 5.12a illustrates the relative increase in the weighted average $\langle\eta\rangle$, which is more than doubled at a pressure of 100 GPa.

Before moving on and presenting the effect of pressure on the electron-phonon coupling constant λ, we explain how the pressure evolution of the phonon contribution was simulated. As for the ambient pressure, Eq. (5.16) with the average mass and Debye approximation was used, but additionally $\theta_D(P)$ (or equivalent $\theta_D(V)$, as the $V(P)$ curve was measured experimentally [53]) had to be modeled. We determined that relationship on the basis of the Grüneisen parameter γ_G [54]

$$\gamma_G(V) = -\frac{\partial \ln \theta_D}{\partial \ln V}. \tag{5.19}$$

and second-order Grüneisen parameter $q(V)$

$$q(V) = \frac{\partial \ln \gamma_G(V)}{\partial \ln V}. \tag{5.20}$$

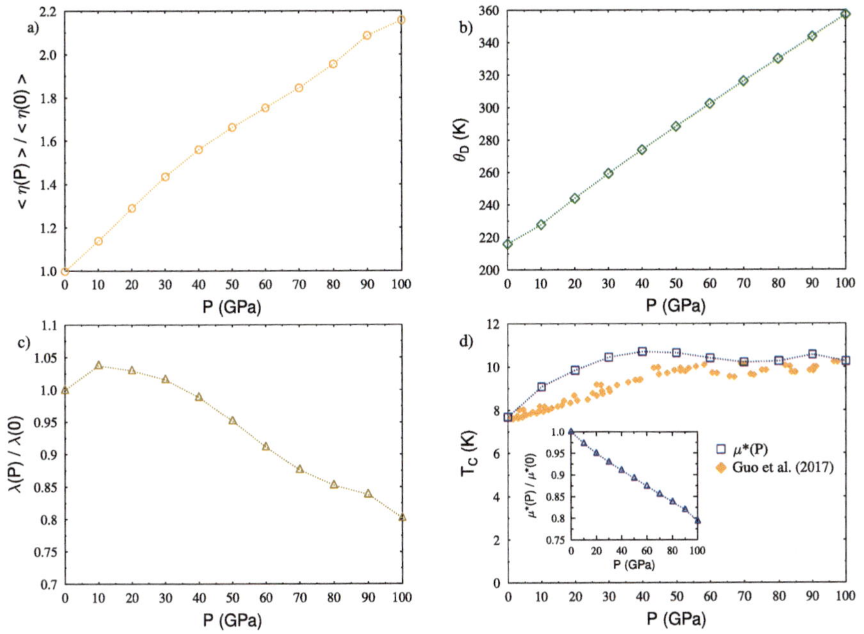

Fig. 5.12 Effect of hydrostatic pressure on: **a** $\langle \eta \rangle$; **b** Debye temperature; **c** electron-phonon coupling constant, normalized with the ambient pressure value λ_0; **d** the critical temperature of the TNHZT-2 alloy. The inset in panel **d** shows the relative change in the Coulomb pseudopotential, which was predicted using Eq. (5.24) and used in calculations of T_c

This approach worked well for Ta and Nb examples, providing results consistent with direct phonon calculations and literature data. For a more detailed discussion see [4] and references therein. The final formula for $\theta_D(V)$ is the following:

$$\theta_D(V) = \theta_D^0 \, (\zeta)^{\gamma_G^0 exp\left[\frac{q_0}{n}(\zeta^n - 1)\right]}, \tag{5.21}$$

where γ_G^0 and q_0 are the Grüneisen parameters at ambient pressure. Parameter n is a positive integer, characteristic for alloys. There was no sufficient data to estimate n; thus, we used the value $n = 16$ which ensured the compatibility of the results for $\theta_D(P)$ obtained from the model and from the direct phonon calculations for the two constituent atoms, Ta and Nb.

The Grüneisen parameter at P = 0 GPa was determined on the basis of measurements of the thermal expansion coefficient, as

$$\gamma_G^0 = \frac{\alpha B V_0 N_A}{C_V}. \tag{5.22}$$

In the above equation N_A is the Avogadro number, C_V the molar constant-volume heat capacity, α_V is the volume thermal expansion coefficient, B denotes the bulk modulus of the crystal and V_0 represents the primitive cell volume. The values of the bulk modulus $B = 177.35$ GPa and $B' = 2.87$ were obtained by fitting the third-order Birch-Murnaghan equation of state [55] to the experimental $V(P)$ data, obtained via the X-ray diffraction (XRD) measurements [53]). The volume thermal expansion coefficient α_V was also measured using XRD [4] and a value of $\alpha_V = 2.07\text{K}^{-1}$ was obtained. This value is slightly lower than α_V of similar refractory HEAs, being in the range from 2.30 to 3.60 K^{-1} [4]. The bulk moduli of several other HEAs listed in [4] do not exceed 125 GPa; only NbTaMoW exhibits B of 162.50 GPa. The Grüneisen parameter of TNHZT-2 (1.62) is of the same order as γ_G^0 of similar HEAs [56],[57]. The second-order Grüneisen parameter at an ambient pressure q_0 is easily calculated with the Grüneisen parameter and the pressure derivative of the bulk modulus:

$$q_0 = 1 + 2\gamma_G^0 - B'. \tag{5.23}$$

For TNHZT-2 we obtained a value of $q = 1.36$. The relationship between the pressure and Debye temperature, obtained with the described method, is shown in Fig. 5.12b. The value of θ_D increases monotonically, from 216 K to 357 K at 100 GPa. Based on Eq. (5.16) the evolution of the electron-phonon coupling constant with pressure, $\lambda(P)$, was determined and is plotted in Fig. 5.12c. Interestingly, after a small initial increase between 0 and 10 GPa, we observe a gradual decrease of λ. There are two competitive factors that determine the pressure evolution of λ: the increase in η, due to the decrease in the unit cell volume, and the increase in θ_D, due to the stiffening of the crystal lattice. As can be shown [4] by calculating a logarithmic derivative of λ with respect to pressure, the increasing or decreasing behavior of λ is determined by the sign of the factor $\left(\frac{d\ln\langle\eta\rangle}{d\ln V} + 2\gamma_G(V)\right)$.

The first term in brackets is negative, since McMillan-Hopfield parameters increase with decreasing unit cell volume; generally the logarithmic volume derivative of η is between –1.0 and –3.0 [30, 58, 59]. The second term is positive, being twice the Grüneisen parameter, and in our case $2\gamma_G \simeq 3.0$. For a λ to increase with pressure, $-\frac{d\ln\langle\eta\rangle}{d\ln V} > 2\gamma_G(V)$. In the case of TNHZT-2, this condition is fulfilled only for pressures below 10 GPa, where the increase in $\langle\eta\rangle$ with pressure is the steepest, see Fig. 5.12c. For higher pressures, we have $\frac{d\ln\langle\eta\rangle}{d\ln V} < 2\gamma_G(V)$ and λ decreases by 20% for $P = 100$ GPa, see Fig. 5.12c.

However, the superconducting critical temperature in the McMillan formula (5.17) also depends on two other factors, the Debye temperature θ_D and the Coulomb pseudopotential μ^*. As we already know, θ_D increases with pressure. To simulate the possible behavior of μ^* with pressure, we used the Benneman and Garland formula [17, 60]

$$\mu^* = \frac{AN(E_F)}{1 + N(E_F)}, \tag{5.24}$$

which relates μ^* to the total density of states at the Fermi level, $N(E_F)$. The parameter A was fixed to reproduce the critical temperature at ambient pressure, and $\mu^*(P)$ evolution was then simulated according to the calculated $N(E_F)(P)$, see [4]. Due to the decrease in $N(E_F)$, μ^* decreases, as shown in the inset in Fig. 5.12d. The final $T_c(P)$ function is presented in Fig. 5.12d, where it is compared with the experimental data. As one can see, we were able to achieve a fairly good description of the pressure changes in T_c, confirming its electron-phonon origin.

5.2.5 *Effect of the Local Lattice Distortions*

In real samples of highly disordered materials built of many atoms of different sizes, because of the relaxation process, the crystal structure will exhibit local distortions from the perfect atomic arrangement. We studied the effects of local distortions on the electronic structure and electron-phonon coupling of TNHZT-2 in [61] and revieweed them briefly here.

First, a set of twelve 3×3×1 supercells (containing 18 atoms) with different atom arrangements was generated, removing all symmetry constraints. This choice of multiplying the TNZHZT-2 unit cell allowed the atomic concentrations of the alloy to remain unchanged. Then, the atomic positions in the supercells were relaxed until the forces acting on all the atoms were less than 2 mRy/a_B. For this purpose, the Wien2k package [62] was used. Next, for the unrelaxed and relaxed supercells, the electronic structure and McMillan-Hopfield parameters were calculated using the same methods as in Sect. 5.2.1; however, here each atomic site was occupied by only one atom. All the results obtained for relaxed supercells are denoted here as D(+) and compared to the results of the calculations for supercells without distortions, denoted as D(–), as well as for the primitive cell with full disorder, obtained using the CPA method. In the calculations here, we set the identical radius of the muffin-tin spheres of all atoms at 2.5 a_B. Such a value does not provide maximal unit cell filling, but was necessary to avoid the overlap of the muffin-tin spheres during relaxation.

Due to relaxation of the atomic positions, the interatomic distances in distorted supercells were changed up to ± 8% from the values in undistorted supercells. The largest atomic displacements are observed in areas with three Hf and/or Zr atoms being the nearest neighbors. These two elements have similar, large atomic radii. Ta and Nb tend to decrease the distance between each other. Notably, bulk Hf, Zr, and Ti, as well as their binary alloys crystallize in the *hcp* structure. Local instability due to large distortion might explain reports of hexagonal phase precipitation in a similar TaNbHfZrTi alloy [63, 64].

Figure 5.13a presents a comparison of the density of states, $N(E)$, of representative supercells, both distorted and undistorted. The density of states for the fully disordered variant (scaled to match the same number of atoms) is also included. The DOS of this particular undistorted supercell (model no. 1) is quite similar to that obtained via the CPA approach. The difference between $N(E_F)$ is only 0.3%, which shows that the minimal 3×3×1 supercell model is a good starting point for studies of structural distortions in TNHZT-2. Due to atomic displacements in the relaxation process, noticeable differences between the DOS of D(–) and D(+) supercells appear, starting at 2 eV below E_F and reaching the vicinity of the Fermi level. The distor-

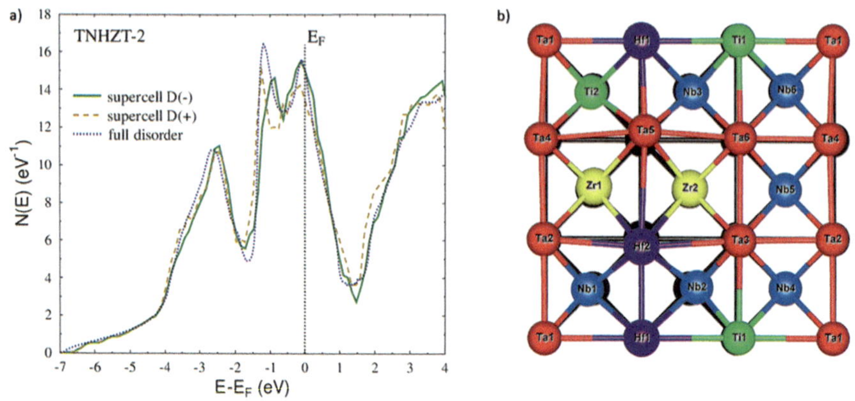

Fig. 5.13 **a** Electronic DOS of TNHZT-2 obtained for the three structural variants: one of the supercell models (no. 1) without distortions, labeled D(–); the same supercell with distortions, labeled D(+); and the KKR-CPA result with random atom occupation, labeled full disorder. **b** Supercell no. 1 with distortions, projected on the xy plane; in the background, the undistorted positions of the atoms are shown in black

tions in this particular supercell result in a decrease in $N(E_F)$ by 10.1%. A similar decrease of $N(E_F)$ was observed for the other supercells, and the average $N(E_F)$ of the 12 models decreased by 11.6% due to relaxation.

To fully analyze the effect of local distortions on the electron-phonon coupling constant, it would be necessary to perform phonon calculations for all supercells (without symmetry). We have not been able to do this thus far, so we have to limit ourselves to a discussion based on changes in the electronic part (McMillan-Hopfield parameters), leaving the Debye temperature and average mass as the phonon contribution in Eq. (5.16).

Figure 5.14 presents the relative changes in the McMillan-Hopfield parameters for each type of atom in different supercells, before and after relaxation. As all of the atoms were nonequivalent in these calculations, the points refer to the average value for each type of atom in the given supercell. The CPA values are taken as a reference (continuous $y = 1.0$ line in the plot). Additionally, the values averaged over all the supercell models are plotted with dashed (no distortions) and dash-dotted (with distortion) lines. First, we note that the supercell η_i D(–) values (i.e., before relaxation, without distortions) averaged over the different models are very close to the CPA values. The difference between the "total" η ($\sum_i \eta_i$ for the supercell and $\sum_i c_i \eta_i$ for the CPA) is on the order of 1%. This confirms that a set of twelve $3\times3\times1$ supercells is a statistically sufficient sample to predict the electronic properties of this alloy.

Taking distortion into account results on average in a decrease in the McMillan-Hopfield parameters. The η_i parameters for Hf and Zr are the most sensitive to distortions, since the decreases are approximately 33% and 31%, respectively. On average, the summed electronic contribution to λ of the TNHZT-2 supercell decreases by 18%

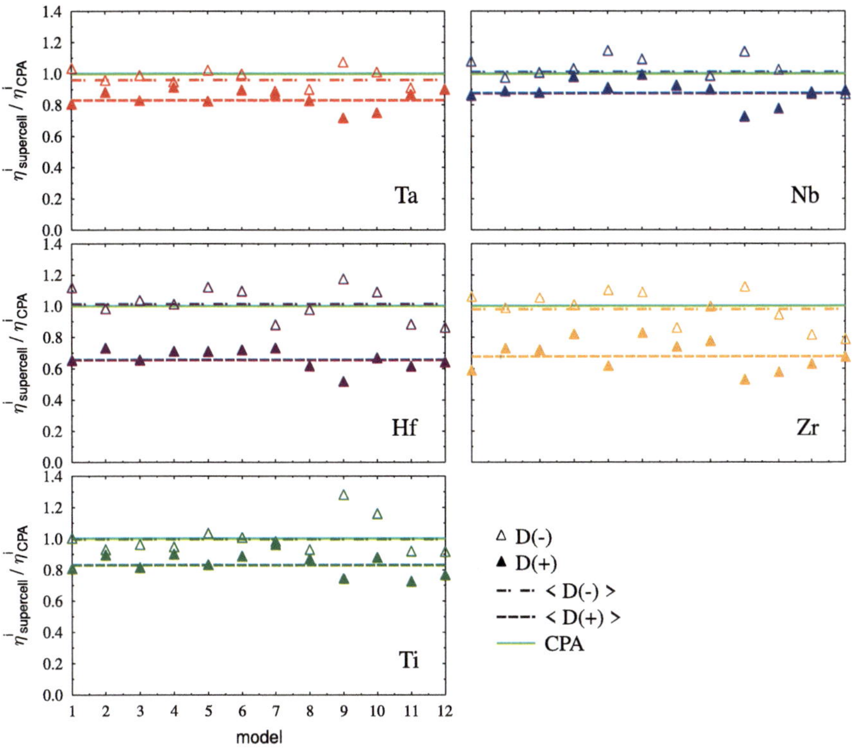

Fig. 5.14 Relative change in the McMillan-Hopfield parameters for all types of atoms in the system in different variants of the supercells of TNHZT-2, computed with distortions, D(+) (full symbols), without distortions, D(–) (empty symbols) with respect to the CPA values calculated for a fully disordered system (continuous reference line at 1.0). Dashed and dash-dotted lines denote the McMillan-Hopfield parameter values for a given atom averaged over all 12 supercell models

as a result of local atomic distortions. This reduces the λ calculated for TNHZT-2 based on McMillan-Hopfield parameters from $\sim$1.2 to $\lambda \sim 1.0$, as presented in Fig. 5.15a. As a consequence, reproducing the experimental critical temperature $T_c = 7.7$ K now requires taking a less-enhanced value of the Coulomb pseudopotential, equal to $\mu^* \approx 0.18$, as shown in Fig. 5.15b.

It is worth noting that, by taking into account the local distortions in the calculations, we also bring the values of λ for TNHZT-2 obtained from the McMillan-Hopfield parameters and as a renormalization of the electronic specific heat closer. According to Table 5.2 the difference between them was greater than 30% (1.19 versus 0.87). The lower average theoretical value of $N(E_F)$ increases λ_γ to 1.09, while the value based on the McMillan-Hopfield parameters decreases to $\lambda \simeq 0.98$, so the difference is reduced to approximately 11%, providing a more consistent prediction of the electron-phonon coupling constant of TNHZT-2, with $\lambda \simeq 1$. These results

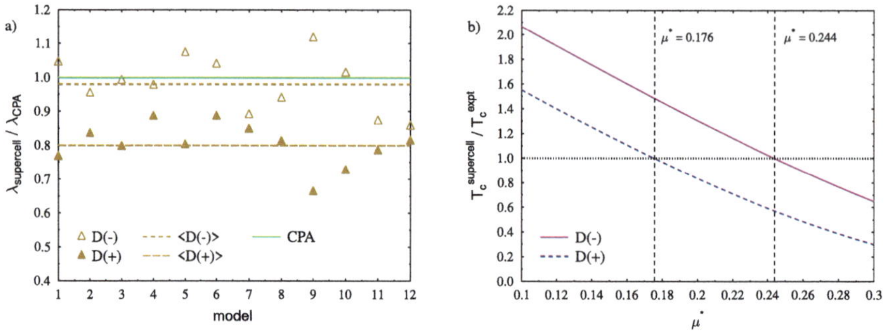

Fig. 5.15 **a** Electron-phonon coupling constant values calculated for twelve supercell models, normalized to the values obtained from the CPA calculations. **b** Critical temperatures obtained for supercell models, normalized with experimental T_C, as a function of the Coulomb pseudopotential parameter

show that local structural distortions should be considered to provide an accurate theoretical description of high entropy alloys.

5.2.6 *Summary*

Our theoretical results shed some light on the influence of disorder on the electronic, phonon, and superconducting properties of TNHZT. The most important findings are as follows:

- KKR-CPA calculations revealed a sharp Fermi surface and well-defined electronic dispersion relations with a relatively weak band-smearing effect.
- The calculated values of $N(E_F)$ and the experimental measurements of the electronic specific heat coefficient predict that the electron-phonon coupling parameter λ is $\sim 0.9 - 1.0$.
- The computed McMillan-Hopfield parameters, through Eq. 5.16 with the average mass and experimental Debye temperature give $\lambda \sim 1.2$ for the ideal *bcc* structure.
- The critical temperature, calculated using the McMillan formula for the computed $\lambda \sim 1.2$ is overestimated when $\mu^* = 0.13$ is used.
- The virtual crystal approximation (VCA) reproduces the KKR-CPA electronic structure well in terms of the density of states and shape of the Fermi surface.
- The VCA used in the phonon calculations predicts a too soft phonon spectrum, which is evidenced by a too large computed lattice specific heat and too low average phonon frequencies, leading to a strongly overestimated electron-phonon coupling parameter $\lambda_{VCA} = 1.70$.
- The pressure effect on T_c is well reproduced by the KKR-CPA calculations and the Debye-Grüneisen model of the phonon spectrum, if the experimental Debye temperature is used as a starting point for ambient conditions. The electron-phonon

coupling parameter λ is predicted to decrease with pressure, and the increase in T_c is mainly caused by an increase in θ_D.

- Local lattice distortions change interatomic distances up to 8% and tend to decrease $N(E_F)$ and the electronic contribution to λ, bringing the computed results closer to the experimental ones.

The presented results show that the chemical (substitutional) disorder in TNHZT has a relatively weak effect on the electronic structure of this system. On the other hand, structural disorder arising from local distortions of the crystalline lattice appears to be an important factor influencing the electronic structure. Both types of disorder are expected to have a strong effect on the phonon spectrum, which cannot be accurately described using virtual crystal and average mass approximations. The question remains as to what the effect of disorder is on the thermodynamics of the superconducting phase. From the present results, if we accept the $\lambda \simeq 1$ value, a slightly enhanced value of the Coulomb pseudopotential μ^* is necessary to reproduce the experimental T_c, suggesting some suppression of superconductivity by disorder. However, as mentioned above, enhanced values of μ^* were needed to reproduce T_c in some well-ordered materials, even in elemental Nb (note that spin fluctuations were also suggested to be responsible for this discrepancy for niobium). Thus, a conclusion on this point is not possible without the full *ab initio* electron-phonon calculations performed for the realistic model of this high-entropy alloy, where both chemical and structural disorders have to be taken into account. Present results suggest that in TNHZT the formation of the superconducting phase is only slightly suppressed by disorder. This issue should be the subject of future theoretical studies of the superconductivity of high-entropy alloys.

Acknowledgements K.J., S.G. and B.W. were supported by the National Science Centre (Poland), project No. 2017/26/E/ST3/00119.

References

1. P. Koželj, S. Vrtnik, A. Jelen, S. Jazbec, Z. Jagličić, S. Maiti, M. Feuerbacher, W. Steurer, J. Dolinšek, Discovery of a superconducting high-entropy alloy. Phys. Rev. Lett. **113**, 107001 (2014)
2. F. von Rohr, M.J. Winiarski, J. Tao, T. Klimczuk, R.J. Cava, Effect of electron count and chemical complexity in the Ta-Nb-Hf-Zr-Ti high-entropy alloy superconductor. Proceed. Nat. Acad. Sci. **113**(46), E7144–E7150 (2016)
3. K. Jasiewicz, B. Wiendlocha, P. Korbeń, S. Kaprzyk, J. Tobola, Superconductivity of $Ta_{34}Nb_{33}Hf_8Zr_{14}Ti_{11}$ high entropy alloy from first principles calculations. Phys. Status Solidi (RRL)–Rapid Res. Lett. **10**(5), 415–419 (2016)
4. K. Jasiewicz, B. Wiendlocha, K. Górnicka, K. Gofryk, M. Gazda, T. Klimczuk, J. Tobola, Pressure effects on the electronic structure and superconductivity of $(TaNb)_{0.67}(HfZrTi)_{0.33}$ high entropy alloy. Phys. Rev. B **100**, 184503 (2019)
5. P. Soven, Coherent-potential model of substitutional disordered alloys. Phys. Rev. **156**, 809–813 (1967)

6. S. Kaprzyk, A. Bansil, Green's function and a generalized lloyd formula for the density of states in disordered muffin-tin alloys. Phys. Rev. B **42**, 7358–7362 (1990)
7. A. Bansil, S. Kaprzyk, P.E. Mijnarends, J. Toboła, Electronic structure and magnetism of $Fe_{3-x}V_xX$ (X = Si, Ga, and Al) alloys by the KKR-CPA method. Phys. Rev. B **60**, 13396–13412 (1999)
8. T. Stopa, S. Kaprzyk, J. Tobola, Linear aspects of the Korringa-Kohn-Rostoker formalism. J. Phys. Condens. Matt. **16**, 4921–4933 (2004)
9. H. Ebert, D. Ködderitzsch, J. Minár, Calculating condensed matter properties using the KKR-Green's function method-recent developments and applications. Rep. Progr. Phys. **74**, 096501 (2011)
10. W.H. Butler, Theory of electronic transport in random alloys: Korringa-Kohn-Rostoker coherent-potential approximation. Phys. Rev. B **31**, 3260–3277 (1985)
11. J.S. Faulkner, G.M. Stocks, Calculating properties with the coherent-potential approximation. Phys. Rev. B **21**, 3222–3244 (1980)
12. B. Wiendlocha, S. Misra, A. Dauscher, B. Lenoir, C. Candolfi, Residual resistivity as an independent indicator of resonant levels in semiconductors. Mater. Horiz. **8**, 1735–1743 (2021)
13. K. Kutorasiński, J. Tobola, S. Kaprzyk, Calculating electron transport coefficients of disordered alloys using the KKR-CPA method and Boltzmann approach: Application to Mg_2 Si_{1-x} Sn_x thermoelectrics. Phys. Rev. B **87**, 195205 (2013)
14. B.E.A. Gordon, W.E. Temmerman, B.L. GyÖrffy, On the fermi surfaces of paramagnetic Cu_cNi_{1-c} alloys. J. Phys. F. Metal Phys. **11**(4), 821 (1981)
15. B. Wiendlocha, Fermi surface and electron dispersion of PbTe doped with resonant Tl impurity from KKR-CPA calculations. Phys. Rev. B **88**, 205205 (2013)
16. B. Wiendlocha, Thermopower of thermoelectric materials with resonant levels: PbTe:Tl versus PbTe:Na and $Cu_{1-x}Ni_x$. Phys. Rev. B **97**, 205203 (2018)
17. S.S. Rajput, R. Prasad, R.M. Singru, S. Kaprzyk, A. Bansil, Electronic structure of disordered Nb-Mo alloys studied using the charge-self-consistent Korringa-Kohn-Rostoker coherent potential approximation. J. Phy. Condens. Matt. **8**, 2929–2944 (1996)
18. H. Ebert et al. *The Munich SPR-KKR Package, version 7.7.3 and 8.6*, https://www.ebert.cup.uni-muenchen.de/index.php/en/software-en/13-sprkkr (2022)
19. J.P. Perdew, Y. Wang, Accurate and simple analytic representation of the electron-gas correlation energy. Phys. Rev. B **45**, 13244–13249 (1992)
20. U. von Barth, L. Hedin, A local exchange-correlation potential for the spin polarized case. i. J. Phys. C. Solid State Phys. **5**, 1629–1642 (1972)
21. G.D. Gaspari, B.L. GyÖrffy, Electron-phonon interactions, d resonances, and superconductivity in transition metals. Phys. Rev. Lett. **28**, 801–805 (1972)
22. M. Wierzbowska, S. de Gironcoli, P. Giannozzi, Origins of low-and high-pressure discontinuities of T_c in niobium. ArXiv preprint arXiv:cond-mat/0504077 (2005)
23. F. Giustino, Electron-phonon interactions from first principles. Rev. Mod. Phys. **89**, 015003 (2017)
24. P.B. Allen, Neutron spectroscopy of superconductors. Phys. Rev. B **6**, 2577–2579 (1972)
25. W.L. McMillan, Transition temperature of strong-coupled superconductors. Phys. Rev. **167**, 331–344 (1968)
26. J.J. Hopfield, Angular momentum and transition-metal superconductivity. Phys. Rev. **186**, 443–451 (1969)
27. I.R. Gomersall, B.L. GyÖrffy, A simple theory of the electron-phonon mass enhancement in transition metal compounds. J. Phys. F. Metal Phys. **4**, 1204–1221 (1974)
28. B.M. Klein, L.L. Boyer, D.A. Papaconstantopoulos, Superconducting properties of $A15$ compounds derived from band-structure results. Phys. Rev. Lett. **42**, 530–533 (1979)
29. I.I. Mazin, S.N. Rashkeev, S.Y. Savrasov, Nonspherical rigid-muffin-tin calculations of electron-phonon coupling in high-T_c perovskites. Phys. Rev. B **42**, 366–370 (1990)
30. B. Wiendlocha, J. Tobola, S. Kaprzyk, Search for Sc_3XB (X = In, Tl, Ga, Al) perovskites superconductors and proximity of weak ferromagnetism. Phys. Rev. B **73**, 134522 (2006)

31. B. Wiendlocha, J. Tobola, M. Sternik, S. Kaprzyk, K. Parlinski, A.M. Oleś, Superconductivity of Mo_3Sb_7 from first principles. Phys. Rev. B **78**, 060507(R) (2008)
32. B. Wiendlocha, M. Sternik, Effect of the tetragonal distortion on the electronic structure, phonons and superconductivity in the Mo_3Sb_7 superconductor. Intermetallics **53**, 150–156 (2014)
33. S. Gutowska, A. Kawala, B. Wiendlocha, Superconductivity near the mott-ioffe-regel limit in the high-entropy alloy superconductor $(ScZrNb)_{1-x}(RhPd)_x$ with a CsCl-type lattice. Phys. Rev. B **108**, 064507 (2023)
34. H. Fukuyama, H. Ebisawa, S. Maekawa, Bulk superconductivity in weakly localized regime. J. Phys. Soc. Japan **53**(10), 3560–3567 (1984)
35. P. Anderson, Theory of dirty superconductors. J. Phys. Chem. Solids **11**(1), 26–30 (1959)
36. P.W. Anderson, K.A. Muttalib, T.V. Ramakrishnan, Theory of the ""universal" degradation of T_c in high-temperature superconductors. Phys. Rev. B **28**, 117–120 (1983)
37. A.P. Mackenzie, R.K.W. Haselwimmer, A.W. Tyler, G.G. Lonzarich, Y. Mori, S. Nishizaki, Y. Maeno, Extremely strong dependence of superconductivity on disorder in Sr_2RuO_4. Phys. Rev. Lett. **80**, 161–164 (1998)
38. Y. Imry, M. Strongin, Destruction of superconductivity in granular and highly disordered metals. Phys. Rev. B **24**, 6353–6360 (1981)
39. M. Strongin, R.S. Thompson, O.F. Kammerer, J.E. Crow, Destruction of superconductivity in disordered near-monolayer films. Phys. Rev. B **1**, 1078–1091 (1970)
40. T.M. Mishonov, E.S. Penev, J.O. Indekeu, V.L. Pokrovsky, Specific-heat discontinuity in impure two-band superconductors. Phys. Rev. B **68**, 104517 (2003)
41. S.Y. Savrasov, D.Y. Savrasov, Electron-phonon interactions and related physical properties of metals from linear-response theory. Phys. Rev. B **54**, 16487–16501 (1996)
42. J.P. Carbotte, Properties of boson-exchange superconductors. Rev. Mod. Phys. **62**, 1027–1157 (1990)
43. S. Gutowska, B. Wiendlocha, *To be published* (2024)
44. P. Giannozzi, S. Baroni, N. Bonini, M. Calandra, R. Car, C. Cavazzoni, D. Ceresoli, G.L. Chiarotti, M. Cococcioni, I. Dabo, A. Dal Corso, S. de Gironcoli, S. Fabris, G. Fratesi, R. Gebauer, U. Gerstmann, C. Gougoussis, A. Kokalj, M. Lazzeri, L. Martin-Samos, N. Marzari, F. Mauri, R. Mazzarello, S. Paolini, A. Pasquarello, L. Paulatto, C. Sbraccia, S. Scandolo, G. Sclauzero, A.P. Seitsonen, A. Smogunov, P. Umari, and R.M. Wentzcovitch, Quantum Espresso: a modular and open-source software project for quantum simulations of materials. J. Phys. Condens. Matt. **21**(39), 395502 (19pp) (2009)
45. P. Giannozzi, O. Andreussi, T. Brumme, O. Bunau, M.B. Nardelli, M. Calandra, R. Car, C. Cavazzoni, D. Ceresoli, M. Cococcioni, N. Colonna, I. Carnimeo, A.D. Corso, S. de Gironcoli, P. Delugas, R.A.D. Jr, A. Ferretti, A. Floris, G. Fratesi, G. Fugallo, R. Gebauer, U. Gerstmann, F. Giustino, T. Gorni, J. Jia, M. Kawamura, H.-Y. Ko, A. Kokalj, E. Kucukbenli, M. Lazzeri, M. Marsili, N. Marzari, F. Mauri, N.L. Nguyen, H.-V. Nguyen, A.O. de-la Roza, L. Paulatto, S. Ponce, D. Rocca, R. Sabatini, B. Santra, M. Schlipf, A.P. Seitsonen, A. Smogunov, I. Timrov, T. Thonhauser, P. Umari, N. Vast, X. Wu, S. Baroni, Advanced capabilities for materials modelling with Quantum Espresso. J. Phys. Condens. Matt. **29**(46), 465901 (2017)
46. A.D. Corso, Pseudopotentials periodic table: From H to Pu. Comput. Mater. Sci. **95**, 337–350 (2014)
47. J.P. Perdew, K. Burke, M. Ernzerhof, Generalized gradient approximation made simple. Phys. Rev. Lett. **77**, 3865–3868 (1996)
48. F. Körmann, Y. Ikeda, B. Grabowski, M.H.F. Sluiter, Phonon broadening in high entropy alloys. Npj Comput. Math. **3**, 36 (2017)
49. P.G. Klemens, The scattering of low-frequency lattice waves by static imperfections. Proceed. Phys. Soc. Secti. A **68**, 1113 (1955)
50. Y. Ikeda, F. Körmann, B. Dutta, A. Carreras, A. Seko, J. Neugebauer, I. Tanaka, Temperature-dependent phonon spectra of magnetic random solid solutions. Npj Comput. Mater. **4**, 7 (2018)
51. Y. Akahama, S. Kawaguchi, N. Hirao, Y. Ohishi, Observation of high-pressure bcc phase of titanium at 243 GPa. J. Appl. Phys. **128**(3), 035901 (2020)

52. H. Xia, G. Parthasarathy, H. Luo, Y.K. Vohra, A.L. Ruoff, Crystal structures of group IVa metals at ultrahigh pressures. Phys. Rev. B **42**, 6736–6738 (1990)
53. J. Guo, H. Wang, F. von Rohr, Z. Wang, S. Cai, Y. Zhou, K. Yang, A. Li, S. Jiang, Q. Wu, R.J. Cava, L. Sun, Robust zero resistance in a superconducting high-entropy alloy at pressures up to 190 GPa. Proceed. Nat. Acad. Sci. **114**(50), 13144–13147 (2017)
54. G. Grimvall, *Thermophysical Properties of Materials*. (North-Holland Publishing Company, 1986)
55. F. Birch, Finite elastic strain of cubic crystals. Phys. Rev. **71**, 809–824 (1947)
56. X. Li, First-principles study of the third-order elastic constants and related anharmonic properties in refractory high-entropy alloys. Acta Materialia **142**, 29–36 (2018)
57. A.S. Ahmad, Y. Su, S.Y. Liu, K. Ståhl, Y.D. Wu, X.D. Hui, U. Ruett, O. Gutowski, K. Glazyrin, H.P. Liermann, H. Franz, H. Wang, X.D. Wang, Q.P. Cao, D.X. Zhang, J.Z. Jiang, Structural stability of high entropy alloys under pressure and temperature. J. Appl. Phys. **121**(23), 235901 (2017)
58. V.K. Ratti, R. Evans, B.L. Gyorffy, The volume dependence of the electron-phonon mass enhancement and the pressure dependence of tc in transition metals. J. Phys. F. Metal Phys. **4**, 371–379 (1974)
59. B. Wiendlocha, M.J. Winiarski, M. Muras, C. Zvoriste-Walters, J.-C. Griveau, S. Heathman, M. Gazda, T. Klimczuk, Pressure effects on the superconductivity of the $HfPd_2Al$ Heusler compound: Experimental and theoretical study. Phys. Rev. B **91**, 024509 (2015)
60. J.W. Garland, K.H. Bennemann, Theory for the pressure dependence of Tc for narrow-band superconductors. AIP Conf. Proceed. **4**(1), 255–292 (1972)
61. K. Jasiewicz, J. Tobola, B. Wiendlocha, Local distortions of the crystal structure and their influence on the electronic structure and superconductivity of the high-entropy alloy $(TaNb)_{0.67}(HfZrTi)_{0.33}$. Phys. Rev. B **108**, 224505 (2023)
62. P. Blaha, K. Schwarz, G.K. Madsen, D. Kvasnicka, J. Luitz, et al., Wien2k, An Augmented Plane Wave+ Local Orbitals Program for Calculating Crystal Properties. 60 (2001)
63. N. Stepanov, N. Yurchenko, S. Zherebtsov, M. Tikhonovsky, G. Salishchev, Aging behavior of the HfNbTaTiZr high entropy alloy. Mater. Lett. **211**, 87–90 (2018)
64. S. Chen, Y. Tong, K.-K. Tseng, J.-W. Yeh, J. Poplawsky, J. Wen, M. Gao, G. Kim, W. Chen, Y. Ren, R. Feng, W. Li, P. Liaw, Phase transformations of HfNbTaTiZr high entropy alloy at intermediate temperatures. Scripta Materialia **158**, 50–56 (2019)

Chapter 6
BCC and HCP Nb-Re-Hf-Zr-Ti High Entropy Alloy Superconductors

Sonika Jangid, Saurav Marik, and Ravi Prakash Singh

Abstract High-entropy alloys (HEAs) have unique and highly disordered atomic structures that provide exceptional mechanical, physical, and chemical properties. Superconducting HEAs are especially intriguing because they allow the relationship between disorder and superconductivity to be explored. Recently, Nb-Re-Zr-Hf-Ti-based HEAs have gained importance as a new family of superconductors because of their composition-dependent crystal structure, providing an opportunity to investigate the influence of crystal structure and disorder on superconductivity. This chapter details the investigation of the superconducting properties of the $[Nb_{0.67-x}Re_x](TiZrHf)_{0.33}$, $Nb_{0.2}Re_{0.2}Zr_{0.2}Hf_{0.2}Ti_{0.2}$ and $Nb_{0.6}Re_{0.1}Zr_{0.1}Hf_{0.1}Ti_{0.1}$ systems. The material phase composition and crystal structure were confirmed using precession electron diffraction, powder X-ray diffraction, and scanning electron microscopy. These results suggest that an increase in the valence electron count stabilizes the hcp phase. Transport, magnetization, heat capacity, and muon spin rotation/relaxation measurements were used to explore the superconducting properties of the materials, and the results suggested that the Nb-Re-Zr-Hf-Ti-based HEAs are bulk, fully gapped moderately coupled superconductors with preserved time-reversal symmetry in the superconducting ground state.

6.1 Introduction

High-entropy alloys (HEAs) are a novel class of disordered materials composed of multiple elements in equimolar or near-equimolar ratios. They have gained significant attention due to their outstanding mechanical properties, including high fracture

S. Jangid · R. P. Singh (✉)
Department of Physics, Indian Institute of Science Education and Research Bhopal, Bhopal 462066, India
e-mail: rpsingh@iiserb.ac.in

S. Marik
School of Physics and Materials Science, Thapar Institute of Engineering and Technology Patiala, 147004 Patiala, India

J. Kitagawa and Y. Mizuguchi (eds.), *High-Entropy Alloy Superconductors*, Springer Series in Solid-State Sciences 202, https://doi.org/10.1007/978-981-97-4129-8_6

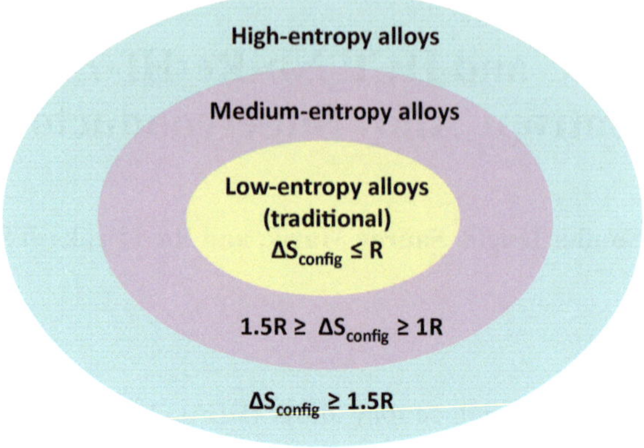

Fig. 6.1 Classification of alloys based on configurational entropy. The figure was taken from [12] with permission

toughness at cryogenic temperatures [1], enhanced ductility by twinning [2], exceptional strength at extremely high pressures [3] and excellent corrosion resistance [4]. These properties make them promising candidates for structural materials. HEAs also exhibit a variety of functionalities, including magnetic refrigeration, energy storage, radiation protection, and biocompatibility [5–8]. HEAs also exhibit fascinating quantum phenomena, such as superconductivity [9] and magnetism [10], which make them attractive for various applications.

HEAs differ from conventional alloys because they have multiple principal alloying elements rather than one or two primary alloying elements. There are two definitions of HEAs: first, they have five or more elements with a concentration between 5% and 35% [11]; second, they exhibit a high configurational entropy, that is, $\Delta S_{\text{config}} \geq 1.5R$ (Fig. 6.1) [12], where R is the gas constant; and $\Delta S_{\text{config}} = \sum_{i=0}^{n} x_i \ln x_i$ [11]. Here, n is the total number of elements and x_i is the mole fraction of the ith element. The dominance of the entropy term in the Gibbs free energy at high temperatures minimizes the Gibbs free energy and stabilizes a solid solution phase over the intermetallic phase [13] in relatively simple crystal structures such as body-centered cubic (bcc), face-centered cubic (fcc), and hexagonal closed-packed (hcp) structures. Atoms of different radii in HEAs are randomly arranged on the crystallographic positions of the crystal lattice, resulting in high configurational entropy and a high degree of disorder [11].

The first HEA superconductor, $Ta_{0.34}Nb_{0.33}Hf_{0.8}Zr_{0.14}Ti_{0.11}$, was discovered in 2014. It is a conventional BCS (Bardeen-Cooper-Schrieffer) superconductor in the weak coupling limit [9]. This was an intriguing finding for physicists, as superconductivity is typically associated with ordered systems, whereas HEAs are known for their high levels of disorder and complex phonon structures. Since then, numerous other HEA superconductors containing 3d, 4d, and 5d elements have been discovered, with various crystal structures, including bcc, fcc, hcp, and CsCl-type structures, and the α -Mn structure.

Many studies have been conducted to uncover the fundamental properties of the superconductive ground state in high-entropy alloys (HEAs) [14–24]. The most remarkable feature of these systems is their superconductivity robustness against pressure [25]. The broadening of the specific heat jump has also been observed in some superconducting high-entropy alloys [26]. Additionally, some HEA superconductors have a Debye temperature in the elemental range regardless of the presence of multiple elements [24, 27].

The bcc Ta-Nb-Hf-Zr-Ti system, which is discussed in detail in Chaps. 2 and 5 of this book, has been extensively studied among all high-entropy alloy superconductors. Several investigations have been conducted on these systems, including the dependence of T_C on VEC, chemical complexity [16], isoelectronic substitution [19], pressure studies [25], first-principle calculations, mechanical properties and thin-film studies [28, 29], to understand the relationships between crystal structure, disorder, and mechanical and electronic properties. However, despite ongoing research, questions still outweigh the answers. Understanding the occurrence of superconductivity and exotic properties is challenging because highly disordered multicomponent alloys such as these are unlikely to have the regular phonons necessary for BCS SC [30, 31]. Although efforts are being made to discover new superconducting high-entropy alloys with different crystal structures and tune their superconducting transition temperature using the combination of 3d, 4d, and 5d elements, the lack of microscopic studies and the unavailability of different crystal structures of HEA superconductors make it difficult to comprehend the superconducting pairing mechanism. There is strong interest in finding new HEA superconductors with different crystal structures and investigating the effect of the presence of different elements in existing HEA compounds. One such study is explored in the Nb-Re-Zr-Hf-Ti system, where Re replaces Ta in the known superconducting HEA Nb-Ta-Zr-Hf-Ti to understand the change in crystal structure with doping and its impact on the superconducting ground state [32–39].

This chapter summarizes the normal and superconducting properties of the Re-Nb-Hf-Zr-Ti system, ranging from the bcc structure to the hcp structure. The chapter is divided into three sections. The first section discusses the experimental details and techniques used. The latter section discusses the results and discussion of the Nb-Re-Zr-Hf-Ti system. The last section concludes with all the findings mentioned in the previous sections.

6.2 Experimental Details

Polycrystalline samples were synthesized using the standard arc-melting technique. Stoichiometric amounts of high-purity Nb, Re, Zr, Hf, and Ti pieces were melted on a water-cooled copper hearth at high currents to form a button. Before melting the samples, a Ti getter was melted to absorb any remaining oxygen. The button was flipped over and remelted multiple times to achieve phase homogeneity, resulting in hard samples with negligible weight loss.

The structural characterization of the Nb-Re-Zr-Hf-Ti HEAs was carried out using powder X-ray diffraction (PXRD) with a PANalytic X′Pert powder diffractometer, precession electron diffraction (PED) and precession electron diffraction (PEDT) techniques. The data were analyzed using PETS and JANA2006 software. The phase compositions of the prepared samples were confirmed using a Zeiss ULTRA Plus high-resolution field emission scanning electron microscope (FESEM) with an energy-dispersive X-ray spectrometer (EDAX).

The magnetic and physical properties of the samples were measured using various techniques. DC magnetization and AC susceptibility measurements were performed using a Quantum Design magnetic property measurement system (MPMS 3). Resistivity and specific heat measurements were performed on a Quantum Design physical property measurement system (PPMS). Muon spin rotation and relaxation (μSR) measurements were performed on a μSR spectrometer at the ISIS Neutron and Muon Pulsed Source, STFC Rutherford Appleton Laboratory, United Kingdom, to investigate the microscopic properties of $Nb_{0.6}Re_{0.1}Zr_{0.1}Hf_{0.1}Ti_{0.1}$ [40, 41]. Transverse-field measurements were utilized to assess the superconducting gap structure. Zero-field (ZF) measurements were conducted to probe local magnetism and the presence of any spontaneous magnetization in the superconducting ground state.

6.3 Results and Discussion

6.3.1 Polymorphic Transition in $[Nb_{0.67-x}Re_x](TiZrHf)_{0.33}$ HEAs and Their Superconducting Behavior

The structural properties of $[Nb_{0.67-x}Re_x](TiZrHf)_{0.33}$ HEAs are studied with increasing Re concentration. The powder XRD patterns at room temperature for various Re concentrations ($x = 0.1, 0.2, 0.25, 0.335$ and 0.56) are presented in Fig. 6.2a. The lower Re content samples exhibit a cubic symmetry with a space group of Im$3\bar{m}$ (229). Samples with higher Re contents exhibit a transition from a cubic structure (bcc) to a hexagonal structure [24]. A sample with a higher Re content ($x = 0.56$, $e/a = 5.79$) can be identified as a pure phase, and its XRD pattern can be well indexed to an hcp structure (space group P6_3/mmc) with lattice parameters $a = b = 5.25(1)$ Å and $c = 8.59(1)$ Å. In conventional alloys, an increase in the electron count (e/a) usually results in a structural transition from bcc to hcp. This polymorphic transition was also observed in $[Nb_{0.67-x}Re_x](TiZrHf)_{0.33}$ HEAs with increased Re content (or e/a ratio). In contrast, when the e/a ratio is decreased, a structural transition starting at $e/a = 5.17$ results in a cubic structure. A similar transition from cubic to hcp has been observed in the Cr-Mn-Fe-Co-Ni system with transition metal only but only under a high external pressure of 14 GPa [42]. Alloying aluminum in HEA superconductors, such as $[TaNb]_{0.67}(ZrHfTi)_{0.33}Al_x$, also leads to complex structural transitions. At $x = 0.4$, the stabilization of a simple crystal structure with high entropy fails, and the material transforms into a nonsuperconducting tetragonal structure (space group P4_2/mnm) structure, causing T_c to decrease gradually with an increasing aluminum content [19]. A phase separation is also observed during long-term annealing of the

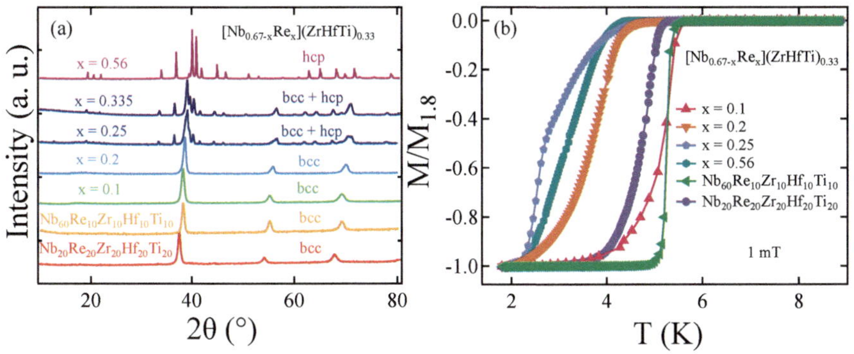

Fig. 6.2 **a** Powder XRD patterns for the equimolar HEA $Nb_{0.2}Re_{0.2}Zr_{0.2}Hf_{0.2}Ti_{0.2}$, Nb rich HEA $Nb_{0.6}Re_{0.1}Zr_{0.1}Hf_{0.1}Ti_{0.1}$ and $[Nb_{0.67-x}Re_x](ZrHfTi)_{0.33}$ with $x = 0.1, 0.2, 0.25, 0.335$ and 0.56. The sample with the highest Re content ($x = 0.56$, $e/a = 5.79$) is a pure phase and crystallizes in an hcp structure (space group $P6_3/mmc$). Samples with reduced Re content (lower e/a ratios) crystallize in a bcc structure (space group $Im3\bar{m}$). **b** Temperature-dependent normalized DC magnetization in a 1 mT field for the equimolar HEA $Nb_{0.2}Re_{0.2}Zr_{0.2}Hf_{0.2}Ti_{0.2}$, the Nb-rich HEA $Nb_{0.6}Re_{0.1}Zr_{0.1}Hf_{0.1}Ti_{0.1}$ and $[Nb_{0.67-x}Re_x](ZrHfTi)_{0.33}$ with $x = 0.1, 0.2, 0.25$ and 0.56. A double transition is detected for the sample with $x = 0.25$, which confirms the presence of a mixed phase (hcp + bcc), as shown in the RT x-ray diffraction (XRD) pattern. The data for both figures were taken from [21, 24, 31] with permission

HEA superconductor $[TaNb]_{1-x}(ZrHfTi)_x$ at high temperatures, affecting the superconducting characteristics depending on the chemical and structural homogeneity level of the sample [43]. Further research on the alloying and thermal treatment of current materials is necessary. In Fig. 6.2b, the temperature variance of the normalized DC magnetization is shown for all $[Nb_{0.67-x}Re_x](ZrHfTi)_{0.33}$ materials. Pure phase materials exhibit an increase in T_c as the e/a ratio decreases.

6.3.2 Nb-Rich bcc HEA Superconductor $Nb_{0.6}Re_{0.1}Zr_{0.1}Hf_{0.1}Ti_{0.1}$

Structural Characterization

According to the powder X-ray diffraction measurements, the $[Nb_{0.6}Re_{0.1}Zr_{0.1}Hf_{0.1}Ti_{0.1}]$ high-entropy alloy (HEA) crystallizes in a centrosymmetric body-centered cubic (bcc) crystal structure with the unit cell parameter $a = b = c = 3.320184(7)$ Å and space group $Im3\bar{m}$, as depicted in the refined XRD pattern shown in Fig. 6.3a. The peak broadening observed in the pattern may be due to the disorder caused by the varying radii of the constituent elements. The homogeneity of the phase was confirmed by performing energy-dispersive X-ray spectroscopy at five different locations in the sample, with one of the spectra shown in the inset of Fig. 6.3a.

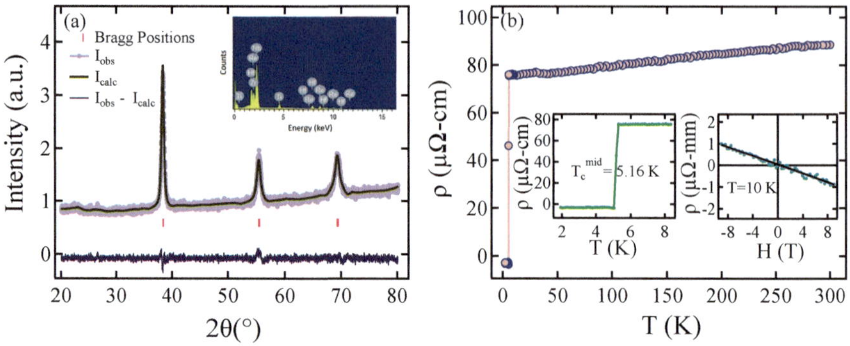

Fig. 6.3 **a** Room-temperature X-ray diffraction pattern of $Nb_{0.6}Re_{0.1}Zr_{0.1}Hf_{0.1}Ti_{0.1}$ obtained using CuK_{α} radiation. The inset shows the energy-dispersive X-ray spectra. **b** Resistivity versus temperature curve in the temperature range from 1.9 K to 300 K at zero fields. The left inset shows the sharp superconducting transition at 5.2(1) K, while the right inset shows the field-dependent Hall resistivity from –9 T to 9 T with a negative slope. The figures are taken from [31] with permission

Electrical Resistivity

The transition of electrical resistivity in zero fields from the normal state to the superconducting state was observed at $T_c = 5.16(5)$ K, as shown in the left inset of Fig. 6.3b. Above the transition temperature, the normal-state resistivity was nearly independent of temperature, as shown in Fig. 6.3b. The residual resistivity ratio (RRR) of the sample was 1.2, which is similar to that of other HEA superconductors [9, 16, 18, 44]. The Hall resistivity was measured to understand the nature and density of the charge carriers in the normal state. The field-dependent Hall resistivity measurement at 10 K is illustrated in the right inset of Fig. 6.3b. The observed Hall resistivity data were linearly fit in the magnetic field range of ±9 T, and the slope yielded the Hall coefficient $R_{\rm H} = -1.08(2) \times 10^{-10}\ \Omega{\rm m}{\rm T}^{-1}$. A negative $R_{\rm H}$ indicates that the charge carriers are electrons. Using the expression $R_{\rm H} = -1/ne$, where n is the quasiparticle carrier density, and e is the electronic charge. The quasiparticle carrier density for this material was calculated to be $n = 6.1(1) \times 10^{28}$ m^{-3}, which is comparable to that of the noncentrosymmetric $(HfNb)_{0.10}(MoReRu)_{0.90}$, and $(ZrNb)_{0.10}(MoReRu)_{0.90}$ HEA superconductors [27].

Magnetization

At an applied magnetic field of 1 mT, temperature-dependent DC magnetization was measured in both zero field cooled warming (ZFCW) and field cooled cooling (FCC) modes. Figure 6.4a displays a diamagnetic signal that corresponds to the onset of the superconducting transition at $T_c = 5.60(4)$ K. The FCC data reveal significant flux pinning in the $Nb_{0.6}Re_{0.1}Zr_{0.1}Hf_{0.1}Ti_{0.1}$ alloy. To calculate the lower critical field H_{c1}, field-dependent magnetization data were collected in the temperature range of 1.8 K to T_c, as illustrated in the inset of Fig. 6.4b. The magnetization displays a linear

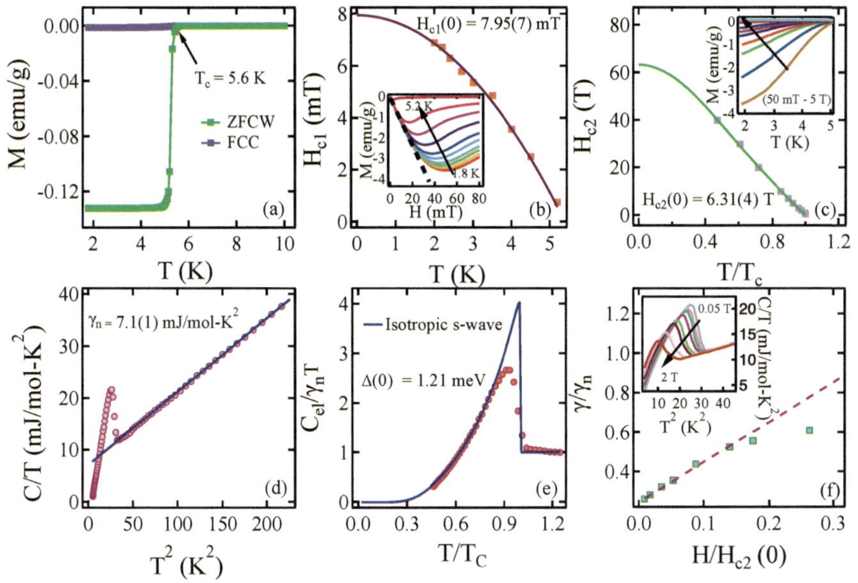

Fig. 6.4 **a** Temperature-dependent DC magnetization in a 1 mT field for $Nb_{0.6}Re_{0.1}Zr_{0.1}Hf_{0.1}Ti_{0.1}$ in both ZFCW and FCC modes. **b** Lower critical field dependence on temperature fitted with Eq. (6.1), where the inset shows the magnetization versus magnetic field curves at different temperatures. **c** Temperature-dependent upper critical field fitted with Eq. (6.2), where the inset shows the magnetization versus temperature curves in different fields. **d** Curve C/T versus T^2 in zero field where the normal state data points are fitted to Eq. (6.6). **e** The electronic-specific heat well fitted with the BCS isotropic s-wave gap. **f** Normalized Sommerfeld coefficient plotted against the normalized upper critical field. The dashed line is the linear fit and the inset shows the C/T versus T^2 curves in different fields. The figures have been adopted from [31] with permission

variation up to a particular field value, after which it deviates from linearity or the Meissner line. The critical field for a given temperature is determined by the field value at which it deviates. After fitting the data points using the Ginzburg-Landau formula,

$$H_{c1}(T) = H_{c1}(0)\left[1 - \left(\frac{T}{T_c}\right)^2\right]. \tag{6.1}$$

The lower critical field H_{c1} at 0 K was calculated by extrapolating $H_{c1}(T)$ to 0 K, which results in $H_{c1}(0) = 7.95(7)$ mT [shown in Fig. 6.4b]. Temperature-dependent magnetization was performed at different magnetic fields to estimate the upper critical field, as shown in the inset of Fig. 6.4c. The initiation of the superconducting transition was used as the upper critical field requirement, and the transition temperature decreased as the applied external magnetic field increased, as shown in Fig. 6.4c. The relationship between H_{c2} and temperature is linear near T_c and may be represented using the Ginzburg-Landau equation:

$$H_{c2}(T) = H_{c2}(0)\left[\frac{(1-t^2)}{(1+t^2)}\right], \tag{6.2}$$

where $t = T/T_c$ is the reduced temperature. Fitting the $H_{c2}(T)$ data with Eq. (6.2) reveals that the upper critical field at 0 K, $H_{c2}(0)$, is 6.31(4) T. The Ginzburg-Landau coherence length can be estimated using the expression

$$H_{c2}(0) = \frac{\Phi_0}{2\pi\xi_{GL}^2}, \tag{6.3}$$

where Φ_0 (= 2.07 × 10^{-15} Tm^2) is the magnetic flux quantum and by substituting $H_{c2}(0) = 6.31(4)$ T, the coherence length $\xi_{GL}(0)$ is obtained as 7.2(1) nm. According to the expression,

$$H_{c1}(0) = \frac{\Phi_0}{4\pi\lambda_{GL}^2(0)}\left(\ln\frac{\lambda_{GL}(0)}{\xi_{GL}(0)} + 0.12\right), \tag{6.4}$$

the magnetic penetration depth $\lambda_{GL}(0)$ is proportional to the lower critical field $H_{c1}(0)$ and $\xi_{GL}(0)$. By replacing the computed parameters $H_{c1}(0) = 7.95(7)$ mT and $\xi_{GL}(0) = 7.2(1)$ nm, $\lambda_{GL}(0)$ is found to be 279(2) nm. The Ginzburg-Landau parameter κ_{GL} differentiates the type of superconductivity and is expressed as $\kappa_{GL} = \frac{\lambda_{GL}(0)}{\xi_{GL}(0)}$. The value of $\kappa_{GL}(0)$ is calculated as 39(1) >> $\frac{1}{\sqrt{2}}$, showing that the $Nb_{0.6}Re_{0.1}Zr_{0.1}Hf_{0.1}Ti_{0.1}$ HEA is a strong type-II superconductor. In type-II superconductors, the Cooper pair is destroyed by applying an external magnetic field by two different kinds of mechanisms, namely, orbital and Pauli limiting effects [33]. The Werthamar-Helfand-Hohenberg-model,

$$H_{c2}^{orb}(0) = -\alpha T_c \left.\frac{dH_{c2}(T)}{dT}\right|_{T=T_c} \tag{6.5}$$

is used to calculate the orbital limiting field $H_{c2}^{orb}(0)$ for BCS superconductors, and in the dirty limit, α can take a value of 0.693. The ratio of the BCS coherence length and the electronic mean free path, $\frac{\xi_0}{l}$ is less than 1, suggesting the dirty limit superconductivity of $Nb_{0.6}Re_{0.1}Zr_{0.1}Hf_{0.1}Ti_{0.1}$ (Sect. 6.3.5). The slope $\frac{dH_{c2}}{dT}$ near T_c is calculated to be -1.08 (3) T, resulting in $H_{c2}^{orb}(0) = 4.2(1)$ T. The latter effect of the Pauli limiting field $H_{c2}^{P}(0)$ can be described using the formula $H_{c2}^{P}(0) = 1.84T_c$ for BCS theory, and the estimated value of $H_{c2}^{P}(0)$ is 10.3(1) T. The Maki parameter (α_M), which is used to determine the relative strength of the orbital and Pauli limiting fields, and is calculated using the formula $\alpha_M = \sqrt{2}H_{c2}^{orb}(0)/H_{c2}^{P}(0)$. $\alpha_M = 0.57$, so the Cooper pair breaking is mainly caused by the orbital limiting field with only a small contribution of the paramagnetic effect.

Specific Heat

The specific heat was measured for the $Nb_{0.6}Re_{0.1}Zr_{0.1}Hf_{0.1}Ti_{0.1}$ alloy in zero fields. As indicated in Fig. 6.4d, a substantial jump in heat capacity was observed at a transition temperature $T_c = 5.4(1)$ K, which is comparable to the magnetization and resistivity. Normal state-specific heat data were well fitted with the following expression:

$$\frac{C}{T} = \gamma_n + \beta_3 T^2. \tag{6.6}$$

The Sommerfeld coefficient $\gamma_n = 7.1(1)$ mJmol^{-1}K^{-2} and $\beta_3 = 0.140(2)$ mJmol^{-1}K^{-4} are obtained from this fitting, which are the measurements of the electronic contribution and the lattice contribution to the total specific heat, respectively. These two parameters have substantial significance in calculating the density of states [$D_c(E_F)$], Debye temperature (θ_D), and electron-phonon coupling constant (λ_{e-ph}). The estimated value of [$D_c(E_F)$] is 3.01(4) $\frac{states}{eV f.u.}$ using the relation

$$\gamma_n = \left(\frac{\pi^2 k_B^2}{3}\right) D_c(E_F), \tag{6.7}$$

where $k_B = 1.38 \times 10^{-23}$ JK^{-1}. The Debye temperature, θ_D, is estimated to be 240(1) K, which is related to β_3 by the expression:

$$\theta_D = \left(\frac{12\pi^4 RN}{5\beta_3}\right)^{\frac{1}{3}}. \tag{6.8}$$

The electron-phonon coupling constant, λ_{e-ph}, which is the measure of the strength of the attractive force between the electron and the phonon, is obtained to be 0.70 using the McMillan equation:

$$\lambda_{e-ph} = \frac{1.04 + \mu^* \ln(\theta_D/1.45T_c)}{(1 - 0.62\mu^*)\ln(\theta_D/1.45T_c) - 1.04}, \tag{6.9}$$

where μ^* is the screened Coulomb repulsion, which is taken to be 0.13 for intermetallic superconductors. The electron-phonon coupling constant, λ_{e-ph} for $Nb_{0.6}Re_{0.1}Zr_{0.1}Hf_{0.1}Ti_{0.1}$ is similar to that of other HEA superconductors [45, 46], indicating moderately coupled superconductivity.

To fully understand the characteristics of the superconducting ground state, the electronic specific heat (C_{el}) in the superconducting state can be determined by subtracting the phononic part from the total specific heat using the formula $C_{el} = C(T) - \beta_3 T^3$ as shown in Fig. 6.4e. The normalized entropy in the superconducting state for a single gap BCS superconductor is given by the expression

$$\frac{S}{\gamma_n T_c} = -\frac{6}{\pi^2}\left(\frac{\Delta(0)}{k_B T_c}\right)\int_0^{\infty}[f\ln(f)+(1-f)\ln(1-f)]dy, \tag{6.10}$$

where $f(\xi) = [\exp(E(\xi)/k_B T) + 1]^{-1}$ is the Fermi function, $E(\xi) = \sqrt{\xi^2 + \Delta^2(t)}$ is the energy of normal electrons relative to the Fermi energy, $y = \xi/\Delta(0), t = T/T_c$ and $\Delta(t) = \tanh[1.82(1.018(1/t) - 1))^{0.51}]$ is the estimation of the temperature-dependent energy gap. The first derivative of the normalized entropy is related to the normalized electronic specific heat below T_c as follows:

$$\frac{C_{el}}{\gamma_n T_c} = t\frac{d(S/\gamma_n T_c)}{dt}. \tag{6.11}$$

The fitting yields a superconducting gap value of $\alpha = \Delta(0)/k_B T_c = 2.57$, which is considerably greater than the predicted BCS value $\alpha = 1.764$ and is also more significant than the equimolar $Nb_{0.2}Re_{0.2}Zr_{0.2}Hf_{0.2}Ti_{0.2}$ HEA [21]. Electronic heat capacity measurements, in conjunction with the electron-phonon coupling constant, demonstrate moderately coupled isotropic s-wave superconductivity in $Nb_{0.6}Re_{0.1}Zr_{0.1}Hf_{0.1}Ti_{0.1}$. The field-dependent specific heat was also examined to obtain more information about the structure of the superconducting gap, as demonstrated in the inset of Fig. 6.4f. The electronic specific heat in different fields is determined by removing the corresponding phononic component ($\beta_3 T^3$) from the overall specific heat. The following equation was used to fit the field-dependent electronic specific heat in the superconducting state.

$$\frac{C_{el}}{T} = \gamma + \frac{A}{T}\exp\left(\frac{-bT_c}{T}\right). \tag{6.12}$$

In this equation, A is a material-dependent parameter, and $b = \Delta(0)/k_B T_c$ is the superconducting gap. Fig. 6.4f shows that the normalized γ increases linearly with the applied external magnetic field H normalized by H_{c2}. The linear increase in γ is due to the confinement of the quasiparticle excitation within the vortex cores. Several vortices move with field H for an isotropic superconducting gap structure [47, 48]. Therefore, at low fields, the linear behavior of $\gamma(H)$ in field H demonstrates the fully isotropic superconducting gap of the s-wave. If any nodes are in the superconducting gap, $\gamma(H)$ is expected to change as the square root of the applied field.

Muon Spin Rotation and Relaxation Spectroscopy

To obtain more information on the superconducting ground-state properties of $Nb_{0.6}Re_{0.1}Zr_{0.1}Hf_{0.1}$ $Ti_{0.1}$, muon spin rotation and relaxation measurements were performed in both the zero-field and transverse-field configurations.

1. Zero-field μSR: Zero-field μSR measurements in the superconducting state can be used to determine the occurrence of spontaneous magnetization caused by broken time-reversal symmetry (TRS). Relaxation spectra above and far below the transition

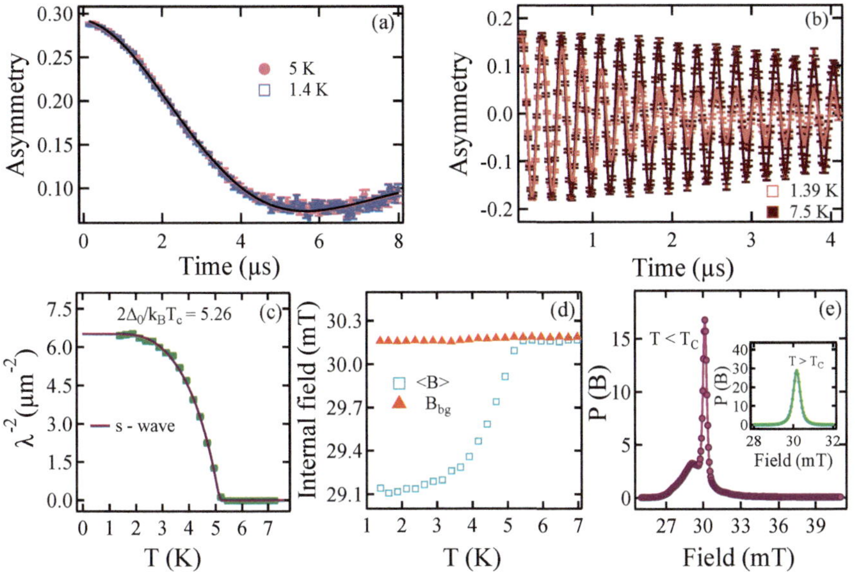

Fig. 6.5 **a** Zero-field μSR asymmetry spectra for above and below T_c fitted with Kubo-Toyabe (black line). **b** Transverse field μSR asymmetry spectra at 1.39 K and 7.5 K in a 30 mT applied field fitted using Eq. (6.15). **c** Temperature dependence of λ^{-2} fitted with an isotropic BCS s-wave superconductor. **d** Temperature variation of the internal magnetic field within the sample. The observed field from the sample is represented by blue squares, while orange triangles show the background field contribution from the sample holder. **e** Field distribution in the vortex state and the normal state (inset). The figures are derived from [31] with permission

temperature were recorded and are presented in Fig. 6.5a. The decaying relaxation behavior suggests that there is no oscillation above or below T_c, and the relaxation is generated by randomly oriented nuclear moments. The Kubo-Toyabe equation [49]:

$$G_{\mathrm{KT}}(t) = \frac{1}{3} + \frac{2}{3}(1 - \sigma_{ZF}^2 t^2)\exp\left(\frac{-\sigma_{ZF}^2 t^2}{2}\right), \tag{6.13}$$

yields the best fit to the ZF data above and below the transition temperature, where σ_{ZF} is the relaxation rate of the muon spin. The expression in the following is useful for explaining the asymmetry spectrum.

$$A(t) = A_1 G_{\mathrm{KT}}(t)\exp(-\Lambda t) + A_{\mathrm{BG}} \tag{6.14}$$

Here, A_1 represents the asymmetry of the sample, A_{BG} is the background asymmetry signal caused by muons resting in the sample holder, and Λ is the electronic relaxation rate. Fitting the data with Eq. (6.14) above and below T_c does not result in any apparent change in the relaxation rate within the detection limit of the μSR instrument (10 μT). The difference corresponding to both relaxations is determined to be $\frac{\sqrt{\Delta}}{\gamma_\mu} = 0.01$

Oe and $\frac{\Lambda}{\gamma_\mu} = 0.04$ Oe, both of which fall outside the detection limit of the instrument. As a result, the time-reversal symmetry in $Nb_{0.6}Re_{0.1}Zr_{0.1}Hf_{0.1}Ti_{0.1}$ is conserved.

2. Transverse field μSR: Transverse field μSR experiments on $Nb_{0.6}Re_{0.1}Zr_{0.1}Hf_{0.1}Ti_{0.1}$ high-entropy alloy were carried out to acquire insights into the superconducting gap symmetry and field distribution of the mixed state of the superconductor. To construct a well-ordered flux line lattice, the sample was first cooled in an applied magnetic field intermediate to H_{c1} and H_{c2}, i.e., 30 mT. TF asymmetry spectra were obtained from 1.39 to 7.5 K, but only two spectra at 1.39 K and above T_c are shown in Fig. 6.5b. The maximum entropy approach was used to acquire the field distribution throughout the temperature scale from the TF spectra [50]. Figure 6.5c provides strong evidence of depolarization below T_c, while the inset shows a uniform field above T_c. The interaction of muon spin with the internal field distribution in a sample leads to the alteration of muon precession in asymmetry spectra. The Muon spin relaxes owing to the existence of random orientations of nuclear dipole moments and due to the nonuniform magnetic field caused by the creation of a flux line lattice in the mixed state of the superconductor together with the nuclear contribution, above and below T_c, respectively. The Gaussian oscillatory decaying function provides the best fit to the TF asymmetry spectra for the entire temperature range and is presented below as [51, 52]:

$$A(t) = \sum_{i=1}^{N} A_i \exp\left(-\frac{1}{2}\sigma_i^2 t^2\right) \cos(\gamma_\mu B_i t + \phi) + A_{bg} \cos(\gamma_\mu B_{bg} t + \phi). \tag{6.15}$$

Here A_i represents the initial asymmetry, σ_i denotes the relaxation rate, $\gamma_\mu/2\pi = 135.5$ MHzT^{-1} is the muon gyromagnetic ratio, B_i represents the mean field of the i_{th} component of the Gaussian distribution, and ϕ is the initial phase. A_{bg} and B_{bg} are the temperature-independent background contributions to the asymmetry and field, respectively. A continuous relaxation rate arises because of the randomly oriented nuclear spin magnetic moments above T_c, which are independent of temperature. The temperature dependence of the internal field is illustrated in Fig. 6.5d, which also demonstrates a continuous background magnetic field contribution over the temperature range studied. The internal magnetic field distribution in the mixed state is represented by $\langle B \rangle$, which is less than the applied field for $T < T_c$, demonstrating the Meissner effect in the superconducting state. The method retrieves the applied field value for temperatures $T < T_c$ and overlaps with the background signal. The overall depolarization rate is caused by the nuclear magnetic moment (σ_n), which is temperature independent, and the flux lattice (σ_{FLL}). The contribution of σ_n was calculated using the above mentioned T_c data points, and the estimated value of σ_{FLL} was obtained by eliminating the contribution of σ_n from the general depolarization rate (σ) as $\sigma_{FLL}^2 = \sigma^2 - \sigma_n^2$. $\sigma_{FLL} \propto \frac{1}{\lambda^{+2}}$ relates the σ_{FLL} to the London penetration depth and the inverse square of the London penetration depth for low applied fields with $H << H_{c2}$ may be estimated using the equation

$$\frac{\sigma_{\text{FLL}}^2(T)}{\gamma_\mu^2} = \frac{0.00371\Phi_0^2}{\lambda^4(T)}, \tag{6.16}$$

where ϕ_0 is the magnetic flux quanta and $\gamma_\mu/2\pi = 135.5$ MHzT^{-1} is the gyromagnetic ratio of the muon. The London penetration depth, λ^μ at 0 K, is calculated to be 391(18) nm using the London approximation, which is slightly greather than the values derived from magnetization. The following expression gives the relationship between the penetration depth and the temperature for the BCS superconducting gap of the s-wave [53–55]:

$$\frac{\lambda^{-2}(T)}{\lambda^{-2}(0)} = 1 + 2\int_{\Delta(T)}^{\infty} \frac{\partial f}{\partial E} \frac{EdE}{\sqrt{E^2 - \Delta(T)^2}} \tag{6.17}$$

where $\lambda(0)$, f, and $\Delta(T)$ are the London penetration depth at 0 K and the Fermi and BCS superconducting gap functions, respectively. The BCS s-wave mode provided by Eq. (6.17) provides the best fit to the experimental data of $\lambda^{-2}(T)$. The resultant $T_c = 5.2(2)$ K also agrees well with the values determined from the magnetization and specific heat data for $Nb_{0.6}Re_{0.1}Zr_{0.1}Hf_{0.1}Ti_{0.1}$. The superconducting gap ratio at zero Kelvin, $\frac{\Delta(0)}{k_B T_c} = 2.65(5)$, which deviates significantly from the BCS value (1.73), and is similar to the estimated value via specific heat measurements [2.57(8)].

6.3.3 Equimolar bcc Nb-Re-Hf-Zr-Ti Superconductor

Structural Characterization

This is the first equimolar HEA superconductor with a bcc structure. The powder XRD pattern confirmed that the material crystallizes in a single bcc phase with a lattice parameter of 3.38 (1) Å [21], as shown in Fig. 6.6a. The broadening of the XRD peaks could be due to a high degree of disorder or challenges in the preparation of fine powder, which may have affected the diffraction experiments. To determine the composition of the material, SEM images were taken from various locations in the sample, revealing an average composition of $Nb_{0.21}Re_{0.16}Zr_{0.20}Hf_{0.23}Ti_{0.20}$ with less than 1% error for each element. The RT-EDX pattern at one of the locations of this material is shown in the inset of Fig. 6.6a.

Magnetization

The temperature dependence of the magnetic susceptibility of DC and AC in both the ZFCW and FCC modes is shown in Fig. 6.6b. A superconducting transition was observed at $T_c = 5.3$ K, confirming bulk superconductivity in this system. The field-dependent magnetization at 1.8 K (up to ± 7 T) is depicted in Fig. 6.6c, suggesting type-II superconductivity. According to the Ginzburg-Landau approximation [Eq.

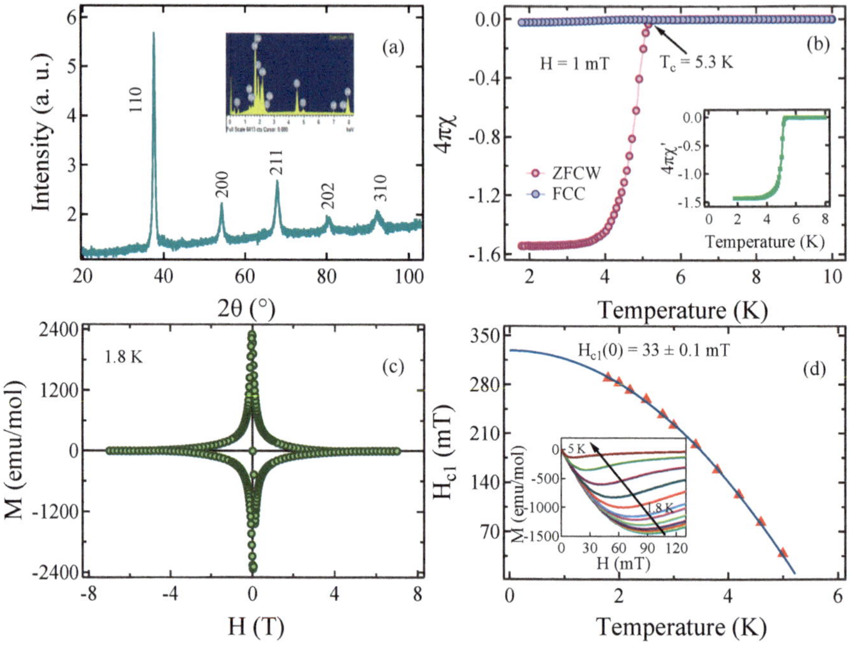

Fig. 6.6 **a** Room temperature X-ray diffraction pattern for $Nb_{0.2}Re_{0.2}Zr_{0.2}Hf_{0.2}Ti_{0.2}$ using CuK_α radiation. The inset shows the energy-dispersive X-ray spectra at room temperature. **b** Temperature-dependent DC and AC (inset) susceptibility for $Nb_{0.2}Re_{0.2}Zr_{0.2}Hf_{0.2}Ti_{0.2}$ in a 1 mT field. **c** Magnetic field-dependent magnetization curve at $T = 1.8$ K in the field $\pm$ 7 T. **d** The temperature-dependent lower critical field H_{c1} fitted using Eq. (6.1). The inset shows the field-dependent magnetization curves in the temperature range 1.8 K − 5 K. The figures are adopted from [21] with permission

(6.1)], the lower critical field $H_{c1}(T)$, defined as the field at which the $M - H$ curve starts deviating from the straight line, is estimated to be 33 mT at 0 K [6.6d].

Electrical Resistivity

The temperature dependence of the resistivity in zero fields is shown in Fig. 6.7a, which shows an abrupt decrease to zero resistivity, illustrating a superconducting transition at 5.3 K. The sluggish increase in resistivity indicates that the normal-state resistivity exhibits weak metallic behavior. The low residual resistivity ratio (RRR) obtained for this system, $\rho(290)/\rho(10) = 1.3$ suggested the occurrence of atomic-scale disorder in the sample, which is comparable to the findings of previous reports [9, 16, 18, 44]. The resistivity measurements at different magnetic fields are shown in the inset of Fig. 6.7a, which reveals a gradual decrease in T_c as the magnetic field increases. The critical temperatures are determined using the mid-values of the superconducting transition. Figure 6.7b shows the temperature-dependent upper critical field (H_{c2}) derived from the resistivity measurements. The upper critical field was determined to be 8.88 T by fitting the $H_{c2}(T)$ curve with Eq. (6.2) and extrap-

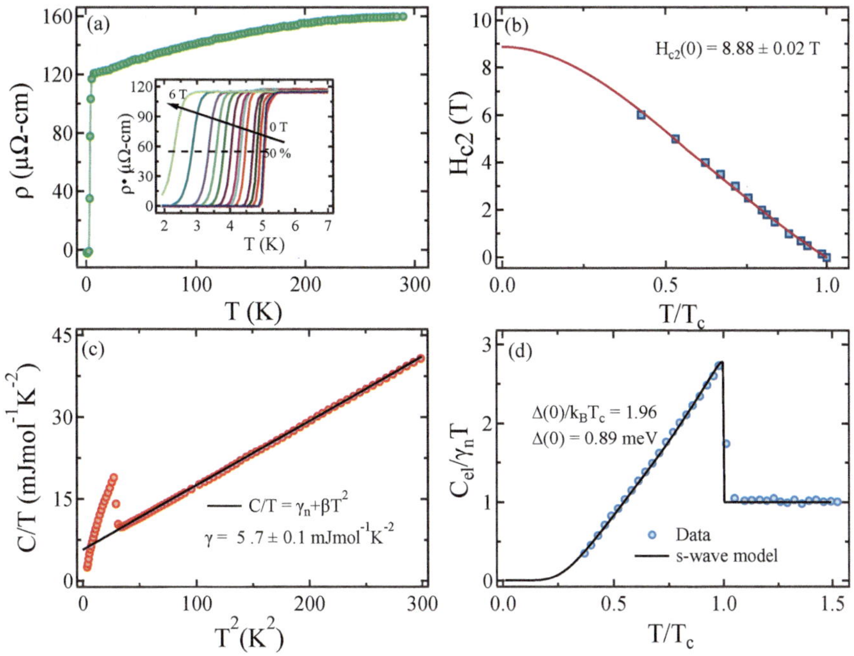

Fig. 6.7 **a** Zero-field resistivity measurement for $Nb_{0.2}Re_{0.2}Zr_{0.2}Hf_{0.2}Ti_{0.2}$ in the temperature range from 1.85 K to 290 K. The inset shows the resistivity versus temperature curves for different magnetic fields. **b** The temperature-dependent upper critical field H_{c2} obtained from resistivity measurements and fitted with Eq. (6.2). **c** The C/T versus T^2 curve in zero fields fitted to Eq. (6.6). **d** The low-temperature electronic specific heat data fitted with the single-gap s-wave superconducting gap using Eq. (6.10). The figures have been taken from [21] with permission

olating the fit to 0 K, which is close to the Pauli limit [$H^{\mathrm{P}}_{c2}(0) = 1.84T_c$, $T_c = 5.3$ K gives $H^{\mathrm{P}}_{c2}(0) = 9.7$ T]. The equivalent bcc-structured $[TaNb]_{0:67}(HfZrTi)_{0:33}$ has $H_{c2}(0) = 8.3$ T with $T_c = 7.3$ K, which is significantly lower than the Pauli limit for weakly coupled BCS superconductors. The increase in H_{c2} (0) for the current system might be due to the increased spin-orbit coupling (SOC) caused by the addition of Re into the structure. However, additional research is needed to demonstrate a link between H_{c2} and the strength of the spin-orbit coupling. The Ginzburg-Landau coherence length $\xi_{GL}(0)$ was determined to be 6.1 nm from Eq. (6.3).

Specific Heat

Specific heat measurements were performed for the $Nb_{0.2}Re_{0.2}Zr_{0.2}Hf_{0.2}Ti_{0.2}$ sample to examine the phononic characteristics and electronic density of the states. Figure 6.7c shows the curve C/T versus T^2, and the bulk superconductivity is confirmed by a substantial increase in the specific heat at 5.3 K. Fitting this curve for the normal region with Eq. (6.6) yields a Sommerfeld coefficient of $\gamma_n = 5.7$ mJmol^{-1}K^{-2} and a lattice contribution of $\beta_3 = 0.1178$ mJmol^{-1}K^{-4}. The Debye temperature θ_D is calculated to be 254.6 K using Eq. (6.8). The density of states at the Fermi energy $D_c(E_F)$ is computed to be 2.42 $\frac{states}{eV f.u.}$ by Eq. (6.7). The McMillan formula (6.9) is used to calculate the electron-phonon coupling constant $\lambda_{e-ph} = 0.69$, which is a measure of the strength of the attractive interaction between the electron and the phonon. The normalized specific heat jump $\frac{C_{el}}{\gamma_n T_c}$ for $Nb_{0.2}Re_{0.2}Zr_{0.2}Hf_{0.2}Ti_{0.2}$ is 1.81, which is higher than the BCS value (1.43) in the weak coupling limit and indicates moderately coupled superconductivity in the sample. The superconducting gap was also evaluated using low-temperature electronic specific heat data (total specific heat-phononic contribution), as shown in Fig. 6.7d. In the superconducting state, the temperature dependence of the electronic-specific heat was best described by a single-gap BCS s-wave superconductor [Eq. (6.10)]. Eq. (6.11) was used to fit the low-temperature electronic specific heat data and yields $\Delta(0)/k_B T_c = 1.96 \pm 0.03$ which exceeds the BCS (1.764) in the weak coupling limit, implying moderately coupled superconductivity in $Nb_{0.2}Re_{0.2}Zr_{0.2}Hf_{0.2}Ti_{0.2}$. The obtained values of the normalized specific heat jump and electron-phonon coupling constant also confirm the moderate coupling in this system.

6.3.4 Re-Rich hcp HEA Superconductor $Re_{0.56}Nb_{0.11}Ti_{0.11}Zr_{0.11}Hf_{0.11}$

Structural Characterization

This is the first hcp-high entropy alloy discussed in Chap. 7 of this book. The PED patterns obtained for $Re_{0.56}Nb_{0.11}Zr_{0.11}Hf_{0.11}Ti_{0.11}$ are displayed in Fig. 6.8. These results are consistent with the reflection conditions obtained for the $P6_3/mmc$ space group (hhl: l = 2n, 00l: l = 2n) [Fig. 6.8a, b]. The structural solution based on the PEDT data (not displayed; JANA2006) straightforwardly verifies the hcp structure of the material. Figure 6.8c shows the potential map and the anticipated atomic locations resulting from this structural solution. The powder XRD pattern acquired at RT was refined to further infer the crystal structure and lattice parameters using the Rietveld refinement technique [Fig. 6.8d]. The element with the lowest covalent radii, that is, Re, preferentially occupies the Wyckoff positions 2a (0,0,0) and 6 h (x, 2x, 1/4) [S (1) and S (2) in Fig. 6.8c and Table 6.1]. In contrast, Hf and Zr have higher covalent radii at position 4f (1/3, 2/3, z) [L in 6.8c and Table 6.1]. However, the medium radii of Nb and Ti might occupy any of the three atomic sites. Considering the probability that Nb and Re share atomic locations [S(1) and S(2)] results in an extremely good refinement. The structural parameters are given in Table 6.1.

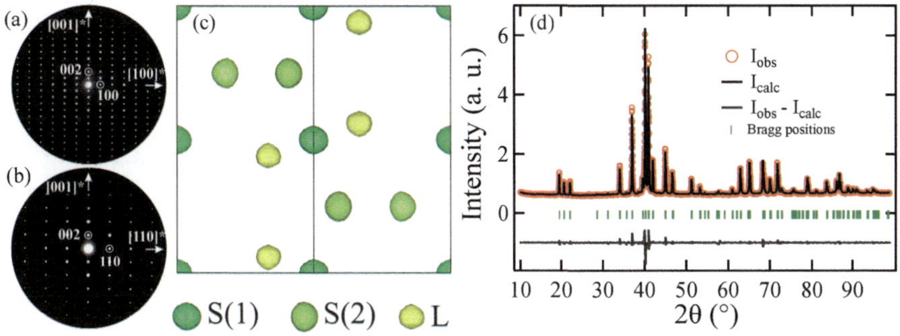

Fig. 6.8 **a** and **b** PED zone axis patterns for $Re_{0.56}Nb_{0.11}Zr_{0.11}Hf_{0.11}Ti_{0.11}$ **a** [010] and **b** [1$\bar{1}$0]. The condition hhl: l = 2n is observed in **b**. **c** [110] Projection of the density map produced by the structural solution technique, which shows the presence of three atomic locations, S(1), S(2), and L. (see Table 6.1). **d** Rietveld refinement of the room-temperature XRD pattern of $Re_{0.56}Nb_{0.11}Zr_{0.11}Hf_{0.11}Ti_{0.11}$. The figures are taken from [24] with permission

Table 6.1 Structural parameters of the hcp superconducting HEA $Re_{0.56}Nb_{0.11}Zr_{0.11}Hf_{0.11}Ti_{0.11}$. Re and Nb occupy the S(1) and S(2) sites, while Zr, Ti, and Hf occupy the L sites. $R_P = 1.48$, $R_{WP} = 2.04$, Goodness of fit (GOF) = 1.75

Structure	Hexagonal
Space group	$P6_3/mmc$
lattice parameters	
$a = b$ (Å)	5.2554(1)
c (Å)	8.5934(1)

	Wyckoff position	x	y	z	Uiso (Å^2)
S(1)	2*a*	0	0	0	0.0038(2)
S(2)	6 *h*	0.7233(4)	0.34466(8)	1/4	0.01154(17)
L	4 *f*	2/3	1/3	0.063(7)	0.0083(2)

Magnetization

The superconducting property of $Re_{0.56}Nb_{0.11}Zr_{0.11}Hf_{0.11}Ti_{0.11}$ was validated by DC and AC susceptibility measurements [Fig. 6.9a]. The bulk superconductivity is confirmed by a diamagnetic signal with an onset of 4.3 K, called the transition temperature, with a superconducting volume fraction greater than 100%. The lower critical field was deduced from the field-dependent magnetization at different

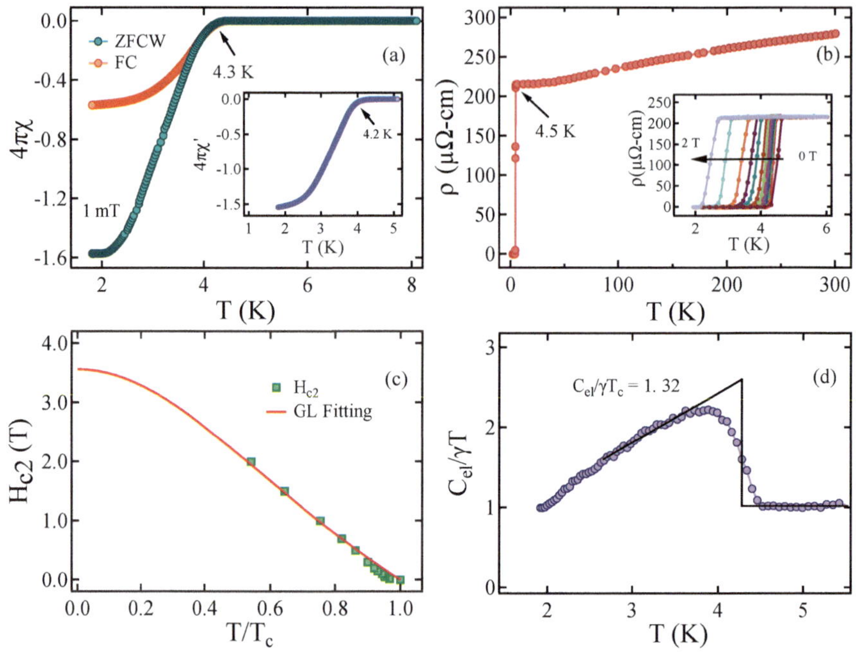

Fig. 6.9 **a** The temperature variation of DC and AC (inset) magnetization for $Re_{0.56}Nb_{0.11}Zr_{0.11}Hf_{0.11}Ti_{0.11}$ at 1 mT applied field. **b** Temperature-dependent resistivity in the zero field for $Re_{0.56}Nb_{0.11}Zr_{0.11}Hf_{0.11}Ti_{0.11}$. The inset shows the resistivity measurements at various fields. **c** Temperature variation of the upper critical field H_{c2} using resistivity measurements for $Re_{0.56}Nb_{0.11}Zr_{0.11}Hf_{0.11}Ti_{0.11}$. The solid line represents the fit using Eq. (6.2). **d** The temperature-dependent normalized electronic specific heat for $Re_{0.56}Nb_{0.11}Zr_{0.11}Hf_{0.11}Ti_{0.11}$ at low temperatures. The figures are derived from [24] with permission

temperatures $(M - H)$ using the Ginzburg-Landau approximation method (Eq. (6.1)), which is found to be 2.3 mT at zero temperature.

Electrical Resistivity

The temperature dependence of the zero-field resistivity is displayed in Fig. 6.9b, which demonstrates a sudden decrease to a zero resistive state (superconducting state) below 4.5 K. The normal state resistivity slowly increases with increasing temperature, indicating weak metallic behavior. The residual resistivity ratio (RRR) is determined to be $\rho(300)/\rho(10) = 1.2$, suggesting the occurrence of atomic-scale disorder in the HEA. However, the low RRR value contrasts with previous reports [9, 16, 18, 44]. Additionally, resistivity measurements were carried out at various magnetic fields to calculate the upper critical field, as shown in the inset of Fig. 6.9b. This shows that T_c is gradually suppressed as the magnetic field increases. The transition temperatures for the respective fields were considered the midpoint values of the superconducting transition, which decreased as the applied magnetic field increasesd. The temperature dependence of the upper critical field was well fitted to Eq. (6.2), as shown in Fig. 6.9c, and the upper critical field at 0 K was determined to be 3.6 T by extrapolating the $H_{c2}(T)$ curve. This value is far below the Pauli limiting field given as $H_{c2}^{\mathrm{P}}(0) = 1.84T_c = 8.3$ T. The Ginzburg-Landau coherence length $\xi_{\mathrm{GL}}(0)$ is estimated to be 11 nm according to Eq. (6.2).

Specific Heat

The specific heat was measured to investigate the characteristics of the phonons, the electronic density of states, and the nature of the superconductivity (BCS or unconventional) in this system. A pronounced increase in the specific heat was observed at approximate 4.4 K, further verifying the bulk superconductivity in this system. The normal state-specific heat data were well fitted with Eq. (6.6), which yields the Sommerfeld coefficient, $\gamma_n = 2.77$ mJmol^{-1}K^{-2} and the lattice contribution to the specific heat, $\beta_3 = 0.0648$ mJmol^{-1}K^{-4}. The Debye temperature, θ_D, was estimated to be 310 K using Eq. (6.8). The density of states at the Fermi level is determined to be 1.17 $\frac{states}{eV f.u.}$ using Eq. (6.7). The McMillan formula (6.9) was used to evaluate the strength of the attractive interaction between the electron and phonon, that is, the electron-phonon coupling constant (λ_{e-ph}). The λ_{e-ph} is found to be 0.57, suggesting weakly coupled superconductivity in $Re_{0.56}Nb_{0.11}Zr_{0.11}Hf_{0.11}Ti_{0.11}$. The electronic specific heat was evaluated by removing the phononic contribution from the total specific heat (Fig. 6.9d), and the normalized specific heat jump $\frac{C_{el}}{\gamma_n T_c}$ was found to be 1.32, which is slightly smaller than the BCS value in the weak coupling limit (1.43), further confirming the weak coupling superconductivity in this sample. Table 6.2 summarizes the parameters of the normal and superconducting states for hcp $Re_{0.56}Nb_{0.11}Zr_{0.11}Hf_{0.11}Ti_{0.11}$.

6.3.5 Electronic Properties and Uemura Plot

To support the experimental findings, various electronic parameters were computed, including the effective mass of the electron, the carrier density of the quasiparticle, the Fermi velocity, the mean free path of the electronic signal and the coherence length of the BCS. Experimentally determined parameters such as the quasiparticle carrier density (n, obtained from Hall measurements), the Sommerfeld coefficient (γ_n, obtained from normal heat capacity), and the residual resistivity (ρ_0, obtained from the resistivity above the transition temperature) were used to extract other electronic parameters. The Sommerfeld coefficient is related to the effective mass (m^*) and the carrier density of the quasiparticle (n) via the following equation:

$$\gamma_n = \left(\frac{\pi}{3}\right)^{2/3} \frac{k_{\mathrm{B}}^2 m^* n^{1/3}}{\hbar^2}, \tag{6.18}$$

where k_B is the Boltzmann constant. The Fermi velocity (v_F) is expressed in terms of the quasi-particle carrier density and effective mass and may be written as:

$$n = \frac{1}{3\pi^2}\left(\frac{m^* v_F}{\hbar}\right)^3. \tag{6.19}$$

The electronic mean-free path l is associated with the Fermi velocity, effective mass, and residual resistivity through the following expression:

$$l = \frac{3\pi^2\hbar^3}{e^2\rho_0 m^{*2} v_F^2}. \tag{6.20}$$

The BCS coherence length (ξ_0) can be determined using the following relation between the Fermi velocity and the transition temperature:

$$\xi_0 = \frac{0.18\hbar v_F}{k_B T_c}. \tag{6.21}$$

In the case of a dirty limit superconductor, the GL penetration depth is related to the BCS coherence length and the electronic mean free path through the following expression:

$$\lambda_{GL}(0) = \left(\frac{m^*}{\mu_0 n e^2}\right)^{1/2}\left(1 + \frac{\xi_0}{l}\right)^{1/2}, \tag{6.22}$$

where the first term is the London penetration depth λ_L. The dirty limit also affects the GL coherence length, which is given by:

$$\frac{\xi_{GL}(0)}{\xi_0} = \frac{\pi}{2\sqrt{3}}\left(1 + \frac{\xi_0}{l}\right)^{-1/2}. \tag{6.23}$$

For $Nb_{0.6}Re_{0.1}Zr_{0.1}Hf_{0.1}Ti_{0.1}$ the electronic parameters were obtained by Eqs. 6.18–6.21 using the experimentally calculated values $\gamma_n = 7.1(1)$ mJmol^{-1}K^{-2}, $n = 6.1(1) \times 10^{28}$ m^{-3}, $\rho_0 = 76(1)$ $\mu\Omega$cm. For $Nb_{0.2}Re_{0.2}Zr_{0.2}Hf_{0.2}Ti_{0.2}$ the Eqs. 6.18–6.20, 6.22 and 6.23 that form a system of equations were solved simultaneously using $\gamma_n = 5.7$ mJmol^{-1}K^{-2}, $\xi_{GL}(0) = 6.1(1)$ nm, and $\rho_0 = 120.01$ $\mu\Omega$cm to obtain the parameters [32]. For both samples, the ratio of BCS the coherence length and electronic mean free path, $\frac{\xi_0}{l}$ was found to be greater than 1, suggesting that the dirty limit superconductivity in both samples. The estimated electronic parameters for $Nb_{0.6}Re_{0.1}Zr_{0.1}Hf_{0.1}Ti_{0.1}$ are listed in Table 6.2.
The Fermi temperature, T_F is given as follows:

$$k_B T_F = \frac{\hbar^2}{2}(3\pi^2)^{2/3}\frac{n^{2/3}}{m^*}. \tag{6.24}$$

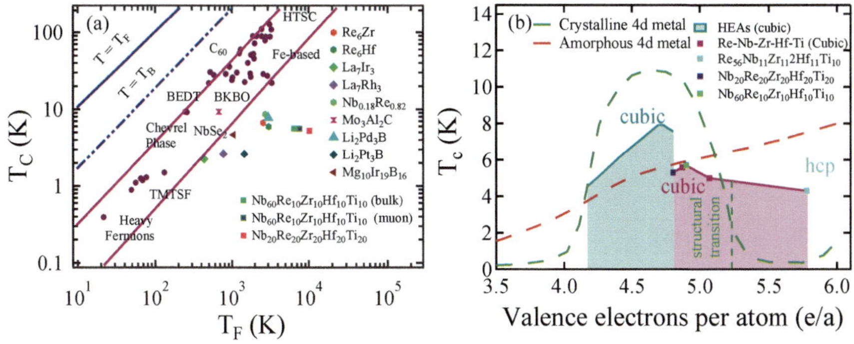

Fig. 6.10 a The Uemura plot displaying the superconducting transition temperature versus the Fermi temperature for various superconducting families. The two solid purple lines represent the unconventional superconductor band. The samples discussed in this chapter ($Nb_{0.6}Re_{0.1}Zr_{0.1}Hf_{0.1}Ti_{0.1}$ and $Nb_{0.2}Re_{0.2}Zr_{0.2}Hf_{0.2}Ti_{0.2}$) are represented by solid green, blue and pink circles that are isolated from the unconventional band. **b** Dependence of the superconducting transition temperature on the valence electron concentration (e/a) for the three samples ($Nb_{0.6}Re_{0.1}Zr_{0.1}Hf_{0.1}Ti_{0.1}$, $Nb_{0.2}Re_{0.2}Zr_{0.2}Hf_{0.2}Ti_{0.2}$ and $Re_{0.56}Nb_{0.11}Zr_{0.11}Hf_{0.11}Ti_{0.11}$). For comparison, the dependence of T_c on the valence electron concentration for crystalline 4d metals [56], amorphous 4d metals [57], and the Ta-Nb-Hf-Zr-Ti system [16] are also included. The data for both figures were taken from [21, 24, 31] with permission

T_F was calculated to be 6480(236) K and 10091 K using the penetration depth by H_{c1} for $Nb_{0.6}Re_{0.1}Zr_{0.1}Hf_{0.1}Ti_{0.1}$ and values of m^* and n for $Nb_{0.2}Re_{0.2}Zr_{0.2}Hf_{0.2}Ti_{0.2}$ respectively. The Fermi temperature, $T_{F_{\mu SR}} = 7590(600)$ K, was also estimated by measuring the penetration depth using μSR measurement (detailed calculations can be found in [58, 59]).

Uemura et al. [60–62] employed the T_c/T_F ratio to distinguish between conventional and unconventional superconductors. Conventional superconductors occupy the region where T_c/T_F is less than 0.01, while unconventional superconductors occupy the range $0.01 \leq T_c/T_F \leq 0.1$. $Nb_{0.6}Re_{0.1}Zr_{0.1}Hf_{0.1}Ti_{0.1}$ and $Nb_{0.2}Re_{0.2}Zr_{0.2}Hf_{0.2}Ti_{0.2}$ fall within the conventional superconductor range, as confirmed by their T_c/T_F ratios determined via H_{c1} and μSR in Fig. 6.10a.

6.3.6 VEC Dependence of T_c

The variation in the transition temperature (T_c) with respect to the valence electron concentration (e/a) is illustrated in Fig. 6.10b. The red dashed line shows T_c dependence on the e/a ratio for transition metal crystalline alloys, whereas the black dashed line shows the T_c dependence on the e/a ratio for amorphous superconductors. The effect of e/a on T_c for the Ta-Nb-Hf-Zr-Ti system is also included (green line). It is evident from the curve that T_c increases with decreasing e/a ratio. The

Table 6.2 Normal and superconducting state parameters of Nb-Re-Zr-Hf-Ti HEA superconductors

Parameter	Unit	Nb-rich bcc HEA	Equimolar bcc HEA	Re-rich hcp HEA
e/a	–	4.9	4.8	5.79
T_c	K	5.7	5.3	4.4
RRR	–	1.2	1.3	1.2
$H_{c1}(0)$	mT	7.75	33	2.3
$H_{c2}(0)$	T	6.31	8.88	3.6
$H_{c2}^{P}(0)$	T	10.3	9.7	8.3
ξ_{GL}	Å	7.2(1)	6.1	11
λ_{GL}	Å	279(2)	–	–
θ_D	K	240	256.6	310
λ_{e-ph}	–	0.70	0.69	0.57
$\frac{\Delta(0)}{k_B T_c}$	–	2.57	1.764	–
$\frac{\Delta C_{el}}{\gamma T_c}$	–	–	1.81	1.32
m^*	m_e	9.9(2)	–	–
ξ_0/l	–	38(5)	–	–
v_F	10^5 ms^{-1}	1.50(4)	–	–
n	10^{28} m^{-3}	6.1(1)	–	–
T_F	K	6480(236)	10091	–

$[Nb_{0.67-x}Re_x](TiZrHf)_{0.33}$ materials exhibit crystalline-like behavior corresponding to the electron count dependency of T_c as demonstrated in Fig. 6.10b.

6.4 Conclusion

The superconducting properties of the Nb-Re-Zr-Hf-Ti system have been investigated in detail for various compositions using different techniques, including XRD, PED, magnetization, resistivity, specific heat, and μSR measurements. The $[Nb_{0.67-x}Re_x](ZrHfTi)_{0.33}$ sample underwent a polymorphic transition from the bcc to the hcp structure with increasing VEC, with $Re_{0.56}Nb_{0.11}Zr_{0.11}Hf_{0.11}Ti_{0.11}$ stabilizing in the pure hcp phase. All the samples exhibit diamagnetic signals below their T_c, indicating bulk superconductivity. Magnetization, resistivity, and specific heat measurements revealed that $Re_{0.56}Nb_{0.11}Zr_{0.11}Hf_{0.11}Ti_{0.11}$ is a weakly coupled BCS s-wave superconductor, while $Nb_{0.6}Re_{0.1}Zr_{0.1}Hf_{0.1}Ti_{0.1}$ and $Nb_{0.2}Re_{0.2}Zr_{0.2}Hf_{0.2}Ti_{0.2}$ are moderately coupled BCS s-wave superconductors in the dirty limit. Specific heat measurements in different fields reveal a nodeless gap in $Nb_{0.6}Re_{0.1}Zr_{0.1}Hf_{0.1}Ti_{0.1}$, while the time-reversal symmetry remains preserved in this sample. The electronic parameters of the samples indicate similarity with those of binary and ternary superconductors, and the Uemura plot classifies $Nb_{0.6}Re_{0.1}Zr_{0.1}Hf_{0.1}Ti_{0.1}$ and $Nb_{0.2}Re_{0.2}Zr_{0.2}Hf_{0.2}Ti_{0.2}$ as conventional superconductors. The upper critical field for $Nb_{0.2}Re_{0.2}Zr_{0.2}Hf_{0.2}Ti_{0.2}$ is very close to the Pauli limiting field, possibly due to enhanced spin-orbit coupling by the introduction of Re. However, further investigations are necessary to explore the pairing mechanism and role of SOC in this sample. Microscopic studies could be useful for understanding the superconducting ground state and pairing mechanism, and searching for time-reversal symmetry breaking in samples with higher Re concentrations, such as $Nb_{0.2}Re_{0.2}Zr_{0.2}Hf_{0.2}Ti_{0.2}$ and $Re_{0.56}Nb_{0.11}Zr_{0.11}Hf_{0.11}Ti_{0.11}$ would be intriguing in solving the mystery surrounding TRS breaking in Re-based systems.

Acknowledgements R. P. S. acknowledges the Science and Engineering Research Board, Government of India, for the CRG/2019/001028 Core Research Grant. The Crystal Growth Lab at IISER Bhopal group members are acknowledged for their valuable assistance.

References

1. B. Gludovatz, A. Hohenwarter, D. Catoor, E.H. Chang, E.P. George, R.O. Ritchie, Science **345**, 1153 (2014)
2. S. Huang, H. Huang, W. Li, D. Kim, S. Lu, X. Li, E. Holmström, S.K. Kwon, L. Vitos, Nat. comm. **9**, 1 (2018)
3. Y. Zou, H. Ma, R. Spolenak, Nat. comm. **6**, 1 (2015)
4. C.P. Lee, Y.Y. Chen, C.Y. Hsu, J.W. Yeh, H.C. Shih, J. Electrochem. Soc. **154**, C424 (2007)
5. X. Wang, W. Guo, Y. Fu, J. Mater. Chem. A **9**, 663 (2021)
6. F. Marques, M. Balcerzak, F. Winkelmann, G. Zepon, M. Felderhoff, Energy Environ. Sci. **14**, 5191 (2021)
7. E.J. Pickering, A.W. Carruthers, P.J. Barron, S.C. Middleburgh, D. Armstrong, A.S. Gandy, Entropy **23**, 98 (2021)
8. D. Castro, P. Jaeger, A.C. Baptista, J.P. Oliveira, Metals **11**, 648 (2021)
9. P. Koželj, S. Vrtnik, A. Jelen, S. Jazbec, Z. Jagličić, S. Maiti, M. Feuerbacher, W. Steurer, J. Dolinšek, Phys. Rev. Lett. **113**, 107001 (2014)
10. Y.F. Kao, S.K. Chen, T.J. Chen, P.C. Chu, J.W. Yeh, S.J. Lin, J. Alloys Compd. **509**, 1607 (2011)
11. J.W. Yeh, S.K. Chen, S.J. Lin, J.Y. Gan, T.S. Chin, T.T. Shun, C.H. Tsau, S.Y. Chang, Adv. Eng. Mater. **6**, 299 (2004)
12. J.W. Yeh, Jom **65**, 1759 (2013)
13. J.W. Yeh, Ann. Chim. Sci. Mat. **31**, 633 (2006)
14. L. Sun, R.J. Cava, Phys. Rev. Mater. **3**, 090301 (2019)
15. J. Kitagawa, S. Hamamoto, N. Ishizu, Metals **10**, 1078 (2020)
16. F.V. Rohr, M.J. Winiarski, J. Tao, T. Klimczuk, R.J. Cava, Proc. Natl. Acad. Sci. USA **113**, E7144 (2016)
17. D. Strong, R.J. Cava, arXiv preprint arXiv:2301.09678 (2023)
18. K. Stolze, F.A. Cevallos, T. Kong, R.J. Cava, J. Mater. Chem. C **6**, 10441 (2018)
19. Fabian O. von Rohr, Robert J. Cava, Phys. Rev. Mater. **2**, 034801 (2018)
20. R. Sogabe, Y. Goto, Y. Mizuguchi, Appl. Phys. Expr. **11**, 053102 (2018)
21. S. Marik, M. Varghese, K.P. Sajilesh, D. Singh, R.P. Singh, J. Alloys Compd. **769**, 1059 (2018)
22. R. Sogabe, Y. Goto, T. Abe, C. Moriyoshi, Y. Kuroiwa, A. Miura, K. Tadanaga, Y. Mizuguchi, Solid State Commun. **295**, 43 (2019)

23. N. Ishizu, J. Kitagawa, Results Phys. **13**, 102275 (2019)
24. S. Marik, K. Motla, M. Varghese, K.P. Sajilesh, D. Singh, Y. Breard, P. Boullay, R.P. Singh, Phys. Rev. Mater. **3**, 060602(R) (2019)
25. J. Guo, H. Wang, F.V. Rohr, Z. Wang, S. Cai, Y. Zhou, K. Yang, A. Li, S. Jiang, Q. Wu, R.J. Cava, L. Sun, Proceed. Nat. Acad. Sci. **114**, 13144 (2017)
26. M.R. Kasem, A. Yamashita, T. Hatano, K. Sakurai, N.O. Hori, Y. Goto, O. Miura, Y. Mizuguchi, Supercond. Sci. Technol. **34**, 125001 (2021)
27. K. Motla, Arushi, P. K. Meena, D. Singh, P. K. Biswas, A. D. Hillier, R. P. Singh, Phys. Rev. B 104, 094515 (2021)
28. S.G. Jung, Y. Han, J.H. Kim, R. Hidayati, J.S. Rhyee, J.M. Lee, W.N. Kang, W.S. Choi, H.R. Jeon, J. Suk, T. Park, Nat. Comm. **13**, 3373 (2022)
29. C.K.W. Leung, X. Zhang, F. von Rohr, R. Lortz, B. Jäck, Sci. Rep. **12**, 12773 (2022)
30. K. Motla, Arushi, V. Soni, P. K. Meena, and R.P. Singh, Supercond. Sci. Technol. 35, 074002 (2022)
31. K. Motla, P. K. Meena, Arushi, D. Singh, P. K. Biswas, A. D. Hillier, and R. P. Singh, Phys. Rev. B 105, 144501 (2022)
32. R.P. Singh, A.D. Hillier, B. Mazidian, J. Quintanilla, J.F. Annett, D.M. Paul, G. Balakrishnan, M.R. Lees, Phys. Rev. Lett. **112**, 107002 (2014)
33. D. Singh, J.A.T. Barker, A. Thamizhavel, D.M. Paul, A.D. Hillier, R.P. Singh, Phys. Rev. B **96**, 180501(R) (2017)
34. D. Singh, K.P. Sajilesh, J.A.T. Barker, D.M. Paul, A.D. Hillier, R.P. Singh, Phys. Rev. B **97**, 100505(R) (2018)
35. P.K. Biswas, M.R. Lees, A.D. Hillier, R.I. Smith, W.G. Marshall, D. McK Paul, Phys. Rev. B **84**, 184529 (2011)
36. J.A.T. Barker, B.D. Breen, R. Hanson, A.D. Hillier, M.R. Lees, G. Balakrishnan, D. McK Paul, R.P. Singh, Phys. Rev. B **98**, 104506 (2018)
37. S. Sharma, K. Motla, J. Beare, M. Nugent, M. Pula, T.J. Munsie, A.D. Hillier, R.P. Singh, G.M. Luke, Phys. Rev. B **103**, 104507 (2021)
38. T. Shang, M. Smidman, S.K. Ghosh, C. Baines, L.J. Chang, D.J. Gawryluk, J.A.T. Barker, R.P. Singh, D. McK, G. Paul, E. Balakrishnan, M. Pomjakushina, M. Shi, A.D. Medarde, H.Q. Hillier, J. Yuan, J. Mesot. Quintanilla, T. Shiroka, Phys. Rev. Lett. **121**, 257002 (2018)
39. D.G.C. Jonas, P.K. Biswas, A.D. Hillier, D.A. Mayoh, M.R. Lees, Phys. Rev. B **105**, L020503 (2022)
40. A.D. Hillier, J.S. Lord, K. Ishida, C. Rogers, Phil. Trans. R. Soc. A **377**, 20180064 (2019)
41. A.D. Hillier, S.J. Blundell, I. McKenzie, I. Umegaki, L. Shu, J.A. Wright, T. Prokscha, F. Bert, K. Shimomura, A. Berlie et al., Nat. Rev. Methods Prim. **2**, 1 (2022)
42. C.L. Tracy, S. Park, D.R. Tittman, S.J. Zinkle, H. Bei, M. Lang, R.C. Ewing, W.L. Mao, Nat. Commun. **8**, 15634 (2017)
43. S. Vrtnik, P. Koželj, A. Meden, S. Maiti, W. Steurer, M. Feuerbacher, J. Dolinsek, J. Alloys Compd. **695**, 3530 (2017)
44. K. Stolze, J. Tao, F.O. von Rohr, T. Kong, R.J. Cava, Chem. Mater. **30**, 906 (2018)
45. B. Liu, J. Wu, Y. Cui, Q. Zhu, G. Xiao, H. Wang, S. Wu, G. Cao, Z. Ren, A.C.S. Appl, Electron. Mater. **2**, 1130 (2020)
46. J. Wu, B. Liu, Y. Cui, Q. Zhu, G. Xiao, H. Wang, S. Wu, G. Cao, Z. Ren, Sci. China Mater. **63**, 823 (2020)
47. M. Isobe, M. Arai, N. Shirakawa, Phys. Rev. B **93**, 054519 (2016)
48. N. Nakai, P. Miranovic, M. Ichioka, K. Machida, Phys. Rev. B **70**, 100503(R) (2004)
49. R.S. Hayano, Y.J. Uemura, J. Imazato, N. Nishida, T. Yamazaki, R. Kubo, Phys. Rev. B **20**, 850 (1979)
50. B.D. Rainford, G.J. Daniell, Hyperfine Interact. **87**, 1129 (1994)
51. M. Weber, A. Amato, F.N. Gygax, A. Schenck, H. Maletta, V.N. Duginov, V.G. Grebinnik, A.B. Lazarev, V.G. Olshevsky, VYu. Pomjakushin, S.N. Shilov, V.A. Zhukov, B.F. Kirillov, A.V. Pirogov, A.N. Ponomarev, V.G. Storchak, S. Kapusta, J. Bock, Phys. Rev. B **48**, 13022 (1993)

52. A. Maisuradze, R. Khasanov, A. Shengelaya, H. Keller, J. Phys.: Condens. Matter **21**, 075701 (2009)
53. B.S. Chandrasekhar, D. Einzel, Ann. Phys. **505**, 535 (1993)
54. R. Prozorov, R.W. Giannetta, Supercond. Sci. Technol. **19**, R41 (2006)
55. D.T. Adroja, A. Bhattacharyya, M. Telling, Y. Feng, M. Smidman, B. Pan, J. Zhao, A.D. Hillier, F.L. Pratt, A.M. Strydom, Phys. Rev. B **92**, 134505 (2015)
56. B.T. Matthias, Phys. Rev. **97**, 74 (1955)
57. M.M. Collver, R.H. Hammond, Phys. Rev. Lett. **30**, 92 (1973)
58. A.D. Hillier, R. Cywinski, Appl. Magn. Reson. **13**, 95 (1997)
59. D. Singh, J.A.T. Barker, A. Thamizhavel, D. McK, Paul, and R. P. Singh, J. Phys.: Condens. Matter **30**, 075601 (2018)
60. Y.J. Uemura, V.J. Emery, A.R. Moodenbaugh, M. Suenaga, D.C. Johnston, A.J. Jacobson, J.T. Lewandowski, J.H. Brewer, R.F. Kiefl, S.R. Kreitzman, G.M. Luke, T. Riseman, C.E. Stronach, W.J. Kossler, J.R. Kempton, X.H. Yu, D. Opie, H.E. Schone, Phys. Rev. B **38**, 909(R) (1988)
61. Y.J. Uemura, G.M. Luke, B.J. Sternlieb, J.H. Brewer, J.F. Carolan, W.N. Hardy, R. Kadono, J.R. Kempton, R.F. Kiefl, S.R. Kreitzman, P. Mulhern, T.M. Riseman, D.L. Williams, B.X. Yang, S. Uchida, H. Takagi, J. Gopalakrishnan, A.W. Sleight, M.A. Subramanian, C.L. Chien, M.Z. Cieplak, G. Xiao, V.Y. Lee, B.W. Statt, C.E. Stronach, W.J. Kossler, X.H. Yu, Phys. Rev. Lett. **62**, 2317 (1989)
62. Y.J. Uemura, L.P. Le, G.M. Luke, B.J. Sternlieb, W.D. Wu, J.H. Brewer, T.M. Riseman, C.L. Seaman, M.B. Maple, M. Ishikawa, D.G. Hinks, J.D. Jorgensen, G. Saito, H. Yamochi, Phys. Rev. Lett. **66**, 2665 (1991)

Chapter 7
Hexagonal Close-Packed HEA Superconductors

Alexander J. Browne

Abstract By drawing together individual reports and supplementing them with new collective insights, this Chapter provides the first comprehensive review of the state of the art of superconducting hexagonal close-packed high-entropy alloys. It is divided into five Sections. Sections 7.1 and 7.2 overview hcp HEAs in general, then specifically the superconducting hcp HEAs that have been reported to date. This leads into Sect. 7.3, where the dependence of the superconducting critical temperature on valence electron count in these materials is evaluated. Then, in Sect. 7.4, the challenges and opportunities for the discovery of new hcp HEAs are discussed. Finally, in Sect. 7.5, superconductivity-motivated future prospects are linked to wider research efforts in the HEA field.

7.1 Overview of hcp HEAs

The hexagonal close-packed structure (Fig. 7.1) is one of the three simple crystallographic structures that can be adopted by HEAs. To start, I will briefly overview the current state of hcp HEAs in general. It is immediately apparent that, in regard to superconductivity or otherwise, hcp HEAs have been less well studied to date than their fcc and bcc analogues. I would encourage any readers interested in materials design and discovery to consider this; as will be illustrated, the chemical space in which they may be found is larger than often thought, so the scope for finding low-hanging fruit is likely to be high.

Whilst the first HEAs were reported in 2004, the first unambiguous report of an hcp HEA, equimolar HoDyYGdTb, came more than a decade later [1]. A number of five- and six-element hcp HEAs, in which rare earth elements are the sole or majority components, have since been reported [2]. The formation of single-phase high-entropy solid solutions, nominally driven by the large mixing entropy, can be prevented if the mixing enthalpy is also large [3]. The mixing enthalpy is smaller, and

A. J. Browne (✉)
School of Chemistry, University of St Andrews, Purdie Building, North Haugh, St Andrews KY16 9ST, UK
e-mail: ajb58@st-andrews.ac.uk

J. Kitagawa and Y. Mizuguchi (eds.), *High-Entropy Alloy Superconductors*, Springer Series in Solid-State Sciences 202, https://doi.org/10.1007/978-981-97-4129-8_7

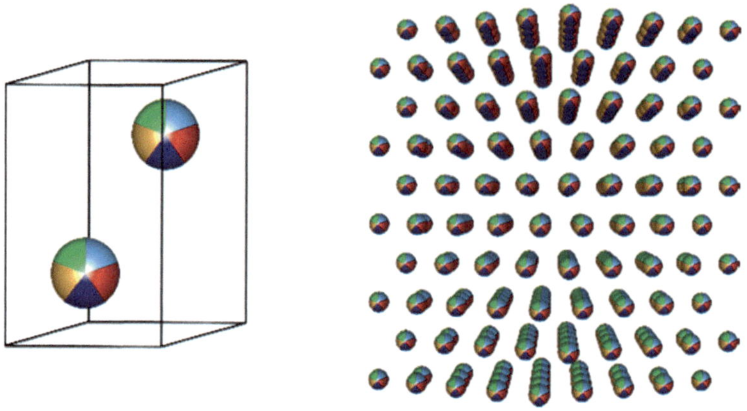

Fig. 7.1 The unit cell of the hcp structure (here with the single crystallographic site occupied by five equimolar species), and a projection of the resulting lattice

hence solid solution formation more likely, if the constituent elements are chemically similar—and given that the rare earth elements themselves often crystallise as hcp or double-hcp, it is unsurprising that their mixing enthalpies are negligible and therefore that they form HE solid solutions.

This bias of hcp HEAs towards compositions dominated by rare earth elements, which have a Valence Electron Count (VEC) of 3, has led to an assumption that HEAs will only crystallise as hcp if their VEC is small [4]. Furthermore, the relatively small number of transition metals the crystallise with the hcp structure seems to have led to an assumption that transition metal HEAs are simply more likely to crystallise with fcc or bcc structures than hcp [5]. Fortunately, chemical intuition and pioneering studies have shown that both these assumptions are false, and various hcp transition metal HEAs have now been reported. There are two regions of the d block in which hcp elements are found. First, the Group 3 (Sc, Y, La) and Group 4 (Ti, Zr, Hf) elements—and combining all six of these in an equimolar HEA does indeed result in a (double-)hcp HEA [6]. The second region is occupied by the Group 7 (Tc, Re) and Group 8 (Ru, Os) 4d and 5d elements. Emphasising the valence electron count here is important, as binary phases with VECs of 7–8 also often crystallise as hcp, even if none of the constituent elements themselves adopt this structure. For example, binary alloys of Mo (bcc) with Rh, Ir, Pd or Pt (all fcc) that have a VEC around 8 crystallise as hcp. The same is found for multi-component alloys. The hcp quaternary Mo-Ru-Rh-Pd was synthesised 30 years before the rare earth HEA HoDyYGdTb was discovered [7]. Pentenary Mo-Ru-Pd-Tc-Rh hcp alloys were even been found in naturally-occurring minerals long before the targeted study of HEAs came into vogue [8]. As will become apparent, a key takeaway from this Chapter should be optimism for the discovery of new hcp HEAs—even though they currently seem more uncommon than fcc or bcc ones.

7.2 Timeline of Superconducting hcp HEAs

In this section I summarise all the reports of superconducting hcp HEAs that have been made to date. Any quantitative parameters related to superconductivity determined in these reports are collated for comparison in Table 7.1.

7.2.1 $Re_{0.56}Nb_{0.11}Ti_{0.11}Zr_{0.11}Hf_{0.11}$

In parallel with the relatively late development of hexagonal HEAs in general compared to cubic ones, the first hexagonal superconducting HEA was only reported in 2019, five years after the first (bcc) superconducting HEA was found [9]. This hexagonal HEA was $Re_{0.56}Nb_{0.11}Ti_{0.11}Zr_{0.11}Hf_{0.11}$, reported by Marik et al. of the Indian Institute of Science Education and Research Bhopal, India, and the Laboratoire CRISMAT, France [10]. The discovery of this material followed earlier work by the same authors that discovered superconductivity in the equimolar HEA of the same five elements [11]. Note how these materials nicely illustrate the possibility of chemical tuning of HEA structure. The lead authors of Refs. 10 and 11 discuss their work more extensively in Chap. 6; discussion of $Re_{0.56}Nb_{0.11}Ti_{0.11}Zr_{0.11}Hf_{0.11}$ is included again here for completeness in the story of superconducting hcp HEAs.

Strictly speaking the structure of is of the hexagonal Laves (C14, $MgZn_2$-type) type, analogous to the Re-rich binary alloys $ZrRe_2$ and $HfRe_2$, rather than simple hcp. The hexagonal Laves structure is an hcp superstructure with three distinct crystallographic sites. In Ref. [10] (see also Chap. 6, Sect. 6.3.4) the authors clearly determine that in $Re_{0.56}Nb_{0.11}Ti_{0.11}Zr_{0.11}Hf_{0.11}$ there are three sites, each of which is occupied by a distinct combination of the five elements present (Fig. 7.2). The unit cell volume is six times that of a true hcp unit cell, which, as shown in Fig. 7.1, only has a single crystallographic site.

Nonetheless, superconductivity had been found in a hexagonal HEA. This was characterised using electrical resistivity, magnetic susceptibility and heat capacity measurements (Chap. 6, Sect. 6.3.4). The critical temperature T_c is 4.4 K, evident in a sharp decrease of the resistivity to zero (Fig. 7.3). Coincident is the onset of diamagnetism in the dc and ac susceptibilities, at 4.3 K and 4.2 K respectively. The lower and upper critical fields, $H_{c1}(0)$ and $H_{c2}(0)$, were determined to be 2.3 mT and 3.6 T, respectively. The latter is well below the value for a BCS superconductor in the weak-coupling Pauli limit, which would be 8.3 T for the observed T_c.

Heat capacity measurement verifies the bulk nature of the superconducting transition. This measurement can also be used to determine the electron–phonon coupling constant $\lambda_{e\text{-}p}$, and ratio $\Delta C_e/\gamma.T_c$, where ΔC_e is the change of the heat capacity through the transition and γ is the Sommerfeld coefficient. For $Re_{0.56}Nb_{0.11}Ti_{0.11}Zr_{0.11}Hf_{0.11}$, $\lambda_{e\text{-}p}$ is 0.57 and $\Delta C_e/\gamma.T_c$ is 1.32. Both of these values indicate weak electron–phonon coupling, with the latter being slightly below the theoretical BCS value of 1.43.

Table 7.1 Compositions, valence electron counts and any reported superconducting parameters for the six families of superconducting hcp HEAs that have been reported to date. Please refer to Sect. 7.2 and the references therein for details on how these parameters were determined (unfortunately, different methods are used in different reports—for example, T_c is sometimes determined from fits to heat capacity data, sometimes as the onset temperature of diamagnetic behaviour, and sometimes as the temperature at which the normal-state resistivity has decreased by 50%). * indicates exact compositions determined by EDX; VECs were calculated for these rather than the nominal compositions when given (and in all cases, were calculated for overall compositions i.e. not accounting for compositional segregation between main and secondary phases)

References	Composition	VEC	T_c (K)	$H_{c2}(0)$ (T)	θ_D (K)	γ (mJ mol^{-1} K^{-1})	$\lambda_{e\text{-}p}$	$\Delta C_e/\gamma.T_c$	ζ_{GL} (nm)
[10]	$Re_{0.56}Nb_{0.11}Ti_{0.11}Zr_{0.11}Hf_{0.11}$	5.79	4.4	3.6	310	2.77	0.57	1.32	11 nm
[12]	$Mo_{0.33}Re_{0.33}Ru_{0.33}$	7.00	9.1	–	–	–	–	–	–
	$Mo_{0.25}Re_{0.25}Ru_{0.25}Rh_{0.25}$	7.50	2.5	–	–	–	–	–	–
	$Mo_{0.2375}Re_{0.2375}Ru_{0.2375}Rh_{0.2375}Ti_{0.05}$	7.33	3.6	–	–	–	–	–	–
	$Mo_{0.225}Re_{0.225}Ru_{0.225}Rh_{0.225}Ti_{0.10}$	7.15	4.7	–	–	–	–	–	–
	$Mo_{0.1}Re_{0.1}Ru_{0.55}Rh_{0.1}Ti_{0.15}$	7.20	2.1	–	–	–	–	–	–
	$Mo_{0.105}Re_{0.105}Ru_{0.527}Rh_{0.105}Ti_{0.158}$	7.16	2.2	–	–	–	–	–	–
	$Mo_{0.118}Re_{0.118}Ru_{0.470}Rh_{0.118}Ti_{0.176}$	7.06	2.5	–	–	–	–	–	–
[14]	$Nb_{0.118}Mo_{0.313}Ru_{0.398}Rh_{0.071}Pd_{0.100}$*	7.29	5.58	6.9	348	3.35	0.63	≈1.43	6.9
	$Nb_{0.139}Mo_{0.323}Ru_{0.347}Rh_{0.091}Pd_{0.100}$*	7.23	6.19	8.1	328	3.27	0.66	≈1.43	6.4
	$Nb_{0.203}Mo_{0.282}Ru_{0.336}Rh_{0.076}Pd_{0.103}$*	7.11	6.10	8.3	324	2.68	0.66	≈1.43	6.3
[15]	$Nb_{0.058}Mo_{0.346}Re_{0.139}Ru_{0.380}Rh_{0.077}$*	7.07	7.54	8.9	338	3.70	0.70	<1.43	6.1
	$Nb_{0.055}Mo_{0.301}Re_{0.187}Ru_{0.383}Rh_{0.074}$*	7.12	6.69	7.5	467	3.93	0.62	<1.43	6.6
	$Nb_{0.062}Mo_{0.251}Re_{0.202}Ru_{0.374}Rh_{0.110}$*	7.21	6.51	7.5	393	4.15	0.64	<1.43	6.6
	$Nb_{0.056}Mo_{0.209}Re_{0.294}Ru_{0.344}Rh_{0.097}$*	7.22	5.46	6.1	378	3.32	0.61	<1.43	7.3
[16]	$Mo_{0.298}Re_{0.288}Ru_{0.361}Pd_{0.010}Pt_{0.043}$*	7.22	8.17	–	369	3.55	0.70	–	–
	$Mo_{0.274}Re_{0.253}Ru_{0.335}Pd_{0.051}Pt_{0.087}$*	7.48	4.91	–	393	3.01	0.59	–	–

(continued)

Table 7.1 (continued)

References	Composition	VEC	T_c (K)	$H_{c2}(0)$ (T)	θ_D (K)	γ (mJ mol^{-1} K^{-1})	$\lambda_{e\text{-}p}$	$\Delta C_e/\gamma.T_c$	ζ_{GL} (nm)
	$Mo_{0.240}Re_{0.204}Ru_{0.299}Pd_{0.117}Pt_{0.140}$*	7.83	2.22	–	411	2.77	0.49	–	–
	$Mo_{0.221}Re_{0.222}Ru_{0.278}Pd_{0.102}Pt_{0.183}$*	7.95	1.64	–	398	2.79	0.47	–	–
[17]	$Mo_{0.219}Re_{0.192}Ru_{0.201}Rh_{0.196}Pt_{0.191}$*	7.94	0.82	–	–	–	–	–	–
	$Mo_{0.220}Re_{0.177}Ru_{0.200}Ir_{0.191}Pt_{0.212}$*	8.00	1.06	–	–	–	–	–	–
	$Mo_{0.2}Re_{0.2}Ru_{0.2}Rh_{0.2}Pd_{0.2}$	8.00	1.38	–	–	–	–	–	–
	$Mo_{0.2}Re_{0.2}Ru_{0.2}Ir_{0.2}Pd_{0.2}$	8.00	1.61	–	–	–	–	–	–
	$Nb_{0.167}Mo_{0.167}Re_{0.167}Ru_{0.167}Rh_{0.167}Pt_{0.167}$	7.50	2.96	–	–	–	–	–	–
	$Nb_{0.167}Mo_{0.167}Re_{0.167}Ru_{0.167}Ir_{0.167}Pt_{0.167}$	7.50	2.84	–	–	–	–	–	–
	$Nb_{0.2}Mo_{0.2}Re_{0.2}Ru_{0.2}Rh_{0.2}$	7.00	6.13	–	–	–	–	–	–
	$Nb_{0.2}Mo_{0.2}Re_{0.2}Ru_{0.2}Ir_{0.2}$	7.00	5.51	–	–	–	–	–	–

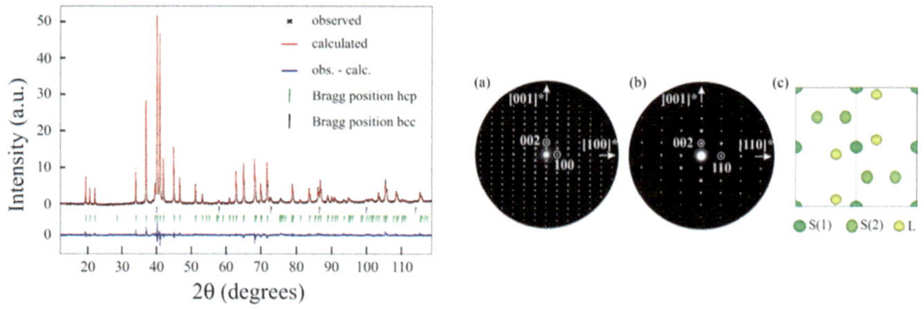

Fig. 7.2 Structural characterisation of $Re_{0.56}Nb_{0.11}Ti_{0.11}Zr_{0.11}Hf_{0.11}$. Powder X-ray diffraction (left) and precession electron diffraction (right, along with the potential map determined by structure solution) show that this HEA has the hexagonal Laves (C14) structure—a supercell of the simple hcp unit cell with three distinct crystallographic sites instead of a single one. Reprinted with permission from Ref. [10]. Copyright (2019) by the American Physical Society

7.2.2 Mo-Re-Ru-Rh-Ti

The first truly-hcp superconducting HEAs were reported shortly after $Re_{0.56}Nb_{0.11}Ti_{0.11}Zr_{0.11}Hf_{0.11}$, by Lee and Cava of Princeton University, USA. Their compositions are based on the equimolar ternary $Mo_{0.33}Re_{0.33}Ru_{0.33}$ [12]. This ternary alloy, and their quaternary $Mo_{0.25}Re_{0.25}Ru_{0.25}Rh_{0.25}$, themselves adopt the hcp structure. For the perspective of materials design this is should be expected—the ternary and quaternary have, respectively, VECs of 7 and 7.5, in the VEC region of the hcp elements Re, Ru and Os. To turn these into HEAs the authors added Ti as a fifth element, thus meeting the HEA criterion of having at least 5 elements in 5–35 at.% concentration [13]. By varying the concentrations of these five elements they produced a series of superconducting HEAs [12].

Powder diffraction patterns of hcp $Mo_{0.33}Re_{0.33}Ru_{0.33}$ and an isostructural HEA are shown in Fig. 7.4. However, the authors found that not all of their synthesised compositions form a single hcp phase. They hypothesise that the hcp phase is destabilised by Ti having a larger atomic radius than the other elements used. In fact, B2 (CsCl-type) phases are found in the binary phase diagrams of Ti-Re, Ti-Ru and Ti-Rh, implying that (because of the difference in ionic radii) they prefer to crystallise with the large and small atoms on different crystallographic sites (i.e. B2) rather than with mixed occupancy on a single site (bcc—the disordered analogue of B2—or hcp, depending on VEC). Electronegativity differences between then constituent elements may also be influential; we will return to these stabilising factors in more detail in Sect. 7.4.2. The existence of hcp and competing phases in similar chemical space will also be returned to in Sect. 7.4.

The superconductivity of these alloys was only characterised using magnetic measurements of the Meissner effect (Fig. 7.4). Whilst ternary $Mo_{0.33}Re_{0.33}Ru_{0.33}$ has a T_c of 9.1 K, those of quaternary $Mo_{0.25}Re_{0.25}Ru_{0.25}Rh_{0.25}$ and the five element hcp (and major-hcp) alloys are much lower, between 2.1 and 4.7 K. The

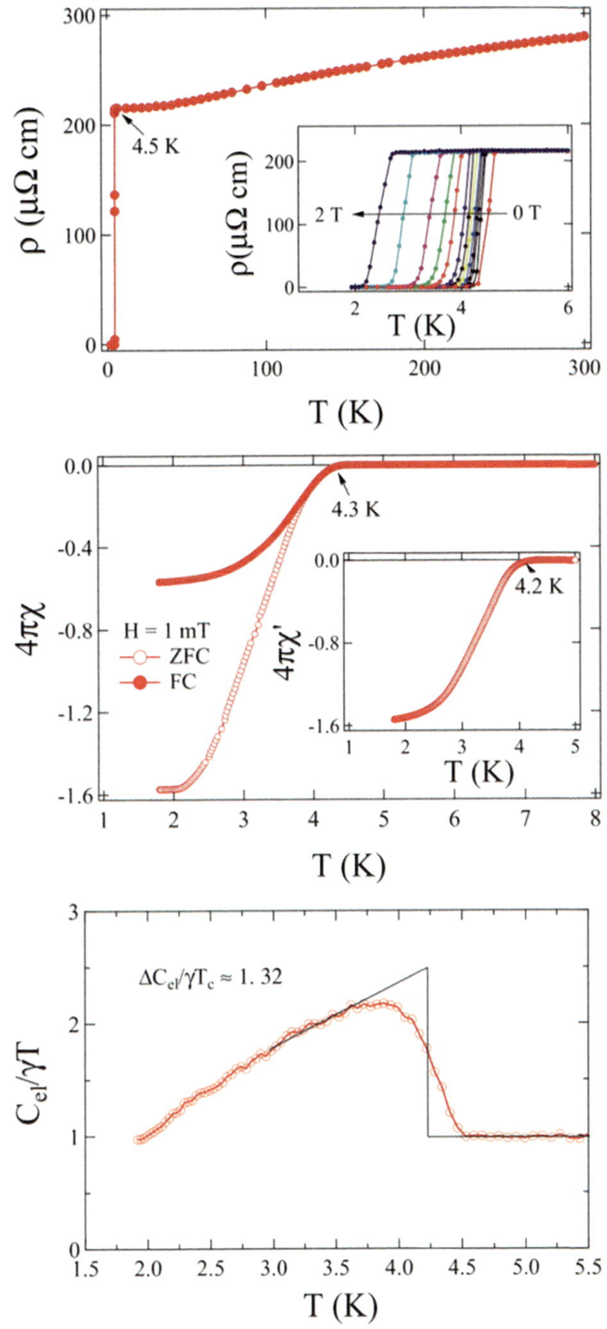

Fig. 7.3 The electrical resistivity of $Re_{0.56}Nb_{0.11}Ti_{0.11}Zr_{0.11}Hf_{0.11}$ shows a superconducting transition with $T_c = 4.4$ K. Reprinted with permission from Ref. [10]. Copyright (2019) by the American Physical Society

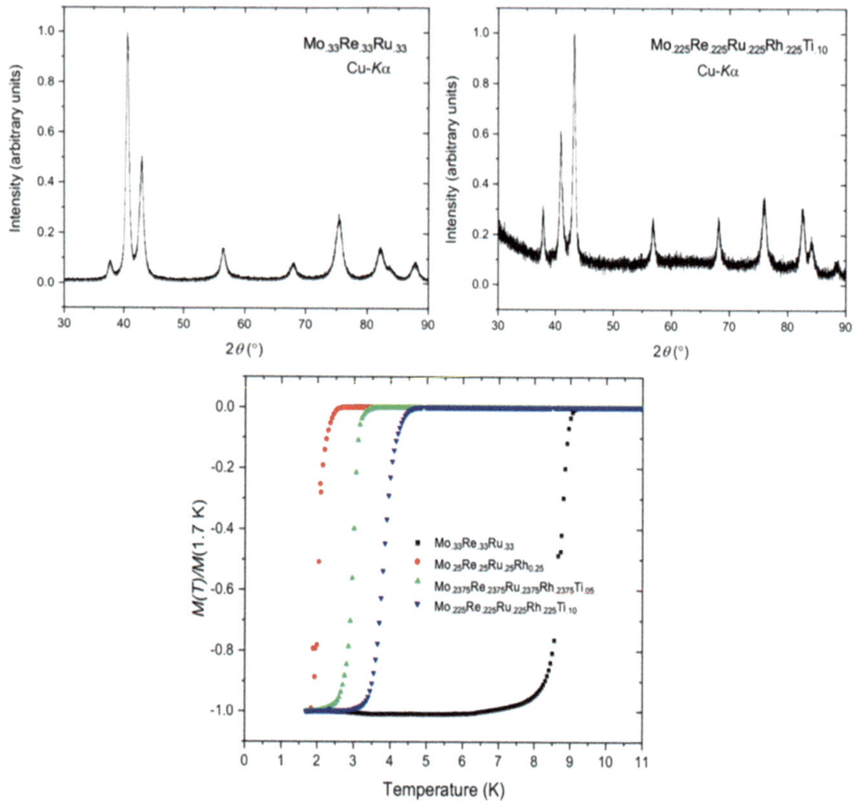

Fig. 7.4 Powder X-ray diffraction patterns of the ternary alloy $Mo_{0.33}Re_{0.33}Ru_{0.33}$, which crystallises with the simple hcp structure, and one of the isostructural five-element HEAs reported in Ref. [12]. Magnetic susceptibility measurements of these and other alloys show a transition to a diamagnetic state, indicating the onset of superconductivity at low temperatures. Reprinted from Ref. [12] with permission from Elsevier

reason for this discrepancy is unclear; a weighted average of the critical temperatures of the constituents in $Mo_{0.33}Re_{0.33}Ru_{0.33}$ is 1.27 K, and its VEC is not significantly different to those of the other alloys. One of the synthesised HEAs, $Mo_{0.167}Re_{0.167}Ru_{0.25}Rh_{0.167}Ti_{0.25}$, crystallises as majority σ-phase and also superconducts. Interestingly its T_c is relatively high, at 5.5 K. This enhancement may be the result of the different structure, and consequent effect on band structure and phonons, in this material relative to the hcp ones even though they are comprised of the same elements.

7.2.3 Nb-Mo-Ru-Rh-Pd

The next family of superconducting hcp HEAs to be found were reported by the group of Zhi Ren of Westlake University, China, who synthesised three materials in the $Nb_{10+2x}Mo_{35-x}Ru_{35-x}Rh_{10}Pd_{10}$ family, nominally with compositions corresponding to $x = 0$, 2.5 and 5 [14]. Powder X-ray diffraction patterns show that they crystallise on simple hcp lattices with only weak additional reflections corresponding to an unidentified trace secondary phase (Fig. 7.5). Unlike for the materials discussed in Sects. 7.2.1 and 7.2.2, the exact compositions of these alloys were determined using EDX spectroscopy. Their VECs fall in the range 7.11–7.29.

Superconducting transitions are evident in resistivity and magnetic susceptibility measurements of all three materials. Heat capacity measurements were also made, and various thermodynamic information extracted. Notably, though exact values are not given inspection shows that the $\Delta C_e/\gamma.T_c$ ratio for all three materials is very close to the ideal BCS value of 1.43 (Fig. 7.5). The reported critical temperatures are 5.58 K for $x = 0$, 6.19 K for $x = 2.5$ and 6.10 K for $x = 5$, though unfortunately it is not made clear exactly how these were determined from the reported property measurements. The upper critical fields show a similar trend, jumping from 6.9 T for $x = 0$ to 8.1 T and 8.3 T for $x = 2.5$ and $x = 5$.

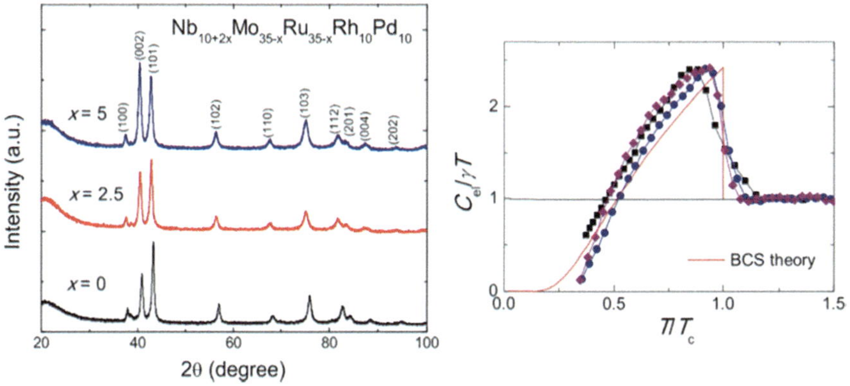

Fig. 7.5 Powder X-ray diffraction patterns and heat capacity measurements of the three $Nb_{10+2x}Mo_{35-x}Ru_{35-x}Rh_{10}Pd_{10}$ HEAs reported in Ref. [14]. These are almost single-phase hcp materials, with just very small impurity peaks visible e.g. between the 100 and 002 reflections of the hcp structure. Nonetheless, the heat capacity data are exceptionally consistent with the prediction from BCS theory. Reprinted from Ref. [14] with permission from Elsevier

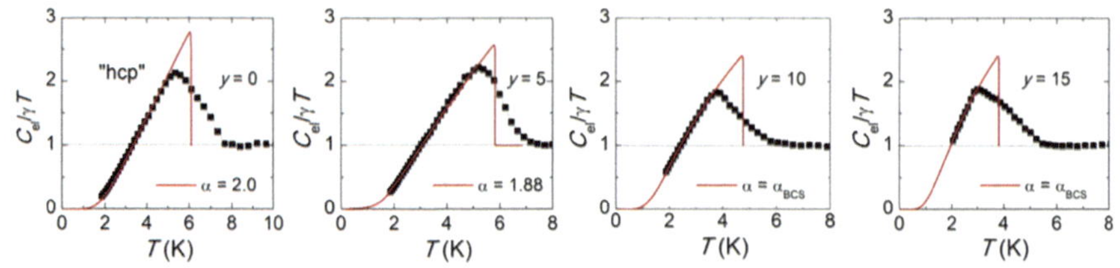

Fig. 7.6 Heat capacity measurements for the $Nb_5Mo_{35\text{-}y}Re_{15+y}Ru_{35}Rh_{10}$ HEAs as reported in Ref. [15]. The width of the transitions (particularly in comparison to the sharp transitions shown in Fig. 7.5) suggest inhomogeneity in these materials that results in an incomplete transition to a superconducting state. α is the superconducting coupling constant and has a value of 1.764 for an ideal BCS superconductor; the larger values of α determined for the $y = 0$ and $y = 5$ phases indicate that the size of the superconducting gap in these materials is larger than ideal. Reprinted from Ref. [15] with permission from Elsevier

7.2.4 Nb-Mo-Re-Ru-Rh

The second family of superconducting hcp HEAs reported by Ren's group is $Nb_5Mo_{35\text{-}y}Re_{15+y}Ru_{35}Rh_{10}$, with nominal $y = 0$, 5, 10 and 15 [15]. Again, these crystallise as simple hcp materials (with powder diffraction showing trace impurities for $y = 0$, 5 and 10) and exact compositions were determined using EDX spectroscopy, giving VECs from 7.07 to 7.22.

Resistivity, magnetic susceptibility and heat capacity measurements were made on all four materials. Using the criterion of T_c being the temperature at which the resistivity has decreased to 50% of its normal-state value, the critical temperatures decrease monotonically from 7.54 K at $y = 0$ to 5.46 K for $y = 15$. A concomitant decrease of $B_{c2}(0)$, from 8.9 T to 6.1 T, also occurs. It should be noted that these changes correlate with increasing VEC. These values of T_c, $B_{c2}(0)$ and the thermodynamic parameters extracted from the heat capacity measurements are broadly similar to those of the Nb-Mo-Ru-Rh-Pd HEAs—as should be expected given their chemical similarity. That said, in the Nb-Mo-Re-Ru-Rh HEAs, the heat capacity increase at T_c appears to be less than the BCS value (Fig. 7.6). An evaluation of the superconducting phase fraction in these HEAs is not reported.

7.2.5 Mo-Re-Ru-Pd-Pt

A third paper from Ren's group reported superconducting hcp HEAs with nominal compositions $[MoReRu]_{(1-2x)/3}[PdPt]_x$ ($x = 0.042$, 0.083, 0.125, 0.167) [16]. To powder X-ray diffraction these all appear to crystallise as single hcp phases (Fig. 7.7). Exact compositions were again determined by EDX spectroscopy, and their VECs range from 7.22 to 7.95.

Resistivity and magnetic susceptibility measurements show superconducting transitions in all four alloys, though it is notable (particularly in the magnetic data) that

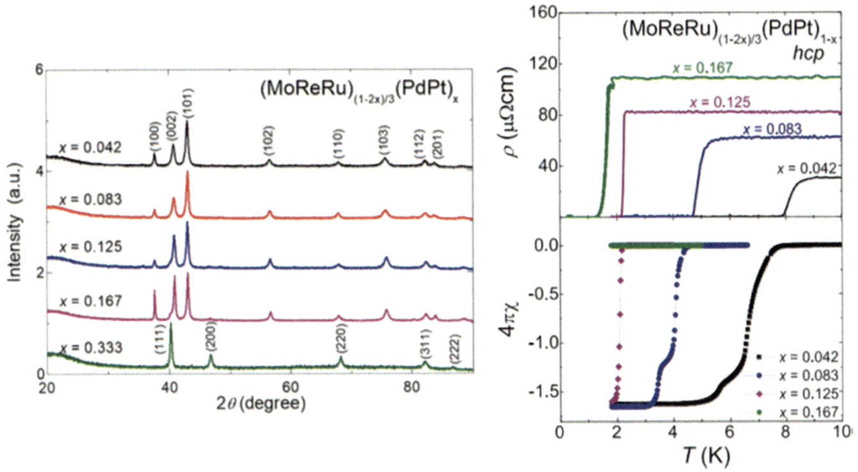

Fig. 7.7 The powder X-ray diffraction patterns of the $[MoReRu]_{(1-2x)/3}[PdPt]_x$ HEAs reported in Ref. [16] show that they crystallise as hcp with trace impurity phases (note that the $x = 0.333$ is fcc not hcp). However, resistivity and magnetic susceptibility measurements of these materials show broad, stepwise transitions, indicative of microstructural inhomogeneity or the presence of strongly superconducting impurity phases that, even if too minor to be visible in diffraction patterns, are sufficient to influence bulk property measurements. Reprinted from Ref. [16] with permission from Elsevier

the transitions in the $x = 0.042$ and $x = 0.083$ systems are broad and stepwise, suggesting some microstructural inhomogeneity or phase impurity (Fig. 7.7). That said, the heat capacity transitions for these two materials are still reasonably sharp (though unfortunately, $\Delta C_e/\gamma.T_c$ values are not reported). Also, due to their low critical temperatures, complete transitions are not observed in the heat capacity data for the $x = 0.125$ and $x = 0.167$ materials.

7.2.6 *Nb-Mo-Re-Ru-Rh/Ir-Pd/Pt*

The most recent discovery of new superconducting hcp HEAs, which are also the first equimolar ones, was made by Browne, Strong and Cava [17]. $Mo_{0.2}Re_{0.2}Ru_{0.2}Rh_{0.2}Pt_{0.2}$ and $Mo_{0.2}Re_{0.2}Ru_{0.2}Ir_{0.2}Pt_{0.2}$, which have VEC = 8, were found to be single-phase hcp alloys with critical temperatures of 0.82 K and 1.06 K, respectively (Fig. 7.8). These comparatively low values of T_c are in line with their comparatively high VECs, as will be elaborated on in the Sect. 7.3. The isoelectronic HEAs $Mo_{0.2}Re_{0.2}Ru_{0.2}Rh_{0.2}Pd_{0.2}$ and $Mo_{0.2}Re_{0.2}Ru_{0.2}Ir_{0.2}Pd_{0.2}$ were also synthesised, though it was found that Pd did not fully dissolve into the HEA mixtures. That said, this has the effect of increasing the critical temperature of the hcp HEA phase—a full explanation of why is provided in Sect. 7.4.4.

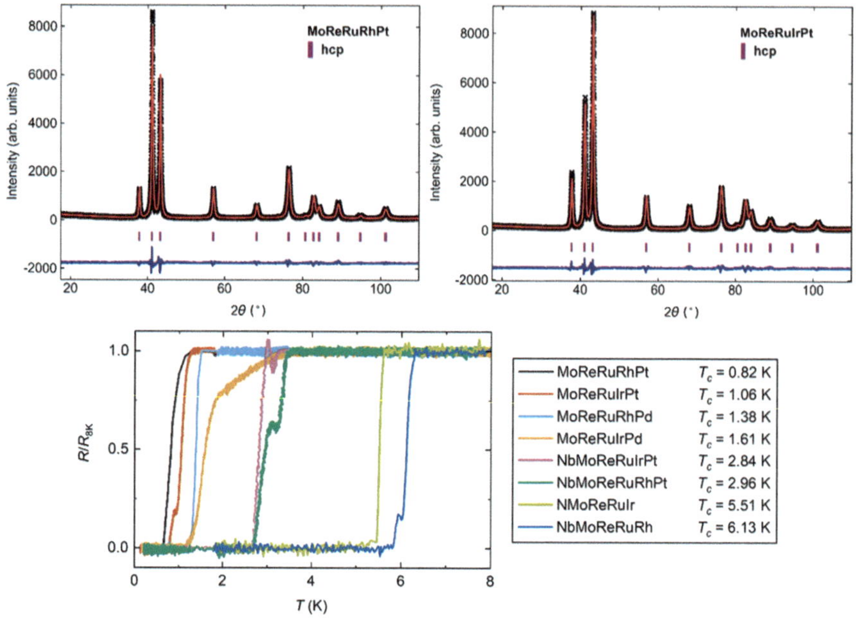

Fig. 7.8 Le Bail profile fitting to powder X-ray diffraction patterns of $Mo_{0.2}Re_{0.2}Ru_{0.2}Rh_{0.2}Pt_{0.2}$ and $Mo_{0.2}Re_{0.2}Ru_{0.2}Ir_{0.2}Pt_{0.2}$ confirms that these equimolar HEAs crystallise as single hcp phases. Resistivity measurements of these and the other HEAs reported in Ref. [17] show their transitions to superconducting ground states, with the temperature at which $R/R_{8K} = 0.5$ taken to be T_c. Reprinted from Ref. [17] with permission from Elsevier

Equimolar VEC = 7 HEAs $Nb_{0.2}Mo_{0.2}Re_{0.2}Ru_{0.2}Rh_{0.2}$ and $Nb_{0.2}Mo_{0.2}Re_{0.2}Ru_{0.2}Ir_{0.2}$ were also synthesised, though these were found to crystallise as mixtures of hcp and σ phases. The authors rationalise this on the basis of the greater differences of the atomic radii and electronegativities of the constituent elements in these HEAs compared to the VEC = 8 ones that do successfully crystallise as single phases. Though $Nb_{0.2}Mo_{0.2}Re_{0.2}Ru_{0.2}Rh_{0.2}$ and $Nb_{0.2}Mo_{0.2}Re_{0.2}Ru_{0.2}Ir_{0.2}$ were found to superconduct, heat capacity measurements show a double transition—in other words, both the hcp and σ phases become superconducting at low temperatures, though their distinct properties could not be disentangled.

Intermediate to these VEC = 8 and VEC = 7 HEAs are the six-element systems $Nb_{0.167}Mo_{0.167}Re_{0.167}Ru_{0.167}Rh_{0.167}Pt_{0.167}$ and $Nb_{0.167}Mo_{0.167}Re_{0.167}Ru_{0.167}Ir_{0.167}Pt_{0.167}$, which have VEC = 7.5. Though the atomic radius and electronegativity differences of the constituent atoms remain large, the increase in entropic stabilisation gained by the addition of the sixth element is sufficient to suppress the formation of the σ phase. Unfortunately this comes with a more stringent requirement for kinetic trapping during the alloy synthesis, and it was found that the samples of both $Nb_{0.167}Mo_{0.167}Re_{0.167}Ru_{0.167}Rh_{0.167}Pt_{0.167}$ and $Nb_{0.167}Mo_{0.167}Re_{0.167}Ru_{0.167}Ir_{0.167}Pt_{0.167}$ comprise two subtly phase-separated high-entropy hcp phases. Nonetheless, these were found to be superconducting, with Tc = 2.96 K for $Nb_{0.167}Mo_{0.167}Re_{0.167}Ru_{0.167}Rh_{0.167}Pt_{0.167}$ and 2.84 K for $Nb_{0.167}Mo_{0.167}Re_{0.167}Ru_{0.167}Ir_{0.167}Pt_{0.167}$. The higher values than those of $Mo_{0.2}Re_{0.2}Ru_{0.2}Rh_{0.2}Pt_{0.2}$ and $Mo_{0.2}Re_{0.2}Ru_{0.2}Ir_{0.2}Pt_{0.2}$ again correlate with the reduction in VEC.

7.3 Trends in T_c

In the literature to date, the most popular way of assessing the phenomenology of superconducting HEAs has been to determine the dependence of T_c on VEC. Of particular interest is comparison to the 'Matthias trend' that is well established for the variation of T_c in the transition metal elements and their simple alloys and compounds. In the 1950s Matthias determined that, in these crystalline systems, the dependence of T_c on VEC shows a double maximum with peaks at VECs around 4.5 and 7.0 (Fig. 7.9) [18]. Later, it was also found that amorphous transition metal alloys can also superconduct—however, a different trend, with a more monotonic dependence of T_c on VEC, was observed [19]. One of the most exciting results in HEA superconductivity to date has been the discovery that the VEC-dependence of T_c in these materials, in which extreme configurational disorder crystallises on a ordered lattice, is intermediate between the ordered and disordered extremes (Fig. 7.9) [20].

That result was demonstrated for bcc superconducting HEAs, which have VECs around the low-VEC maximum of the Matthias trend. With their higher VECs, hcp HEAs afford an opportunity for determining whether that intermediate phenomenology is more generally true—in different structure types, and in the

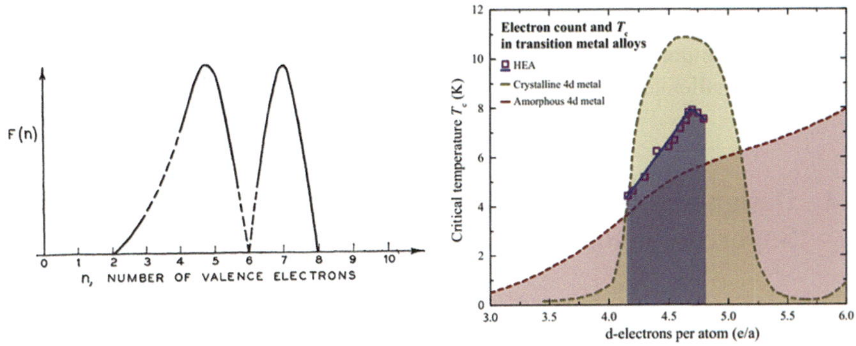

Fig. 7.9 Left—the 'Matthias trend' for the dependence of T_c on VEC in crystalline transition metal-based superconductors. Reprinted with permission from Ref. [18]. Copyright (1955) by the American Physical Society. Right—the dependence of T_c on VEC in bcc HEAs in the regime of the low-VEC Matthias maximum is of intermediate form between distinct trends found in truly crystalline and truly amorphous materials. Reprinted from Ref. [20] with permission from the National Academy of Sciences of the United States of America

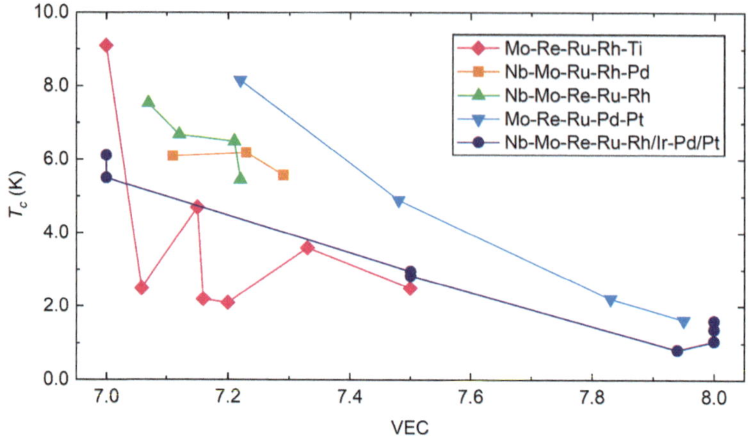

Fig. 7.10 Dependence of T_c on VEC for the hcp HEAs listed in Table 7.1 (excluding $Re_{0.56}Nb_{0.11}Ti_{0.11}Zr_{0.11}Hf_{0.11}$, which not being a true hcp material has an anomalously low VEC). These are all the superconducting hcp HEAs that have so far been reported, and though a Matthias-like decrease of T_c with increasing VEC is evident, more materials—and particularly some that extend the observable range to VEC < 7—are required if a generalised crystalline, amorphous or intermediate phenomenology is to be determined

regime of the high-VEC Matthias maximum. Plotting T_c against VEC for the hcp HEAs listed in Table 7.1 shows in all cases that T_c decreases as the VEC is raised (Fig. 7.10). However, in comparison to the bcc HEAs studied in Ref. [20], there are currently an insufficient number of hcp HEAs to conclusively determine whether the phenomenology of their superconductivity reflects their simultaneously ordered-but-disordered nature, or whether it is closer to one of the crystalline or amorphous limits. To make such a conclusive determination will particularly require new hcp HEAs with VEC < 7; in other words, spanning the complete width of the high-VEC Matthias maximum and not just filling out one side of it. To this end, Sect. 7.4 will explore the challenges and opportunities raised by the quest for new hcp HEAs.

7.4 Design and Optimisation of New hcp HEA Superconductors

The critical bottleneck in the study of hcp HEA superconductors to date is simply the paucity of such systems. That said, every bottleneck presents an opportunity. In this Section I will draw together the state of the art to provide an overview to guide future materials design and discovery. This will encompass both chemical (that is, where in chemical space new hcp phases might be found) and physical (how superconductivity may be stabilised in them) points of view. However, the crucial point to take away is that there is no substitute for simply getting into the lab and trying something. I suspect that the lack of hcp HEAs reported to date is, at least in part, simply the result of a lack of attention, rather than any intrinsic impossibility in finding them. Furthermore, the power of predictive models is increased the more data is available—including negative results. So go, experiment, and see what you find.

7.4.1 Thermodynamic Principles

The stabilisation of HEA phases is principally thermodynamic—that is, it requires finding the right balance of ΔS and ΔH to minimise the Gibbs free energy, ΔG:

$$\Delta G = \Delta H - T\Delta S \tag{7.1}$$

The very name of the high-entropy classification gives away the importance of the former, in the form of the mixing entropy of a solid solution, ΔS_{mix}. However, we will focus extensively on ΔH_{mix} too, because its significance in 'real' materials cannot be understated. Simply mixing a large number of elements together does not guarantee that a single-phase high-entropy solid solution will form—this will only happen if its entropically-dominated ΔG is not outweighed by that of another product (or combination of products) for which ΔG has a large enthalpic driving force [3]. In principle, predicting whether a combination of elements will successfully form a HEA requires ΔG to be calculated for all possible products of that combination, but doing so for a multicomponent mixture is non-trivial. Instead, various empirical proxies have been put forward as guides to help predict whether a particular mixture will form a HEA. The two most well-established are the atomic size difference, δ:

$$\delta = 100 \times \sqrt{\sum_i x_i \left(1 - \frac{r_i}{\sum_i x_i r_i}\right)^2} \tag{7.2}$$

where r_i is the atomic radius of element i [21], and the electronegativity difference, $\Delta\chi$:

$$\Delta\chi = \sqrt{\sum_i x_i \left(\chi_i - \sum_i x_i \chi_i\right)^2} \tag{7.3}$$

where χ_i is the Pauling electronegativity of element i [22]. Single-phase HEAs with simple structures such as hcp, in which there is a single crystallographic site, are more likely to form when all the pairwise ΔH_{mix} interactions in the system are small, hence the different constituent atoms are more comfortable sharing that single crystallographic site and thus forming a high-entropy solid solution. Empirically, that generally correlates with small values of δ and $\Delta\chi$. Conversely, large values of δ and

$\Delta\chi$ suggest non-zero ΔH_{mix} interactions—meaning that the constituent atoms don't want to mix. In this case, high-entropy compositions are more likely to crystallise with complex structures in which there are multiple crystallographic sites, and/or as multiple phases. In other words, enthalpic effects outweigh the nominal entropic driving force of a large ΔS_{mix}.

7.4.2 VEC, δ and Δχ

Finding stability trends in the vast chemical space of high-entropy materials is the key to the effective design of these materials. As the examples given in Sect. 7.2 demonstrate, all current examples of superconducting hcp HEAs are composed principally of 4d and 5d transition metal elements and have VECs around 7–8. It is therefore likely that this will be the chemical space in which new systems can be found most easily. However, it is also evident from the literature that this chemical space is shared with HEAs that adopt other structures, such as σ and α-Mn. It is therefore instructive to look for trends in the enthalpic proxies δ and $\Delta\chi$, as well as the VEC (in addition to its influence on T_c, VEC in effect determines the most stable chemical bonding, and hence the most stable structure, of a system), and locate regions where hcp HEAs are more likely to form.

Table 7.2 collates all the hcp HEAs discussed in Sect. 7.2 along with others that have similar compositions but different structures. Some of these were found in the same compositional series as the hcp phases, whilst others were reported elsewhere [23–25]. For each material, the values of δ (calculated according to Eq. 7.2), $\Delta\chi$ (calculated according to Eq. 7.3) and VEC are provided. The variation of these three parameters is visualised in Fig. 7.11. It is immediately apparent from the first two panels that none of the known superconducting hcp HEAs have VEC < 7. That said, even with the insights we can now gain from quantum chemistry [26], it is not clear to me that this is not a result of the statistics of small numbers rather than an intrinsic physical limit. To access this VEC < 7 space may just require materials design that optimises δ and $\Delta\chi$.

The δ values of the VEC < 7 phases in Fig. 7.11 span a range of values, but none of them has a value of $\Delta\chi$ less than 0.192. The importance of minimising $\Delta\chi$ is highlighted when δ and $\Delta\chi$ are compared without considering VEC, as done in the third panel of Fig. 7.11. There is a concentration of single hcp phases in the chemical space of small δ (<2) and small $\Delta\chi$ (<0.19) values—as should be expected if these really are good proxies for ΔH_{mix}. Interestingly, there is also clear trend of hcp phases forming outside of this regime, when the values of δ and $\Delta\chi$ are increased together at just the right rate. Hence, for example, most of the Mo-Re-Ru-Rh-Ti HEAs reported in Ref. [12] crystallise as single hcp phases despite their $\Delta\chi$ values being as large as those of the σ phase HEAs reported in Ref. [24] or the α-Mn ones reported in Ref. [23]. Targeting HEA compositions that give δ and $\Delta\chi$ values in line with these trends should therefore be a core strategy for the discovery of new hcp HEAs.

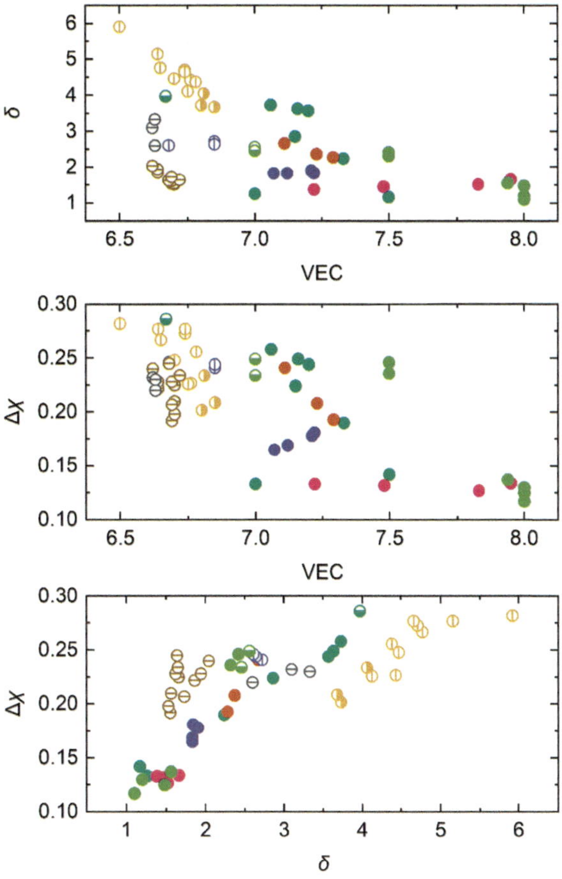

Fig. 7.11 Interdependencies of δ, $\Delta\chi$ and VEC for the HEAs listed in Table 7.2. Filled circles are single-phase hcp HEAs, whilst a horizonal bar indicates a σ phase and a vertical bar indicates an α-Mn one (and half-filling indicates mixed hcp-σ or mixed hcp-α-Mn). Colours group the different phases by source publication

A final point related to VEC is also worth considering. In Sect. 7.1 of this chapter mentioned that the study of hcp HEAs to date has been dominated by rare earth-based systems with VECs of 3–4. Though none of them have yet been shown to exhibit superconductivity, they do show interesting electronic and magnetic phase transitions [27]. Furthermore, many of the rare earth elements themselves have the hcp structure, and some of them become superconducting under high pressures—lanthanum even does so at ambient pressure [28, 29]. The Group 4 transition metals Ti, Zr and Hf are also hcp superconductors. Might this low-VEC region of the Periodic Table provide a whole new chemical space for the discovery of superconducting hcp HEAs that complements the VEC = 7–8 region that to date has been more well-studied?

Table 7.2 Structural assessment of known superconducting hcp HEAs plus other HEAs in similar chemical space. Phase identification follows reported results ('?' indicates that unidentified impurity phases are evident in diffraction patterns). Calculated values of VEC, δ (Eq. 7.2) and $\Delta\chi$ (Eq. 7.3) are provided for each phase. * indicates exact compositions determined by EDX; VEC, δ and $\Delta\chi$ determined for these rather than nominal compositions when appropriate (and in all cases, for overall compositions i.e. not accounting for compositional segregation between main and secondary phases). † indicates materials not included in Table 7.1

References	Composition	Major phase(s)	Minor phase(s)	VEC	δ	$\Delta\chi$
[10]	$Re_{0.56}Nb_{0.11}Ti_{0.11}Zr_{0.11}Hf_{0.11}$	C14 (hcp supercell)	bcc	5.79	6.67	0.243
[12]	$Mo_{0.33}Re_{0.33}Ru_{0.33}$	hcp	None	7.00	1.26	0.133
	$Mo_{0.25}Re_{0.25}Ru_{0.25}Rh_{0.25}$	hcp	None	7.50	1.17	0.142
	$Mo_{0.2375}Re_{0.2375}Ru_{0.2375}Rh_{0.2375}Ti_{0.05}$	hcp	None	7.33	2.24	0.190
	$Mo_{0.225}Re_{0.225}Ru_{0.225}Rh_{0.225}Ti_{0.10}$	hcp	None	7.15	2.86	0.224
	$Mo_{0.1}Re_{0.1}Ru_{0.55}Rh_{0.1}Ti_{0.15}$	hcp	None	7.20	3.57	0.244
	$Mo_{0.105}Re_{0.105}Ru_{0.527}Rh_{0.105}Ti_{0.158}$	hcp	B2	7.16	3.63	0.249
	$Mo_{0.118}Re_{0.118}Ru_{0.470}Rh_{0.118}Ti_{0.176}$	hcp	B2	7.06	3.73	0.258
	$Mo_{0.167}Re_{0.167}Ru_{0.25}Rh_{0.167}Ti_{0.25}$†	σ	hcp, B2	6.67	3.97	0.286
[14]	$Nb_{0.118}Mo_{0.313}Ru_{0.398}Rh_{0.071}Pd_{0.100}$*	hcp	?	7.29	2.28	0.193
	$Nb_{0.139}Mo_{0.323}Ru_{0.347}Rh_{0.091}Pd_{0.100}$*	hcp	?	7.23	2.37	0.208
	$Nb_{0.203}Mo_{0.282}Ru_{0.336}Rh_{0.076}Pd_{0.103}$*	hcp	?	7.11	2.67	0.241
[15]	$Nb_{0.058}Mo_{0.346}Re_{0.139}Ru_{0.380}Rh_{0.077}$*	hcp	?	7.07	1.83	0.165
	$Nb_{0.055}Mo_{0.301}Re_{0.187}Ru_{0.383}Rh_{0.074}$*	hcp	?	7.12	1.83	0.169
	$Nb_{0.062}Mo_{0.251}Re_{0.202}Ru_{0.374}Rh_{0.110}$*	hcp	?	7.21	1.90	0.178
	$Nb_{0.056}Mo_{0.209}Re_{0.294}Ru_{0.344}Rh_{0.097}$*	hcp	?	7.22	1.84	0.181
	$Nb_{0.255}Mo_{0.056}Re_{0.366}Ru_{0.254}Rh_{0.072}$*†	α-Mn	?	6.85	2.72	0.241
	$Nb_{0.239}Mo_{0.100}Re_{0.338}Ru_{0.221}Rh_{0.102}$*†	α-Mn	?	6.85	2.64	0.244
	$Nb_{0.264}Mo_{0.146}Re_{0.329}Ru_{0.166}Rh_{0.095}$*†	α-Mn	?	6.68	2.61	0.246

(continued)

Table 7.2 (continued)

References	Composition	Major phase(s)	Minor phase(s)	VEC	δ	$\Delta\chi$
[16]	$Mo_{0.298}Re_{0.288}Ru_{0.361}Pd_{0.010}Pt_{0.043}$*	hcp	?	7.22	1.38	0.133
	$Mo_{0.274}Re_{0.253}Ru_{0.335}Pd_{0.051}Pt_{0.087}$*	hcp	?	7.48	1.46	0.132
	$Mo_{0.240}Re_{0.204}Ru_{0.299}Pd_{0.117}Pt_{0.140}$*	hcp	?	7.83	1.52	0.127
	$Mo_{0.221}Re_{0.222}Ru_{0.278}Pd_{0.102}Pt_{0.183}$*	hcp	None	7.95	1.66	0.134
[17]	$Mo_{0.219}Re_{0.192}Ru_{0.201}Rh_{0.196}Pt_{0.191}$*	hcp	None	7.94	1.56	0.137
	$Mo_{0.220}Re_{0.177}Ru_{0.200}Ir_{0.191}Pt_{0.212}$*	hcp	None	8.00	1.48	0.125
	$Mo_{0.2}Re_{0.2}Ru_{0.2}Rh_{0.2}Pd_{0.2}$	hcp	Pd	8.00	1.20	0.130
	$Mo_{0.2}Re_{0.2}Ru_{0.2}Ir_{0.2}Pd_{0.2}$	hcp	Pd	8.00	1.10	0.117
	$Nb_{0.167}Mo_{0.167}Re_{0.167}Ru_{0.167}Rh_{0.167}Pt_{0.167}$	hcp	hcp (phase separation)	7.50	2.42	0.246
	$Nb_{0.167}Mo_{0.167}Re_{0.167}Ru_{0.167}Ir_{0.167}Pt_{0.167}$	hcp	hcp (phase separation)	7.50	2.32	0.236
	$Nb_{0.2}Mo_{0.2}Re_{0.2}Ru_{0.2}Rh_{0.2}$	hcp	σ	7.00	2.56	0.249
	$Nb_{0.2}Mo_{0.2}Re_{0.2}Ru_{0.2}Ir_{0.2}$	hcp	σ	7.00	2.46	0.234
[23]	$Zr_{0.10}Nb_{0.10}Mo_{0.2667}Re_{0.2667}Ru_{0.2667}$	α-Mn	None	6.50	5.92	0.282
	$Zr_{0.048}Nb_{0.048}Mo_{0.301}Re_{0.299}Ru_{0.304}$*	α-Mn	None	6.76	4.43	0.227
	$Hf_{0.106}Ta_{0.085}W_{0.081}Ir_{0.099}Re_{0.630}$*†	α-Mn	None	6.64	5.15	0.277
	$Hf_{0.0875}Ta_{0.0875}W_{0.0875}Ir_{0.0875}Re_{0.65}$†	α-Mn	None	6.65	4.77	0.267
	$Hf_{0.075}Ta_{0.075}W_{0.075}Ir_{0.075}Re_{0.70}$†	α-Mn	None	6.70	4.47	0.248
	$Hf_{0.0625}Ta_{0.0625}W_{0.0625}Ir_{0.0625}Re_{0.75}$†	α-Mn	None	6.75	4.12	0.226
	$Hf_{0.05}Ta_{0.05}W_{0.05}Ir_{0.05}Re_{0.80}$†	α-Mn	hcp	6.80	3.73	0.202

(continued)

Table 7.2 (continued)

References	Composition	Major phase(s)	Minor phase(s)	VEC	δ	$\Delta\chi$
	$Hf_{0.090}Ta_{0.080}W_{0.083}Pt_{0.083}Re_{0.665}$*†	α-Mn	None	6.74	4.71	0.273
	$Hf_{0.0875}Ta_{0.0875}W_{0.0875}Pt_{0.0875}Re_{0.65}$†	α-Mn	None	6.74	4.66	0.277
	$Hf_{0.075}Ta_{0.075}W_{0.075}Pt_{0.075}Re_{0.70}$†	α-Mn	None	6.78	4.38	0.256
	$Hf_{0.0625}Ta_{0.0625}W_{0.0625}Pt_{0.0625}Re_{0.75}$†	α-Mn	hcp	6.81	4.06	0.234
	$Hf_{0.05}Ta_{0.05}W_{0.05}Pt_{0.05}Re_{0.80}$†	α-Mn	hcp	6.85	3.68	0.209
[24]	$Ta_{0.048}Mo_{0.342}W_{0.060}Re_{0.358}Ru_{0.192}$*†	σ	None	6.69	1.55	0.192
	$Ta_{0.044}Mo_{0.297}W_{0.108}Re_{0.356}Ru_{0.195}$*†	σ	None	6.70	1.53	0.198
	$Ta_{0.047}Mo_{0.244}W_{0.156}Re_{0.355}Ru_{0.198}$*†	σ	None	6.70	1.56	0.210
	$Ta_{0.058}Mo_{0.188}W_{0.205}Re_{0.342}Ru_{0.207}$*†	σ	None	6.70	1.66	0.225
	$Ta_{0.052}Mo_{0.153}W_{0.262}Re_{0.327}Ru_{0.206}$*†	σ	None	6.69	1.62	0.228
	$Ta_{0.054}Mo_{0.105}W_{0.284}Re_{0.345}Ru_{0.212}$*†	σ	None	6.72	1.65	0.234
	$Ta_{0.057}Mo_{0.051}W_{0.350}Re_{0.344}Ru_{0.198}$*†	σ	None	6.68	1.64	0.245
	$Ta_{0.069}Mo_{0.317}W_{0.057}Re_{0.355}Ru_{0.202}$*†	σ	None	6.69	1.73	0.207
	$Ta_{0.092}Mo_{0.303}W_{0.064}Re_{0.346}Ru_{0.195}$*†	σ	None	6.64	1.86	0.222
	$Ta_{0.103}Mo_{0.295}W_{0.061}Re_{0.339}Ru_{0.202}$*†	σ	None	6.64	1.94	0.228
	$Ta_{0.124}Mo_{0.268}W_{0.065}Re_{0.346}Ru_{0.197}$*†	σ	None	6.62	2.04	0.240
[25]	$Ta_{0.094}Mo_{0.321}Cr_{0.048}Re_{0.349}Ru_{0.188}$*†	σ	None	6.63	2.60	0.220
	$Ta_{0.111}Mo_{0.261}Cr_{0.085}Re_{0.354}Ru_{0.189}$*†	σ	None	6.62	3.10	0.232
	$Ta_{0.106}Mo_{0.222}Cr_{0.112}Re_{0.382}Ru_{0.178}$*†	σ	None	6.63	3.33	0.230

7.4.3 Control Through Entropy

Of course, in multi-principle element systems the mixing entropy can still be highly influential, but its effect needs to be considered in conjunction with enthalpic ones as a contribution to the overall ΔG rather than a singular dominating factor. Take the series of equimolar three- to six-element compositions $Mo_{0.33}Re_{0.33}Ru_{0.33}$, $Mo_{0.25}Re_{0.25}Ru_{0.25}Rh_{0.25}$, $Nb_{0.2}Mo_{0.2}Re_{0.2}Ru_{0.2}Rh_{0.2}$ and $Nb_{0.167}Mo_{0.167}Re_{0.167}Ru_{0.167}Rh_{0.167}Pt_{0.167}$. They are evidently similar in composition, and the series progresses simply by adding one more element each time. Doing so increases ΔS_{mix} increases from 1.10*R* to 1.79*R*. However, increasing the number of components also increases the number of possible pairwise interactions, making it more likely that an enthalpically-stabilised phase (rather than an entropy-stabilised solid solution) is formed [30]. Hence, whilst $Mo_{0.33}Re_{0.33}Ru_{0.33}$ and $Mo_{0.25}Re_{0.25}Ru_{0.25}Rh_{0.25}$ are single-phase hcp systems, $Nb_{0.2}Mo_{0.2}Re_{0.2}Ru_{0.2}Rh_{0.2}$ crystallises as a mixture of hcp and σ phases. However, the mixing entropy gained by adding the sixth element in $Nb_{0.167}Mo_{0.167}Re_{0.167}Ru_{0.167}Rh_{0.167}Pt_{0.167}$ is sufficient to suppress the formation of the σ phase once more (though, as noted in Sect. 7.2.6, suppressing the hcp phase separation in $Nb_{0.167}Mo_{0.167}Re_{0.167}Ru_{0.167}Rh_{0.167}Pt_{0.167}$ also requires control of the synthesis kinetics—this discussion focuses on thermodynamic considerations, but comprehensive chemical consideration is essential).

Even more intriguingly, whilst equimolar $Nb_{0.2}Mo_{0.2}Re_{0.2}Ru_{0.2}Rh_{0.2}$ crystallises as a mixture of hcp and σ phasesΔS_{mix} must be less than that of $Nb_{0.2}Mo_{0.2}Re_{0.2}Ru_{0.2}Rh_{0.2}$. But breaking the equimolarity apparently also alleviates competing enthalpic forces (hence, as given in Table 7.2, their smaller values of δ and $\Delta\chi$) by a significant enough amount that a σ phase no longer forms. Furthermore, these non-equimolar alloys have markedly higher critical temperatures than $Nb_{0.2}Mo_{0.2}Re_{0.2}Ru_{0.2}Rh_{0.2}$ (Fig. 7.10). The design of HEAs is often dominated by the concept of maximising ΔS_{mix}, but equimolar systems need only be the starting point in explorations of as-yet uncharted chemical space.

It also pertinent to remember that ΔS_{mix} is but one contribution to the total ΔS. The well-studied fcc HEA CrMnFeCoNi converts to a hcp polymorph under applied pressure [31, 32]. Calculations suggest that the driving force for this transition is the effect of high pressure on the electronic, magnetic and vibrational contributions to the total entropy [33]. In the search for new hcp HEA superconductors, then, it should not be necessary to focus singularly on materials which crystallise as hcp directly from the melt. A more comprehensive predictive capability, for example to identify systems that transition to hcp polymorphs at high pressures, will widen the scope for materials discovery.

7.4.4 *Materials Design to Optimise* T$_c$

In Sect. 7.3, the dependence of T_c on VEC in hcp HEAs was investigated. However, it is essential to remember that VEC is not the only parameter that influences T_c—and that this has consequences for materials design. For BCS superconductors, which all HEA superconductors so far seem to be, T_c is dependent on electron–phonon interactions and the density of states at the Fermi level, $N(E_F)$. VEC—that is, the filling of the valence band—is evidently one way in which $N(E_F)$ is determined, but it is not the only one. A comparison of the values of T_c for the hcp elements Ru and Os with the hcp HEAs $Mo_{0.219}Re_{0.192}Ru_{0.201}Rh_{0.196}Pt_{0.191}$ and $Mo_{0.220}Re_{0.177}Ru_{0.200}Ir_{0.191}Pt_{0.212}$ illustrates this (Fig. 7.12). These are all isoelectronic systems with VEC $= 8$, so any variation of T_c cannot be the result of varying VEC. But nonetheless there is a variation, as T_c is also dependent on the unit cell volume. Increasing the unit cell volume narrows the valence band, which increases $N(E_F)$ and thus increases T_c for a constant band filling. In the search for new HEA superconductors, a singular focus on VEC cannot give optimised results.

Data for $Mo_{0.2}Re_{0.2}Ru_{0.2}Rh_{0.2}Pd_{0.2}$ and $Mo_{0.2}Re_{0.2}Ru_{0.2}Ir_{0.2}Pd_{0.2}$ are also plotted in Fig. 7.12, showing that their critical temperatures are considerably higher than would be expected from their unit cell volumes. This illustrates a further strategy for optimising functionality through materials design—exploiting the very (in)stability of multi-principle element mixtures. As mentioned in Sect. 7.2.6, $Mo_{0.2}Re_{0.2}Ru_{0.2}Rh_{0.2}Pd_{0.2}$ and $Mo_{0.2}Re_{0.2}Ru_{0.2}Ir_{0.2}Pd_{0.2}$ consist of an hcp HEA phase plus a Pd impurity—in other words, the hcp phases are not equimolar $Mo_{0.2}Re_{0.2}Ru_{0.2}(Rh/Ir)_{0.2}Pd_{0.2}$ but Pd-deficient $[MoReRu(Rh/Ir)]_{0.8+x}Pd_{0.2-x}$. Therefore, the VEC of the hcp phase in these HEAs is actually less than, rather than equal

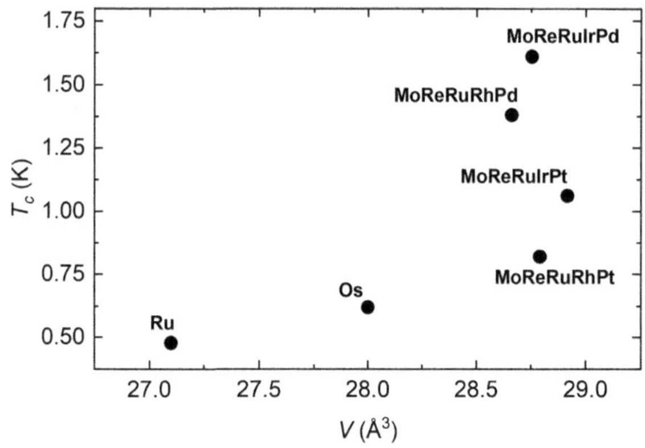

Fig. 7.12 Dependence of T_c on unit cell volume V for the VEC $= 8$ elements Ru and Os (unit cell volumes taken from Ref. [34], T_c values from Ref. [35]) and some (nominally) isoelectronic HEAs. Increasing V increases the density of states at the Fermi level, which in turn increases T_c. The apparent enhancement for the Pd-containing HEAs is due to the hcp phases not fully dissolving the equimolar amount of Pd, meaning that their VECs are actually less than 8

to, 8. In line with the Matthias-like dependence shown in Fig. 7.10, their critical temperatures are therefore higher than VEC = 8 systems with the same unit cell volume. This example should, firstly, reiterate the importance of thorough structural and compositional analysis over a range of length scales of HEAs, and doing so over bulk and microstructural scales. It should also prompt a more open-minded approach to their design. Do superconducting HEAs need to be single phase materials—and might it even be advantageous for application if they are not? For example, might the segregation of a superconducting phase within a non-superconducting bulk lead to a material with multifunctional properties or one that is of improved commercial viability?

7.5 Intersection with Other Areas of HEA Research

The key point of Sects. 7.3 and 7.4 is that understanding of superconductivity in hcp HEAs will intrinsically be limited until new materials are found. In this final section I hope to motivate efforts in materials discovery further by highlighting some opportunities of interdisciplinary interest.

Because there are currently fewer examples all properties of hcp HEAs have been less well-studied than those of their bcc and fcc counterparts. However, they may well have the extraordinary mechanical properties shown by cubic HEAs. The hcp high-entropy titanium aluminide analogue $Al_{0.15}Hf_{0.25}Sc_{0.10}Ti_{0.25}Zr_{0.25}$ has been found to have both higher compression strength and ductility at room temperature than conventional Ti_3Al [36]. There is no reason that HEAs synthesised for mechanical studies can't also be superconductors—or that those synthesised as superconductors can also have useful mechanical properties. The response of HEAs to high pressures is also topical. The hcp HEA $Ir_{0.19}Os_{0.22}Re_{0.21}Rh_{0.20}Ru_{0.19}$, for example, remains mechanically stable to at least 45 GPa [37]. This may offer opportunities for synergistic combination of mechanical and functional properties (as, for example, in the bcc HEA $[TaNb]_{0.67}[HfZrTi]_{0.33}$, for which both the structure and the superconducting state retain their integrity under applied pressures as high as 190 GPa [38]).

$Ir_{0.19}Os_{0.22}Re_{0.21}Rh_{0.20}Ru_{0.19}$ also exhibits excellent catalytic behaviour, whilst other hcp HEAs have been found to show promising shape-memory and magnetocaloric effects [39, 40]. It is not currently known if these materials are also superconductors—low-temperature properties are often simply not measured by researchers in other fields, and personally I suspect that many of these reported materials will show superconductivity if it is measured for. Again, just because a material is synthesised for investigation in one particular area doesn't mean it can't be of interest elsewhere too.

Finally, hcp HEAs with compositions similar to those deliberately synthesised as superconductors are also formed through decay processes in the waste products of nuclear fission reactors [41]. That may not be a practical environment for superconductivity investigations, but it highlights the importance of developing an interdisciplinary understanding of the chemical and physical properties of these materials. In the vast chemical space of HEAs, plenty of interesting materials surely remain to be found, and who knows what exciting properties efforts to make and study them will reveal.

7.6 Conclusion

In this Chapter I have reviewed the state of the art (up to the time of writing in early 2023) of superconducting hcp HEAs, elucidated trends in their stabilities and properties, and mapped out design principles that may guide the discovery of new materials and the enhancement of their properties. I have found superconducting hcp HEAs to be fascinating materials and hope I have illustrated why I believe that their further study is of rich potential. It is therefore my sincere hope that this work will inspire similar fascination and thus motivate readers to join this nascent field.

Acknowledgements This Chapter was written whilst the author was a research fellow at the University of St Andrews, supported by an EPSRC Early Career Fellowship (EP/T011130/1) awarded to Dr A. S. Gibbs. The author is indebted to Dr Gibbs for her support, without which this Chapter could not have been written.

References

1. M. Feuerbacher, M. Heidelmann, C. Thomas, Mater. Res. Lett. **3**, 1 (2015)
2. R.-X. Li, J.-W. Qiao, P.K. Liaw, Y. Zhang, Acta Metall. Sin. (English Lett.) **33**, 1033 (2020)
3. F. Otto, Y. Yang, H. Bei, E.P. George, Acta Mater. **61**, 2628 (2013)
4. Y.F. Ye, Q. Wang, J. Lu, C.T. Liu, Y. Yang, Mater. Today **19**, 349 (2016)
5. M.C. Gao, B. Zhang, S.M. Guo, J.W. Qiao, J.A. Hawk, Metall. Mater. Trans. A **47**, 3322 (2016)
6. A. Takeuchi, K. Amiya, T. Wada, K. Yubuta, Intermetallics **69**, 103 (2016)
7. J.O.A. Paschoal, H. Kleykamp, F. Thümmler, Int. J. Mater. Res. (Z. Metallkd.), **74**, 652 (1983)
8. S. Utsunomiya, R.C. Ewing, Radiochim. Acta **94**, 749 (2006)
9. P. Koželj, S. Vrtnik, A. Jelen, S. Jazbec, Z. Jagličić, S. Maiti, M. Feuerbacher, W. Steurer, J. Dolinšek, Phys. Rev. Lett. **113**, 107001 (2014)
10. S. Marik, K. Motla, M. Varghese, K.P. Sajilesh, D. Singh, Y. Breard, P. Boullay, R.P. Singh, Phys. Rev. Mater. **3**, 060602(R) (2019)
11. S. Marik, M. Varghese, K.P. Sajilesh, D. Singh, R.P. Singh, J. Alloys Compd. **769**, 1059 (2018)
12. Y.-S. Lee, R.J. Cava, Phys. C Supercond. **566**, 1353520 (2019)
13. J.-W. Yeh, S.-K. Chen, S.-J. Lin, J.-Y. Gan, T.-S. Chin, T.-T. Shun, C.-H. Tsau, S.-Y. Chang, Adv. Eng. Mater. **6**, 299 (2004)
14. B. Liu, J. Wu, Y. Cui, Q. Zhu, G. Xiao, S. Wu, G. Cao, Z. Ren, Scr. Mater. **182**, 109 (2020)
15. B. Liu, J. Wu, Y. Cui, Q. Zhu, G. Xiao, S. Wu, G. Cao, Z. Ren, J. Mater. Sci. Technol. **85**, 11 (2021)
16. Q. Zhu, G. Xiao, Y. Cui, W. Yang, S. Song, G.-H. Cao, Z. Ren, J. Alloys Compd. **909**, 164700 (2022)

17. A.J. Browne, D.P. Strong, R.J. Cava, J. Solid State Chem. **321**, 123881 (2023)
18. B.T. Matthias, Phys. Rev. **97**, 74 (1955)
19. M.M. Collver, R.H. Hammond, Phys. Rev. Lett. **30**, 92 (1973)
20. F. von Rohr, M.J. Winiarski, J. Tao, T. Klimczuk, R.J. Cava, Proc. Natl. Acad. Sci. USA **113**, E7144 (2016)
21. Y. Zhang, Y.J. Zhou, J.P. Lin, G.L. Chen, P.K. Liaw, Adv. Eng. Mater. **10**, 534 (2008)
22. S. Guo, C. Ng, J. Lu, C.T. Liu, J. Appl. Phys. **109**, 103505 (2011)
23. K. Stolze, F.A. Cevallos, T. Kong, R.J. Cava, J. Mater. Chem. C **6**, 10441 (2018)
24. B. Liu, J. Wu, Y. Cui, Q. Zhu, G. Xiao, H. Wang, S. Wu, G. Cao, Z. Ren, A.C.S. Appl, Electron. Mater. **2**, 1130 (2020)
25. B. Liu, J. Wu, Y. Cui, Q. Zhu, G. Xiao, S. Wu, G. Cao, Z. Ren, J. Alloys Compd. **869**, 159293 (2021)
26. S.K. Riddle, T.R. Wilson, M. Rajivmoorthy, M.E. Eberhart, Molecules **26**, 5396 (2021)
27. J. Lužnik, P. Koželj, S. Vrtnik, A. Jelen, Z. Jagličić, A. Meden, M. Feuerbacher, J. Dolinsek, Phys. Rev. B **92**, 224201 (2015)
28. W. Chen, D.V. Semenok, I.A. Troyan, A.G. Ivanova, X. Huang, A.R. Oganov, T. Cui, Phys. Rev. B **102**, 134510 (2020)
29. J. Wittig, Phys. Rev. Lett. **21**, 1250 (1968)
30. D.B. Miracle, JOM **69**, 2130 (2017)
31. C.L. Tracy, S. Park, D.R. Rittman, S.J. Zinkle, H. Bei, M. Lang, R.C. Ewing, W.L. Mao, Nat. Commun. **8**, 15634 (2017)
32. F. Zhang, Y. Wu, H. Lou, Z. Zeng, V.B. Prakapenka, E. Greenberg, Y. Ren, J. Yan, J.S. Okasinski, X. Liu, Y. Liu, Q. Zeng, Z. Lu, Nat. Commun. **8**, 15687 (2017)
33. D. Ma, B. Grabowski, F. Körmann, J. Neugebauer, D. Raabe, Acta Mater. **100**, 90 (2015)
34. *ASM Handbook Volume 3: Alloy Phase Diagrams*, ed. H. Okamoto, M.E. Schlesinger, E.M. Mueller (ASM International, Materials Park, Ohio, 2016)
35. C.W. Chu, W.L. McMillan, H.L. Luo, Phys. Rev. B **3**, 3757 (1971)
36. L. Rogal, P. Bobrowski, F. Körmann, S. Divinski, F. Stein, B. Grabowski, Sci. Rep. **7**, 2209 (2017)
37. K.V. Yusenko, S. Riva, P.A. Carvalho, M.V. Yusenko, S. Arnaboldi, A.S. Sukhikh, M. Hanfland, S.A. Gromilov, Scr. Mater. **138**, 22 (2017)
38. J. Guo, H. Wang, F. von Rohr, Z. Wang, S. Cai, Y. Zhou, K. Yang, A. Li, S. Jiang, Q. Wu, R. J. Cava, L. Sun, Proc. Natl. Acad. Sci. USA **114**, 13144 (2017)
39. D. Wu, K. Kusada, T. Yamamoto, T. Toriyama, S. Matsumura, S. Kawaguchi, Y. Kubota, H. Kitagawa, J. Am. Chem. Soc. **142**, 13833 (2020)
40. A.O. Moghaddam, E.A. Trofimov, J. Alloys Compd. **851**, 156838 (2021)
41. R. Devanathan, W. Jiang, K. Kruska, M.A. Conroy, T.C. Droubay, J.M. Schwantes, J. Mater. Res. **34**, 709 (2019)

Part II
The Other Crystal Structures

Chapter 8
Superconducting Properties of High-Entropy Metal Chalcogenides

Md. Riad Kasem and Yoshikazu Mizuguchi

Abstract High-entropy (HE) compounds have been extensively studied in the fields of material science, chemistry, and physics, owing to their unique properties. In this chapter, we review the material development of HE-type metal chalcogenide (*M*Ch) superconductors and discuss the superconductivity of metal tellurides (*M*Te) and *M*Ch with multi-chalcogen compositions. Furthermore, we examined the exotic superconducting property of *M*Te under high pressure, which is one of the most important features of HE-type *M*Te.

8.1 Introduction

Superconductivity is a quantum phenomenon that is observed under cryogenic conditions. In superconducting states, the temperature-dependent electrical resistance decreases to zero at a certain temperature called the critical temperature (T_c). This phenomenon was discovered by Kamerlingh in 1911 [1]. Thereafter, superconducting properties were observed in various materials, such as alloy-type compounds. Nevertheless, several decades passed before researchers developed high-T_c superconductors, which may be useful for practical applications. In 1986, a CuO-layered superconductor was discovered by Bednorz and Müller [2]. With further research on this system [3], its T_c exceeded the boiling point of liquid nitrogen (77 K). In 2008, an Fe-based layered superconductor was discovered [4], and its T_c later increased to 55 K [5]. While exploring the origin of the high T_c of Fe-based layered superconductors, researchers have found a correlation between the T_c and the crystal structure (anion height from the Fe plane) [6]. In 2012, a BiS_2-based layered superconductor was fabricated [7]. In BiS_2-based compounds, the optimization of the in-plane local disorder is essential for bulk superconductivity [8]. These studies indicate that it

Md. R. Kasem (✉) · Y. Mizuguchi
Department of Physics, Tokyo Metropolitan University, 1-1, Minami-osawa, Hachioji, Tokyo 192-0397, Japan
e-mail: riad.phy.du@gmail.com

Y. Mizuguchi
e-mail: mizuguchi-yoshikazu@tmu.ac.jp

J. Kitagawa and Y. Mizuguchi (eds.), *High-Entropy Alloy Superconductors*, Springer Series in Solid-State Sciences 202, https://doi.org/10.1007/978-981-97-4129-8_8

is sometimes necessary to control the local crystal structure and/or local disorder to develop new superconductors. Therefore, there have been very few examples of disordered superconductors. Herein, we focus on HE-type compounds to examine the effect of local disorder on superconductivity. In particular, we review recently developed HE-type metal chalcogenide (*M*Ch, *Ch* = Se, Te) superconductors and discuss the impact of disorder, induced by the increase in the configurational entropy of mixing, on their superconducting properties.

8.1.1 NaCl-Type Metal Chalcogenide Superconductors

NaCl-type materials have unique physical properties that are suitable for superconductors, thermoelectric materials, and topological insulators. Simple alloy-based metal telluride (*M*Te) semiconductors such as SnTe and PbTe have been extensively studied as thermoelectric materials [9–11]. Moreover, SnTe has been shown to be suitable as a topological insulator, which is an unusual quantum state of matter [12–15], and it has been observed that self-doped $Sn_{1-x}Te$ shows superconductivity below 0.3 K [12]. Superconductivity has also been observed in Ag-doped SnTe [16, 17]. Recently, In-doped SnTe superconductors have been studied owing to their controllable carrier concentrations and electronic structure [18, 19]. End-member InTe with a NaCl-type structure can be formed through high-pressure synthesis [20]. The valence-skipping states of In^{+} and In^{3+} play important roles in inducing superconductivity. Another semiconductor, PbTe, has been shown to exhibit superconductivity with the partial substitution of In at the Pb site [21]. A similar valence-skipping state for Tl was observed in a cubic (Tl, Pb)Te superconductor [22]. These results indicate that there is a strong relationship between the superconductivity and valence-skipping states. A report on (Pb, Sn, In)Te superconductors indicated that In doping modified the density of the state at the Fermi level and resulted in superconductivity [23].

Similar to *M*Te, superconductivity was observed in NaCl-type metal selenides (*M*Se) [24–26]. In $Ag_{1-x}Sn_{1+x}Se_{2-y}$ [24], the T_c varied from 4.85 to 6.96 K. In another report by Ren et al. [25], the T_c of $AgSnSe_2$ was found to be 4.6 K. The valence-skipping states of Sn^{2+} and Sn^{4+} thus play an important role in the emergence of superconductivity. In 2017, Hijikata et al. showed that T_c is enhanced by the substitution of Bi in $AgSnSe_2$ [26]. These reports indicate that increasing the disorder in the metal site and/or the valence-skipping state is essential for the superconductivity of NaCl-type *M*Se. To induce further disorder, the concept of high-entropy alloys (HEAs) was developed.

The concept of HEAs was first introduced by Yeh et al. for simple-structure metals and alloys [27]. Generally, HEAs are defined as alloys formed by a combination of five or more elements with concentrations between 5 and 35 at%. The configurational entropy of mixing of that alloy (calculated from Boltzmann's equation, $\Delta S_{\text{mix}} = -R \Sigma_i c_i \ln c_i$, where, c_i and R are the molar fraction and the molar gas constant, 8.314 J/mol·K, respectively) should be greater than $1.5R$. HEAs have been extensively studied in the fields of material science and engineering for their thermostability, strength, and corrosion resistance [28].

In 2014, the first observation of superconductivity in an HEA, specifically $Ta_{34}Nb_{33}Hf_8Zr_{14}Ti_{11}$, was reported by Koželj et al. [29]. It was found that HEAs can exhibit robust superconductivity under extreme conditions, such as high pressure [30]. Recently, the HEA concept has been extended to HEA-type *compounds* in layered superconductors [31, 32]. Here, simple metals (or alloys) are considered to contain *one* crystallographic site, and compounds are considered to contain *two or more* sites. One example is the synthesis of HEA-type $REO_{0.5}F_{0.5}BiS_2$ and RE123 cuprates, wherein the RE site contains five elements. The superconducting properties were improved by increasing ΔS_{mix}. The concept of HEAs has also been applied to metal chalcogenides.

8.2 HEA-Type Metal Tellurides

In 2019, HEA-type $AgInSnPbBiTe_5$ was discovered [33], and its T_c was 2.6 K (Fig. 8.1). The cation sites contained five elements, Ag, In, Sn, Pb, and Bi, at equimolar concentrations. In $AgInSnPbBiTe_5$, Ag and Bi have valence states of +1 and +3, respectively [34, 35]. Moreover, +2 is preferable for Pb and Sn [21, 36]. In exhibits valence-skipping (In^+ and In^{3+}) states. Therefore, the average valence in the metal site is +2, which is neutralized by anionic Te^{2-}. Since this discovery, further investigations have been conducted, and other HEA-type *M*Te superconductors have been synthesized [37]. The temperature dependence of the electrical resistivity of the HEA-type compounds under cryogenic conditions confirms their superconductivity (Fig. 8.2), and the highest T_c^{onset} observed was 1.4 K for $AgInSnSbPbTe_5$. Its residual resistivity ratio was very small, similar to that of other conventional HEA superconductors with evidence of local disorder [29, 33, 38]. The T_c of the *M*Te was successfully tuned by partially substituting the element at the metal site. This indicates that the configurational entropy change at the *M*-site can be a reliable way to tune T_c. To determine the origin of the change in the T_c, the correlation between the T_c and the lattice constant was observed. In Fig. 8.3, *M*Te compounds with higher lattice constants exhibit a higher T_c. However, the T_c values of the HEA-type *M*Te were lower than those of the low- and mid-entropy alloy-type *M*Te. This relationship is useful for changing the entropy of mixing at the *M*-site and tuning the lattice constant to achieve the desired T_c.

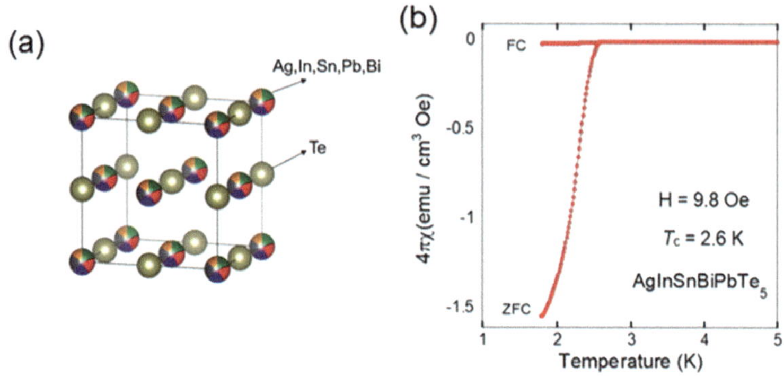

Fig. 8.1 **a** Schematic image of the NaCl-type crystal structure of $AgInSnPbBiTe_5$. **b** Temperature-dependent magnetic susceptibility of $AgInSnPbBiTe_5$. The original data were partly published in Ref. [33]

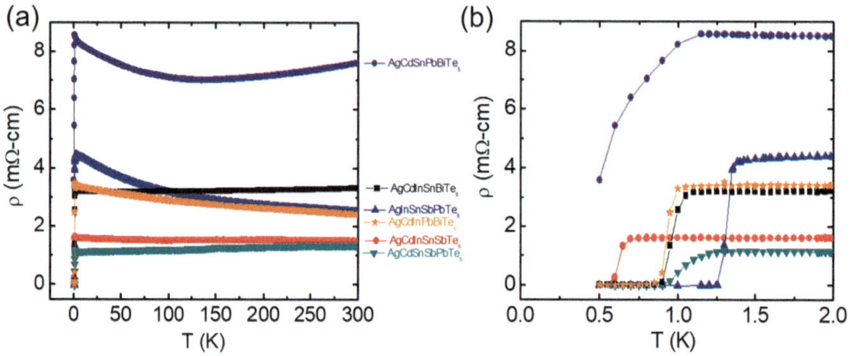

Fig. 8.2 **a** Temperature-dependent electrical resistivity of HEA-type *M*Te. **b** Enlarged view of **a** at low temperatures. The original data were partly published in Ref. [37]

8.3 Effect of Indium Doping on Superconductivity in the *M*Te

In earlier reports, it was found that In doping induces superconductivity in low- and mid-entropy-alloy-based *M*Te [12–14, 18–21] due to the valence-skipping state of In. A gradual increase in In doping is a useful way to determine its impact on superconductivity in disordered systems. In the In-free $AgSnBiTe_3$ system, which is a mid-entropy-alloy-type compound, semiconducting behavior was observed via temperature-dependent electrical resistivity measurements (Fig. 8.4). In addition, band calculations revealed that $AgSnBiTe_3$ was a narrow-gap semiconductor with band inversion, similar to SnTe [39]. However, as In is gradually doped, superconductivity is observed in $(AgSnBi)_{(1-y)/3}In_yTe$. Figure 8.5 displays the temperature dependence of the magnetic susceptibility of $(AgSnBi)_{(1-y)/3}In_yTe$ ($y = 0.2$–0.5) with

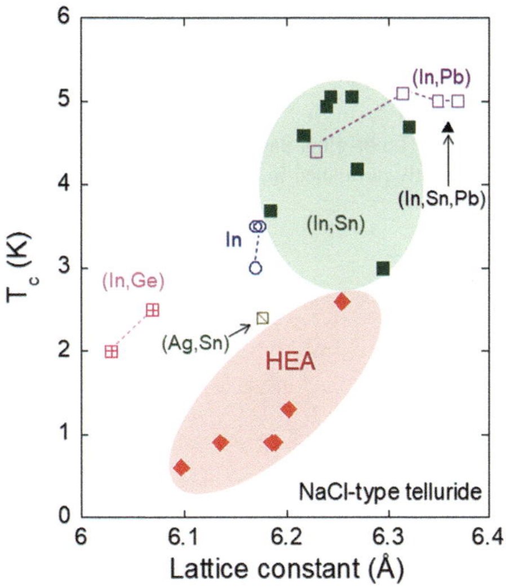

Fig. 8.3 Correlation of the superconducting critical temperature and lattice constant of compounds with NaCl-type structures. The original data were partly published in Ref. [37]

zero-field cooling and field cooling, which confirms the superconductivity. However, the shielding volume fraction is large for y = 0.3 and 0.4 and small for y = 0.2 and 0.5. Bulk superconductivity was observed at y = 0.4.

To investigate the effect of In doping on HEA-type *M*Te, (Ag, Sn, Pb, $Bi)_{(1-x)/4}In_xTe$ (x = 0–0.4) was synthesized [40]. In Fig. 8.6, the temperature dependence of the electrical resistivity of the undoped $AgSnPbBiTe_4$ system (x = 0) shows a semiconductor-like behavior similar to that of $AgSnBiTe_3$ [39]. In contrast, superconductivity was observed when x = 0.1–0.4. The highest T_c^{zero} value was 2.8 K when x = 0.3. Semiconductor-like behavior was observed when x = 0.1 at high temperatures (Fig. 8.6c); meanwhile, metallic behavior was observed when x = 0.2–0.4. This indicates that partial In substitution at the *M*-site generates carriers and converts the system from a semiconductor to a metal. Figure 8.7 shows the

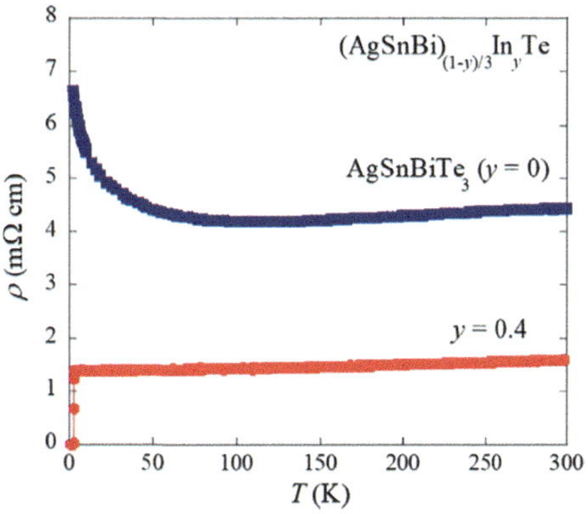

Fig. 8.4 Temperature dependence of the electrical resistivity of $(AgSnBi)_{(1-y)/3}In_yTe$; y = 0 and 0.4. Superconductivity is observed in the In-doped compound (y = 0.4) when T_c = 2.4 K. The original data were partly published in Ref. [39]

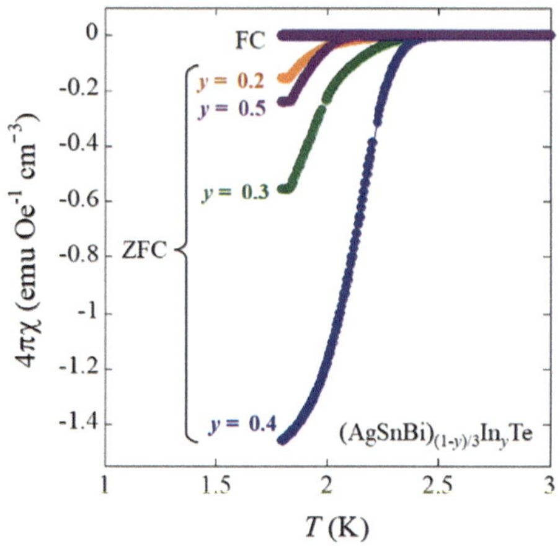

Fig. 8.5 Temperature dependence of the magnetic susceptibility of $(AgSnBi)_{(1\text{-}y)/3}In_yTe$ for y = 0.2–0.5. The original data were partly published in Ref. [39]

temperature-dependent susceptibility of $(Ag, Sn, Pb, Bi)_{(1\text{-}x)/4}In_xTe$. Bulk superconductivity was observed when x = 0.2–0.4, whereas filamentary superconductivity was observed when x = 0.1.

From the above discussion, it can be inferred that the partial substitution of In played a crucial role in the emergence of superconductivity. This highlights the importance of investigating carriers in In-doped superconductors. In pure InTe, both In^{+} and In^{3+} ions were present, and superconductivity was observed. In In-doped PbTe [21], a positive Seebeck coefficient indicates the presence of the In^{+} state, which is a hole carrier. For In-doped SnTe, the carrier switches from holes to electrons as the In concentration changes [13]. However, for $(Ag, Sn, Bi)_{(1\text{-}y)/3}In_yTe$ and $(Ag, Sn, Pb, Bi)_{(1\text{-}x)/4}In_xTe$, the Seebeck coefficients were negative (Fig. 8.8), signifying the presence of an electron carrier in the In^{3+} state.

PbTe exhibits a larger lattice constant. The ionic radius of In^{+} is larger than that of In^{3+}. In^{+} is preferable to In-doped PbTe. Moreover, $(Ag, Sn, Bi)_{(1\text{-}y)/3}In_yTe$ and $(Ag, Sn, Pb, Bi)_{(1\text{-}x)/4}In_xTe$ have smaller lattice constants than PbTe and SnTe. Consequently, In^{3+} ions are preferred in the same crystallographic site. This trend aids in tuning the lattice constant to obtain a desirable T_c for NaCl-type *M*Te.

8.4 Robustness of Superconductivity in $AgInSnPbBiTe_5$ Under High Pressure

In a recent study, robust superconductivity was observed for the HEA $(TaNb)_{0.67}(HfZrTi)_{0.33}$ under extremely high pressure [30]. It was theoretically calculated that the e_g ($d_{x^2\text{-}y^2}$ and d_{z^2}) orbital electrons are almost unchanged at the Fermi level (E_F), which is distinctively different from those of Nb-Ti [41]. An earlier study on 2D HEA-type compounds, such as BiS_2-layered superconductors [31] and high-T_c YBCO compounds [32], revealed their superconducting properties, bulk nature,

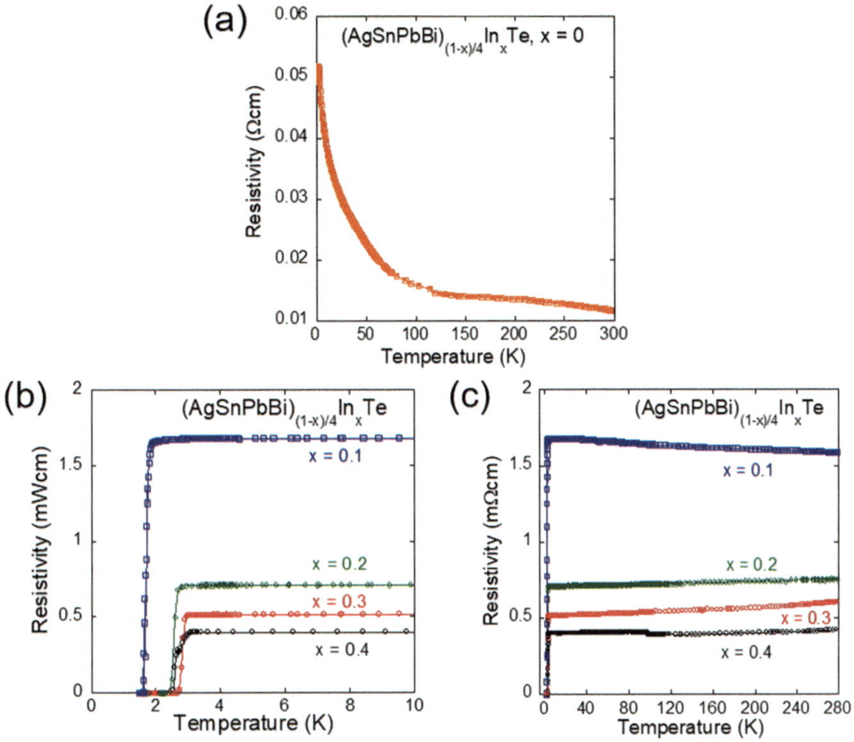

Fig. 8.6 Temperature dependence of the electrical resistivity of (Ag, Sn, Pb, Bi)$_{(1\text{-}x)/4}$In$_x$Te for **a** x = 0 and **b**, **c** x = 0.1–0.4. The original data were partly published in Ref. [40]

Fig. 8.7 Temperature dependence of the magnetic susceptibility of (Ag, Sn, Pb, Bi)$_{(1\text{-}x)/4}$In$_x$Te for x = 0.1–0.4. The original data were partly published in Ref. [40]

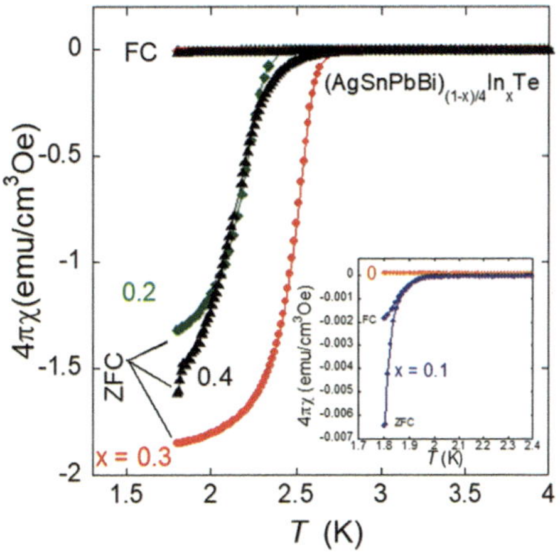

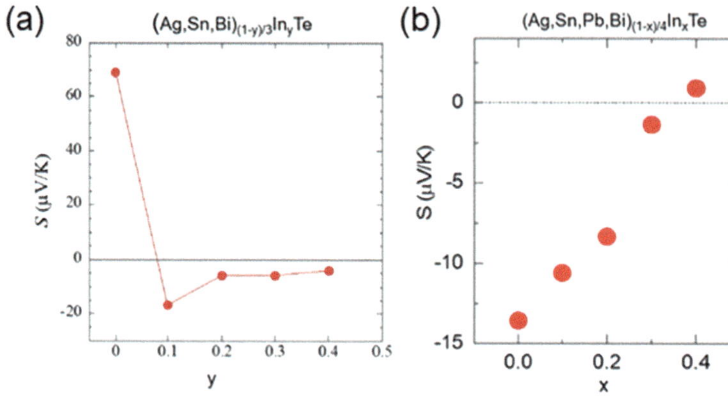

Fig. 8.8 Seebeck coefficient at room temperature for **a** $(Ag, Sn, Bi)_{(1-y)/3}In_yTe$ and **b** $(Ag, Sn, Pb, Bi)_{(1-x)/4}In_xTe$. The original data were partly published in Refs. [39, 40]

upper critical field, and critical current density. However, T_c did not increase significantly for these HEA-type compounds. In contrast, the T_c of 3D-structure-based NaCl-type compounds varied with the change in the configurational entropy of mixing ΔS at the metal site [37]. Three low- to HEA-type *M*Tes-PbTe, $AgPbBiTe_3$, and $AgInSnPbBiTe_5$ exhibited this effect under high pressure [42].

8.4.1 Crystal Structure at Applied Pressure

From the pressure-dependent synchrotron XRD measurements of PbTe, $AgPbBiTe_3$, and $AgInSnPbBiTe_5$, a clear phase separation was observed for all the samples. In Fig. 8.9, the lattice volume per unit formula is displayed against pressure. Three phases, NaCl-type, orthorhombic Pnma, and CsCl-type, were found in the low-, mid-, and high-pressure ranges, respectively. The NaCl-type phase was more stable in the low-pressure range for $AgPbBiTe_3$ and $AgInSnPbBiTe_5$ than for PbTe, whereas the formation of the orthorhombic Pnma phase was suppressed. Moreover, the CsCl-type phase is common for all three *M*Te samples in the high-pressure region, and the lattice volume decreases with increasing pressure. These observations indicate that the configurational entropy of mixing at *the M*-site significantly affected the NaCl and Pnma phases.

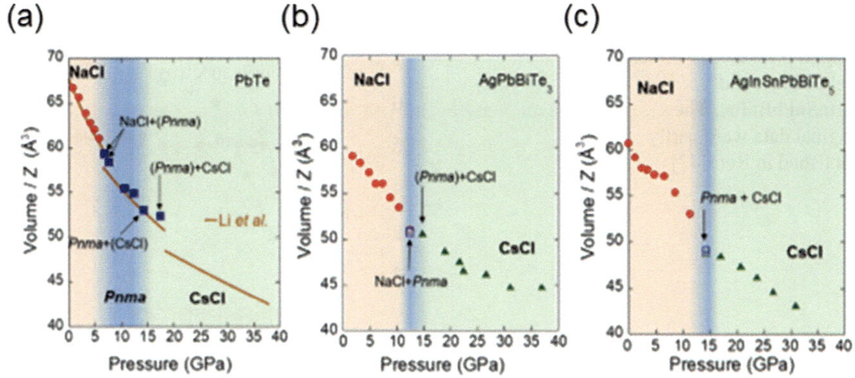

Fig. 8.9 Lattice volume per unit formula against applied pressure for **a** PbTe, **b** $AgPbBiTe_3$, and **c** $AgInSnPbBiTe_5$. An earlier report by Li et al. [43] is shown as the orange line for PbTe. The original data were partly published in Ref. [42]

8.4.2 *Superconducting Properties*

To investigate the superconducting properties, the temperature-dependent electrical resistances of PbTe, $AgPbBiTe_3$, and $AgInSnPbBiTe_5$ were measured under high-pressure and cryogenic conditions. In Fig. 8.10, the pressure dependence of T_c^{onset} is shown. The variation in T_c with structural changes is clearly observed. PbTe with a NaCl-type structure in the low-pressure range exhibits semiconducting behavior. Moreover, superconductivity was observed above 17 GPa for the CsCl-type structure. These results indicate that T_c decreases monotonically with increasing pressure. In contrast, for $AgPbBiTe_3$, superconductivity is observed when the pressure is greater than 2.6 GPa. T_c reaches its maximum value (6.5 K) in the CsCl-type structure but subsequently decreases slightly with increasing pressure. For $AgInSnPbBiTe_5$, superconductivity was observed in the NaCl-type structure, which increased sharply with pressure. These findings show that for $AgInSnPbBiTe_5$ with a CsCl-type structure, T_c is almost independent of pressure, which is very different from the behavior of PbTe and $AgPbBiTe_3$. This indicates that the robustness of the superconductivity under high pressure for HEA-type $AgInSnPbBiTe_5$ is similar to that for $(TaNb)_{0.67}(HfZrTi)_{0.33}$ [30].

8.4.3 *Electronic Structure of* MTe

From a theoretical study, it was found that unchanged d-orbital electrons at the Fermi level are the origin of the robustness of the superconductivity of HEA-type compounds under high pressure [41]. To investigate the electronic structure of MTe under high pressure, X-ray absorption spectroscopy in partial fluorescence mode

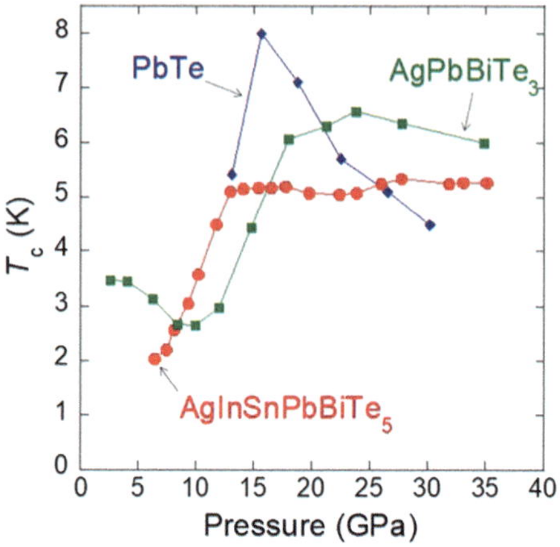

Fig. 8.10 Critical temperature T_c against applied pressure for PbTe, $AgPbBiTe_3$, and $AgInSnPbBiTe_5$. The original data were partly published in Ref. [42]

(PFY-XAS) was performed for PbTe and $AgInSnPbBiTe_5$. PbTe is a low-entropy-alloy-type *M*Te; in contrast, $AgInSnPbBiTe_5$ is an HEA-type compound. The variations in the electronic states of these two compounds are expected to be due to the difference in the entropy of mixing at the *M* site. Hence, for PbTe, the Pb-L_3 absorption spectrum was measured, and for $AgInSnPbBiTe_5$, the Pb-L_3 and Bi-L_3 absorption spectra were obtained. Pb-L_3 and Bi-L_3 were found to be similar. These spectra were fitted using Voigt functions P1, P2, P3, P4, and P5 with an arctan-like background according to an earlier report [44].

In Fig. 8.11, P1, P2, and P3 correspond to the dipole transitions of the $2P_{3/2}$ electron to the 6 s and 6d states of the t_{2g} (d_{xy}, d_{yz}, d_{zx}) and e_g ($d_{x^2-y^2}$ and d_{z^2}) orbital electrons, respectively. For PbTe, the intensity of peak P1 increased continuously up to 5 GPa. The increase in P1 indicated an increase in the unoccupied 6 s states, which signified a change in the density of states (DOS). In the pressure range of 5–15 GPa, P1 remained almost unchanged. However, the intensity of P1 increased again in the high-pressure range. Notably, the superconductivity of PbTe was observed in high-pressure regions with a CsCl-type structure. The increase in P1 in the CsCl-type structure indicated an increase in the unoccupied 6 s state, which might be correlated with the emergence of superconductivity. Moreover, the incident energy of peak P1 decreases as the pressure increases to 5 GPa, remains constant in the middle-pressure range, and decreases again above 18 GPa. The energy of peak P1 indicated an upward shift in E_F, and thus, the DOS changed. The trend of peak P3 was similar to that of P1. In contrast, the intensity of peak P2 decreases with increasing pressure above 17 GPa. Although the electronic structure changes above 20 GPa, the CsCl-type structure remains, and superconductivity is still observed, but it gradually decreases monotonically with increasing pressure. In the case of $AgInSnPbBiTe_5$, the Pb-L_3 absorption edges showed a trend similar to that of PbTe. This indicates that despite increasing the entropy of mixing at the metal site, the electronic structure does not change for *M*Te. Therefore, the robustness of the superconductivity cannot be explained by the pressure-evaluated electronic structure.

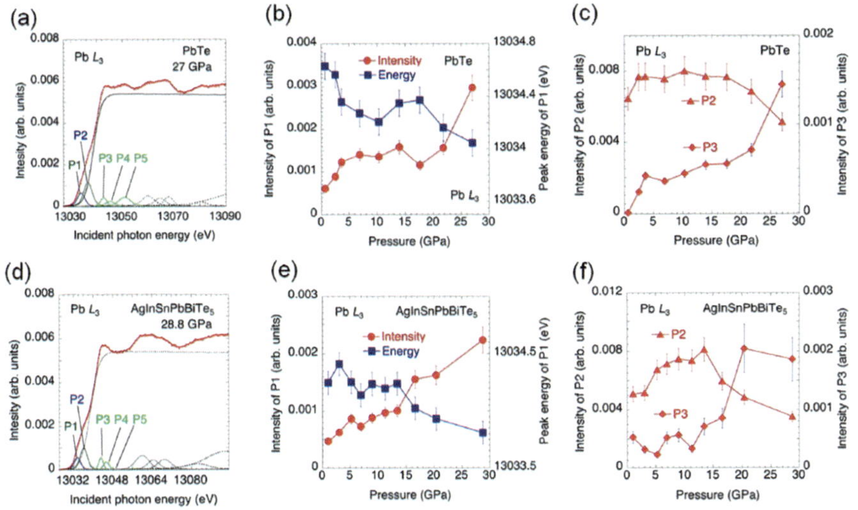

Fig. 8.11 **a, d** PFY-XAS spectra under high pressure, **b, e** pressure dependence intensity (red) and incident energy of the peak (blue) for P1, and **c, f** pressure dependence intensity of the peak for P2 and P3 for PbTe and $AgInSnPbBiTe_5$. The original data were partly published in Ref. [42]

8.5 HEA-Type Metal Chalcogenide Superconductors

In the *M*Te system, the entropy of mixing changes only at the *M*-site. However, it has been observed that two different crystallographic positions possess different entropies of mixing for CsCl-type HEA $(ScZrNb)_{1-x}(RhPd)_x$ [45]. Similarly, in the case of HEA-type *M*Ch (Ag, In, Pb, Bi)$Te_{1-x}Se_x$ (x = 0, 0.25, 0.5), both the *M*- and *Ch*-sites are alloyed [46]. The lattice constants are 6.237, 6.176 and 6.121 Å for x = 0, 0.25 and 0.5, respectively, which indicates the shrinkage of the lattice constant. Few impurity peaks are also observed for x = 0.5 during powder XRD.

Figure 8.12a shows the temperature dependence of the electrical resistivity. T_c^{onset} is 2.7, 2.5, and 2.0 K for x = 0, 0.25, and 0.5, respectively. The decrease in T_c is consistent with the decrease in the lattice constant reported previously [37]. Furthermore, carrier localization in the high-temperature range was observed at x = 0.5. This might be due to an increase in the entropy of mixing at the *Ch*-site, which increases the overall entropy in the unit cell, or the presence of impurities in the crystal structure. Figure 8.12b shows the temperature-dependent susceptibility. Bulk superconductivity was observed at x = 0 and 0.25. The critical current density and critical magnetic field are greater for x = 0.25 than for the Se-free compounds. These results suggest that inducing the entropy of mixing is another effective way to improve the superconducting properties of *M*Te.

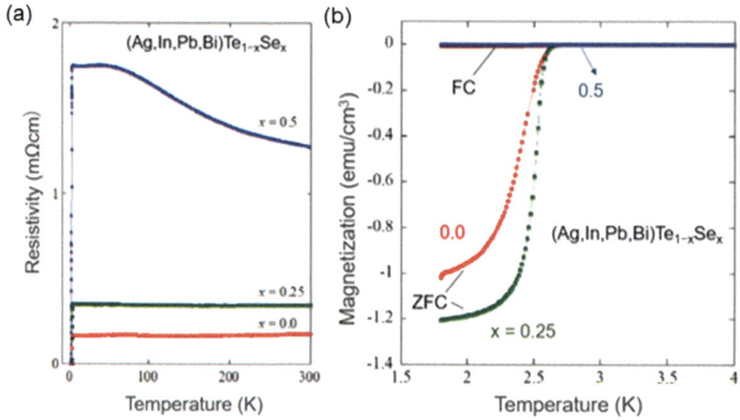

Fig. 8.12 Temperature dependence of **a** electrical resistivity and **b** magnetic susceptibility of (Ag, In, Pb, Bi)$Te_{1-x}Se_x$. The original data were partly published in Ref. [46]

8.6 Conclusion

Superconductivity in *M*Ch is observed even when the configurational entropy of mixing in the *M* and *Ch* sites is increased and when local disorder is introduced. The presence of In at the *M* site was essential for the emergence of metallic and superconducting states in *M*Te and *M*Ch. Although the superconducting properties were improved by increasing the configurational entropy of mixing in layered systems, BiS_2-based systems, and RE123 cuprates, no clear evidence of the improvement in the superconducting properties of NaCl-type *M*Ch was observed. Nevertheless, exotic superconductivity characteristics were observed in the CsCl-type phase of HEA-type *M*Te under high pressures. T_c did not change with pressure for the HEA-type *M*Te. Robust superconductivity with respect to pressure is a common feature of the HEA superconductor Ti–Zr–Hf–Nb–Ta. Recently, we proposed that robust superconductivity is caused by glassy atomic vibrations and electronic states in the *M*Te [47], and a similar glassy nature should exist in all HE superconducting materials. Studies on HE superconductors with local disorder will open up a new field for superconductor science and its applications.

Acknowledgements We are grateful to all the coauthors for their contributions. In addition, special thanks to A. Yamashita and Y. Nakahira for their fruitful suggestion. This work is partially supported by Tokyo Metropolitan Government Advanced Research (No.: H31-1).

References

1. H.K. Onnes, The superconductivity of mercury. Comm. Phys. Lab. Univ. Leiden **122**, 124 (1911)
2. J.G. Bednorz, K.A. Mueller, Possible high Tc superconductivity in the Ba−La−Cu−O system. Z. Phys. B. **64**(2), 189 (1986)
3. M.K. Wu, J.R. Ashburn, C.J. Torng, P.H. Hor, R.L. Meng, L. Gao, Z.J. Huang, Y.Q. Wang, C.W. Chu, Superconductivity at 93 K in a new mixed-phase Y-Ba-Cu-O compound system at ambient pressure. Phys. Rev. Lett. **58**, 908 (1987)
4. Y. Kamihara, T. Watanabe, M. Hirano, H. Hosono, Iron-based layered superconductor $La[O_{1-x}F_x]FeAs$ (x = 0.05−0.12) with $T_c = 26$ K. J. Am. Chem. Soc. **130**(11), 3296–3297 (2008)
5. N.D. Zhigadlo, S. Katrych, S. Weyeneth, R. Puzniak, P.J.W. Moll, Z. Bukowski, J. Karpinski, H. Keller, B. Batlogg, Th-substituted SmFeAsO: structural details and superconductivity with T_c above 50 K. Phys. Rev. B **82**, 064517 (2010)
6. Y. Mizuguchi, Y. Hara, K. Deguchi, S. Tsuda, T. Yamaguchi, K. Takeda, H. Kotegawa, H. Tou, Y. Takano, Anion height dependence of T_c for the Fe-based superconductor. Supercond. Sci. Technol. **23**, 054013 (2010)
7. Y. Mizuguchi, H. Fujihisa, Y. Gotoh, K. Suzuki, H. Usui, K. Kuroki, S. Demura, Y. Takano, H. Izawa, O. Miura, BiS_2-based layered superconductor $Bi_4O_4S_3$. Phys. Rev. B **86**, 220510 (2012)
8. Y. Mizuguchi, Material development and physical properties of BiS_2-based layered compounds. J. Phys. Soc. Jpn. **88**, 041001 (2019)
9. T. Fu, X. Yue, H. Wu, C. Fu, T. Zhu1, X. Liu, L. Hu, P. Ying, J. He, X. Zhao, Enhanced thermoelectric performance of PbTe bulk materials with figure of merit zT > 2 by multi-functional alloying, J. Mater. **2**(2), 141–149 (2016)
10. J.P. Heremans, V. Jovovic, E.S. Toberer, A. Saramat, K. Kurosaki, A. Charoenphakdee, S. Yamanaka, G.J. Snyder, Enhancement of thermoelectric efficiency in PbTe by distortion of the electronic density of states. Science **321**(5888), 554–557 (2008)
11. Q. Zhang, B. Liao, Y. Lan, K. Lukas, W. Liu, K. Esfarjani, C. Opeil D. Broido, and Z. Ren, High thermoelectric performance by resonant dopant indium in nanostructured SnTe. Proc. Natl. Acad. Sci. **110**(33), 13261–13266 (2013)
12. A.S. Erickson, J.-H. Chu, M.F. Toney, T.H. Geballe, I.R. Fisher, Enhanced superconducting pairing interaction in indium doped tin telluride. Phys. Rev. B **79**, 024520 (2009)
13. N. Haldolaarachchige, Q. Gibson, W. Xie, M.B. Nielsen, S. Kushwaha, R.J. Cava, Anomalous composition dependence of the superconductivity in In-doped SnTe. Phys. Rev. B **93**, 024520 (2016)
14. G. Balakrishnan, L. Bawden, S. Cavendish, M.R. Lees, Superconducting properties of the In-substituted topological crystalline insulator SnTe. Phys. Rev. B **87**, 140507(R) (2013)
15. L. P. He, Z. Zhang, J. Pan, X. C. Hong, S. Y. Zhou, S. Y. Li, Full superconducting gap in the doped topological crystalline insulator $Sn_{0.6}In_{0.4}Te$. Phys. Rev. B **88**, 014523 (2013)
16. Y. Mizuguchi, O. Miura, High-pressure synthesis and superconductivity of Ag-doped topological crystalline insulator SnTe ($Sn_{1-x}Ag_xTe$ with x = 0–0.5). J. Phys. Soc. Jpn. **85**, 053702 (2016)
17. Y. Mizuguchi, A. Yamada, R. Higashinaka, T.D. Matsuda, Y. Aoki, O. Miura, M. Nagao, Specific heat and electrical transport properties of $Sn_{0.8}Ag_{0.2}Te$ superconductor. J. Phys. Soc. Jpn. **85**, 103701 (2016)
18. R.D. Zhong, J.A. Schneeloch, X.Y. Shi, Z.J. Xu, C. Zhang, J.M. Tranquada, Q. Li, G.D. Gu, Optimizing the superconducting transition temperature and upper critical field of $Sn_{1-x}In_xTe$. Phys. Rev. B **88**, 020505 (2013)
19. K. Kobayashi, Y. Ai, H. O. Jeschke, J. Akimitsu, Enhanced superconducting transition temperatures in the rocksalt-type superconductors $In_{1-x}Sn_xTe$ ($x \leq 0.5$)

20. M. Kriener, M. Kamitani, T. Koretsune, R. Arita, Y. Taguchi, Y. Tokura, Tailoring band structure and band filling in a simple cubic (IV, III)-VI superconductor. Phys. Rev. Mater. **2**, 044802 (2018)
21. M. Katsuno, R. Jha, K. Hoshi, R. Sogabe, Y. Goto, Y. Mizuguchi, High-pressure synthesis and superconducting properties of NaCl-Type $In_{1-x}Pb_xTe$ (x = 0–0.8). Condens. Matter **5**, 14 (2020)
22. Y. Matsushita, H. Bluhm, T.H. Geballe, I.R. Fisher, Evidence for charge Kondo effect in superconducting Tl-doped PbTe. Phys. Rev. Lett. **94**, 157002 (2005)
23. R. Zhong, J. Schneeloch, Q. Li, W. Ku, J. Tranquada, G. Gu, Indium substitution effect on the topological crystalline insulator family $(Pb_{1-x}Sn_x)_{1-y}In_yTe$: topological and superconducting properties. Crystals **7**, 55 (2017)
24. D.C. Johnston, H. Adrian, Superconducting and normal state properties of $Ag_{1-x}Sn_{1+x}Se_{2-y}$. J. Phys. Chem. Solids **38**, 355 (1977)
25. Z. Ren, M. Kriener, A.A. Taskin, S. Sasaki, K. Segawa, Y. Ando, Anomalous metallic state above the upper critical field of the conventional three-dimensional superconductor $AgSnSe_2$ with strong intrinsic disorder. Phys. Rev. B **87**, 064512 (2013)
26. Y. Hijikata, A. Nishida, K. Nagasaka, O. Miura, A. Miura, C. Moriyoshi, Y. Kuroiwa, Y. Mizuguchi, Bi substitution effects on superconductivity of valence-skip superconductor $AgSnSe_2$. J. Phys. Soc. Jpn. **86**, 054711 (2017)
27. J.W. Yeh, S.K. Chen, S.J. Lin, J.Y. Gan, T.S. Chin, T.T. Shun, C.H. Tsau, S.Y. Chang, Nanostructured high-entropy alloys with multiple principal elements: novel alloy design concepts and outcomes. Adv. Eng. Mater. **6**, 299 (2004)
28. X. Wang, W. Guoa, Y. Fu, High-entropy alloys: emerging materials for advanced functional applications. J. Mater. Chem. A **9**, 663–701 (2021)
29. P. Koželj, S. Vrtnik, A. Jelen, S. Jazbec, Z. Jagličić, S. Maiti, M. Feuerbacher, W. Steurer, J. Dolinšek, Discovery of a superconducting high-entropy alloy. Phys. Rev. Lett. **113**, 107001 (2014)
30. J. Guo, H. Wang, F. von Rohr, Z. Wang, S. Cai, Y. Zhou, K. Yang, A. Li, S. Jiang, Q. Wu, R.J. Cava, L. Sun, Robust zero resistance in a superconducting high-entropy alloy at pressures up to 190 GPa. Proc. Natl. Acad. Sci. USA **114**, 13144 (2017)
31. R. Sogabe, Y. Goto, Y. Mizuguchi, Superconductivity in REO0.5F0.5BiS2 with high-entropy-alloy-type blocking layers. Appl. Phys. Express **11**, 053102 (2018)
32. Y. Shukunami, A. Yamashita, Y. Goto, Y. Mizuguchi, Synthesis of RE123 high-T_c superconductors with a high-entropy-alloy-type RE site. Phys. C **572**,1353623 (2020)
33. Y. Mizuguchi, Superconductivity in high-entropy-alloy telluride $AgInSnPbBiTe_5$. J. Phys. Soc. Jpn. **88**, 124708 (2019)
34. L. Pan, D. Bérardan, N. Dragoe, High thermoelectric properties of n-type $AgBiSe_2$. J. Am. Chem. Soc. **135**, 4914 (2013)
35. Y. Goto, A. Nishida, H. Nishiate, M. Murata, C.H. Lee, A. Miura, C. Moriyoshi, Y. Kuroiwa, Y. Mizuguchi, Effect of Te substitution on crystal structure and transport properties of $AgBiSe_2$ thermoelectric material. Dalton Trans. **47**, 2575 (2018)
36. M.P. Mathur, D.W. Deis, C.K. Jones, W.J. Carr Jr., Superconductivity as a function of carrier density and magnetic spin concentration in the SnTe-MnTe system. J. Phys. Chem. Solids **34**, 183 (1973)
37. M.R. Kasem, K. Hoshi, R. Jha, M. Katsuno, A. Yamashita, Y. Goto, T.D. Matsuda, Y. Aoki, Y. Mizuguchi, Superconducting properties of high-entropy-alloy tellurides M-Te (M: Ag, In, Cd, Sn, Sb, Pb, Bi) with a NaCl-type structure. Appl. Phys. Express **13**, 033001 (2020)
38. S. Marik, K. Motla, M. Varghese, K.P. Sajilesh, D. Singh, Y. Breard, P. Boullay, R.P. Singh, Superconductivity in a new hexagonal high-entropy alloy. Phys. Rev. Mater. **3**, 060602 (2019)
39. T. Mitobe, K. Hoshi, M.R. Kasem, R. Kiyama, H. Usui, A. Yamashita, R. Higashinaka, T.D. Matsuda, Y. Aoki, Y. Goto, Y. Mizuguchi, Superconductivity in In-doped $AgSnBiTe_3$ with possible band inversion. Sci. Rep. **11**, 22885 (2021)
40. M.R. Kasem, R. Ishii, T. Katase, O. Miura, Y. Mizuguchi, Tuning of carrier concentration and superconductivity in high-entropy-alloy-type metal telluride $(AgSnPbBi)_{(1-x)/4}In_xTe$. J. Alloys Compd. **920**, 166013 (2022)

41. C. Huang, J. Guo, J. Zhang, K. Stolze, S. Cai, K. Liu, H. Weng, Z. Lu, Q. Wu, T. Xiang, R.J. Cava, L. Sun, RSAVS superconductors: Materials with a superconducting state that is robust against large volume shrinkage. Phys. Rev. Mater. **4**, 071801(R) (2020)
42. M.R. Kasem, Y. Nakahira, H. Yamaoka, R. Matsumoto, A. Yamashita, H. Ishii, N. Hiraoka, Y. Takano, Y. Goto, Y. Mizuguchi, Robustness of superconductivity XE "Robustness of superconductivity" to external pressure in high-entropy-alloy-type metal telluride AgInSnPbBiTe5. Sci. Rep. **12**, 7789 (2022)
43. Y. Li, C. Lin, H. Li, X. Li, J. Liu, Phase transitions in PbTe under quasi-hydrostatic pressure up to 50 GPa High Press. Res. **33**, 713 (2013)
44. H. Yamaoka, H.O. Jeschke, X. Yang, T. He, H. Goto, N. Hiraoka, H. Ishii, J. Mizuki, Y. Kubozono, Electronic structures of Bi_2Se_3 and $Ag_xBi_2Se_3$ under pressure studied by high-resolution x-ray absorption spectroscopy and density functional theory calculations. Phys. Rev. B **102**, 155118 (2020)
45. K. Stolze, J. Tao, F.O.V. Rohr, T. Kong, R.J. Cava, Sc–Zr–Nb–Rh–Pd and Sc–Zr–Nb–Ta–Rh–Pd high-entropy alloy superconductors on a CsCl-type lattice. Chem. Mater. **30**(3), 906 (2018)
46. A. Yamashita, R. Jha, Y. Goto, T.D. Matsuda, Y. Aoki, Y. Mizuguchi, An efficient way of increasing the total entropy of mixing in high-entropy-alloy compounds: a case of NaCl-type (Ag, In, Pb, Bi)$Te_{1-x}Se_x$ ($x = 0.0, 0.25, 0.5$) superconductors. Dalton Trans. **49**, 9118 (2020)
47. Y. Mizuguchi, H. Usui, R. Kurita, K. Takae, M.R. Kasem, R. Matsumoto, K. Yamane, Y. Takano, Y. Nakahira, A. Yamashita, Y. Goto, A. Miura, C. Moriyoshi, Glassy atomic vibrations XE "Glassy atomic vibrations" and blurry electronic structures created by local structural disorders in high-entropy metal telluride superconductors. Mater. Today Phys. **31**, 101019 (2023)

Chapter 9
High Entropy van der Waals Superconductors

Tongxu Yu and Tianping Ying

Abstract High entropy alloys and van der Waals materials represent two mainstream research directions in condensed matter physics and materials science. The recent combination of these two domains has given rise to a novel class of materials known as high entropy van der Waals materials, amalgamating the advantages of both realms, such as corrosion resistance, exfoliability, and diverse degrees of freedom including composition, spin, charge, intercalation, and twisting. Much like the surprising discovery of superconductivity in high entropy alloys, superconductivity has also been observed in these 2D counterparts. The layered structures not only enhance their geometric tunability but also open avenues for exploring exotic superconducting pairing mechanisms in these highly confined and disordered planar systems. This chapter delves into the superconducting properties of this newly discovered material category, encompassing its historical background, design strategies, characterization, and forecasting future research directions.

9.1 Introduction

Since the discovery of superconductivity in pure mercury metals by Kamerlingh Onnes in 1911, one main theme/impetus for superconductivity research is to find diverse material systems exhibiting superconducting behavior and elucidate the underlying mechanisms. Over the past century, thousands of superconductor materials with distinct pairing rules are discovered, in which the cuprates and iron pnictides have occupied the most conspicuous position. An explicit characteristic of these material categories is the complexity of their elemental constituents. The discovery of superconductivity in high entropy alloy with the composition $Ta_{34}Nb_{33}Hf_8Zr_{14}Ti_{11}$

T. Yu
Suzhou Laboratory, Jiangsu 215123, China

T. Ying (✉)
Beijing National Laboratory for Condensed Matter Physics, Institute of Physics, Chinese Academy of Sciences, Beijing 100190, China
e-mail: ying@iphy.ac.cn

J. Kitagawa and Y. Mizuguchi (eds.), *High-Entropy Alloy Superconductors*, Springer Series in Solid-State Sciences 202, https://doi.org/10.1007/978-981-97-4129-8_9

by P. Koželj et al. in 2014 bring this complexity to a new degree and stimulate the research community to find a variety of superconducting high entropy alloys [1–4].

Up till now, the exploration of high entropy alloy superconductors has extended from the simple crystal structures, i.e., face-centered-cubic (fcc), body-centered-cubic (bcc) and hexagonal-closed packing (hcp) systems, to multisite lattices, e.g., α-Mn-type, A15, σ-phase and layered structures. The metallic nature of these alloys makes certain that the BCS electron–phonon coupling theoretical framework is able to clarify the superconducting mechanism, even though the disordered distribution of multi-components in ordered multisite lattices is daunting if not impossible to deal with from the first-principle perspective.

In the past few years, many attempts have been made to apply the high entropy concept to layered materials in order to lower the crystalline dimension, which not only has practical applications, e.g., in the development of superconducting thin films, but also holds promise for revealing new physics and novel pairing mechanisms. However, efforts to introduce superconductivity in cuprate $(La_{0.2}Pr_{0.2}Nd_{0.2}Sm_{0.2}Eu_{0.2})_2CuO_4$ with a Ruddlesden-Popper structure by annealing the sample to create oxygen vacancies proved unsuccessful, likely due to significant lattice distortion caused by variation in the atomic sizes of rare earth elements [5]. This finding is in contrast to the observation of superconductivity in high-entropy blocking layer BiS-based superconductors, such as $REO_{0.5}F_{0.5}BiS_2$ and $REOBiS_2$, where detailed structural analysis has shown that the high-entropy blocking layer modifies both bond lengths and bond angles, but remains the S site in the BiS plane intact, which is now considered the key ingredient for the realization of superconductivity [6–9]. Optimization of the rare-earth site combination has been found to further enhance the transition temperature and critical current compared to monoelement counterparts. Measurements of the upper critical field in these materials have revealed a value beyond the Pauli limit predicted from the weak-coupling BCS framework, suggesting a potential unconventional pairing mechanism [7, 8]. It is worth noting, however, that the high-entropy approach has thus far only been applied to the blocking layer, and there have been fewer attempts to extend this approach to the functional BiS_2 layer, possibly due to the lack of isostructural alternatives. Furthermore, prominent interlayer interactions between the BiS_2 layer and the REO layer make it infeasible to cleave these materials towards its 2D limit.

Very recently, the concept of high entropy has given rise to a new family of van der Waals materials, coined as high entropy van der Waas materials (HEX), characterized by the presence of high-entropy transition metals (HE) and anion clusters (X) [10, 11]. These materials display a range of intriguing properties, including corrosion resistance, long-range magnetic ordering, and metal–insulator transitions [10–14]. Notably, some HEX materials, such as $HETe_2$ at ambient pressure and $HEPSe_3$ at high pressure, exhibit superconductivity [10, 15]. In these compounds, the *d* orbitals of the transition metals strongly hybridize with the *p* orbitals, contributing to density of states (DOS) at the Fermi surface. This chapter focuses on the superconductivity of these HEX materials. Section 9.1 introduces the superconductivity of $(CoAuIrPdPt)Te_2$ at ambient pressure, while Sect. 9.2 summarizes the superconductivity of medium-entropy materials, including $MnFeCoInPSe_3$,

$Fe_{0.7}(MnCd)_{0.1}In_{0.1}PSe_3$ and $Fe_{0.8}Mn_{0.1}Cd_{0.05}In_{0.05}PSe_3$, which only exhibit superconductivity at high pressure due to entropy-driven P-P dimer breaking. In Sect. 9.3, we provide a perspective on the outstanding challenges and unanswered questions regarding the HEX superconductors, including the underlying mechanism and the ability of superconductivity to survive in a random 2D lattice.

9.2 Discovery of Superconductivity in High-Entropy van der Waals Materials

As mentioned previously, introducing high entropy constituents into the superconducting layer of the BiS-based family is challenging due to the lack of alternative isostructure functional units, unlike in high-entropy alloys where multi-elements favor the closed packing arrangements. However, 2D materials offer a solution to this difficulty. Transition metal chalcogenides (TMD) are widely available and are generally characterized by their structural units and mutual stacking sequences, such as the 1T, 2H, and 3R phases. For instance, multiple TMD compounds exhibit the 1T phase at high temperature, where the MX_6 building block features octahedral units and necessitates a quenching procedure in the synthesis process. This quenching procedure promotes randomization and facilitates a high-entropy single phase. Through further analysis of the acquired sample, we discovered that the synergistic effect of electronegativity and M-X bond length determines whether a single phase can be achieved [11]. Consequently, various properties, including metal–insulator transition, catalysis, corrosion resistance, and superconductivity, have been successfully realized.

As a proof of concept, we start from transition metal tellurides, as superconductivity has been observed in this system, such as Pt-doped $AuTe_2$ [16] and Pd-doped $IrTe_2$ [17]. The Rietveld refinements and element distribution of two typically-synthesized tellurides are shown in Fig. 9.1. It should be emphasized that the maximum T_cs in these ternary systems usually appear at the boundary of structural instability or on the verge of Te_2 dimer breaking, and their superconducting regions are relatively narrow. As shown in Fig. 9.2a, b, $(CoAu)_{0.2}(RhIrPdPt)_{0.8}Te_2$ compound exhibits a superconducting transition temperature (T_c^{onset}) at 4.5 K, with a bulk response at 3.6 K, which is comparable to the highest T_c^{onset} in the transition metal-ditelluride system. The cobalt atoms are intentionally introduced. Although the actual composition is determined to be $Co_{0.03}Au_{0.06}Rh_{0.23}Ir_{0.24}Pd_{0.16}Pt_{0.28}Te_2$, the incorporation of small amount of cobalt is proven to effectively suppress the phase separation of the system. It is quite interesting that superconductivity could survive with a stoichiometric amount of magnetic element incorporating in the crystal structure. Such a robustness and easily accessible superconductivity in $HETe_2$ is quite different from the binary tellurides. The absence of structural transition and the widespread dimer fluctuation play an important role, which can be viewed as a cocktail effect of the high entropy system. By modulating the composition, the T_c can be further enhanced, as

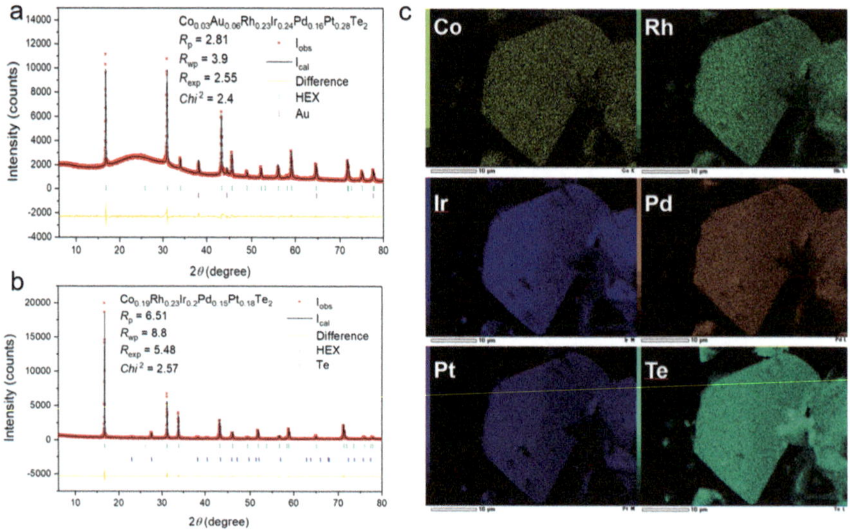

Fig. 9.1 **a**, **b** Rietveld refinements of $(CoAu)_{0.2}(RhIrPdPt)_{0.8}Te_2$ and $(CoRhIrPdPt)Te_2$. **c** EDX mapping of $(CoRhIrPdPt)Te_2$. The actual composition is determined to be $Co_{0.19}Rh_{0.23}Ir_{0.2}Pd_{0.15}Pt_{0.18}Te_2$. (Reprinted with permission from ref 10. Copyright 2021 American Chemical Society)

it directly influences the valence state and the strength of electron–phonon coupling. Nonetheless, an apparent suppression of the superconductivity can be observed by increasing the doping content of Co. In $(CoRhIrPdPt)Te_2$ (with the actual composition of $Co_{0.19}Rh_{0.23}Ir_{0.2}Pd_{0.15}Pt_{0.18}Te_2$, superconductivity cannot be observed in the resistance characteristic down to 2 K. Incorporating elements with local spin state offers a platform to investigate the unconventional pairing mechanism of superconductivity and its ostensible competition with magnetism.

While superconductivity has already been observed in high-entropy van der Waals materials, the full potential and benefits of entropy are yet to be fully explored. On one hand, most noble metal ditellurides, including Rh, Ir, Pd, Pt, and Au, are known to exhibit superconductivity under doping or substitution. On the other hand, the high-entropy tellurides remain in a polycrystalline form due to inhomogeneous precipitation, which makes it challenging to explore their intrinsic properties. In the following section, we will discuss in detail the synergistic effect of high entropy and pressure on medium entropy $MEPSe_3$ (ME = medium entropy), in which the entropy not only induces superconductivity in a randomized system but also leads to an intriguing modification of the crystal structure that cannot be achieved through any other known method.

MPX_3 compounds, where M represents divalent metal elements and X is S or Se, are a broad class of magnetic van der Waals materials that have attracted significant attention in fundamental and applied research. These compounds are characterized by a honeycomb lattice of magnetic M^{2+} ions, leading to anisotropic magnetic properties

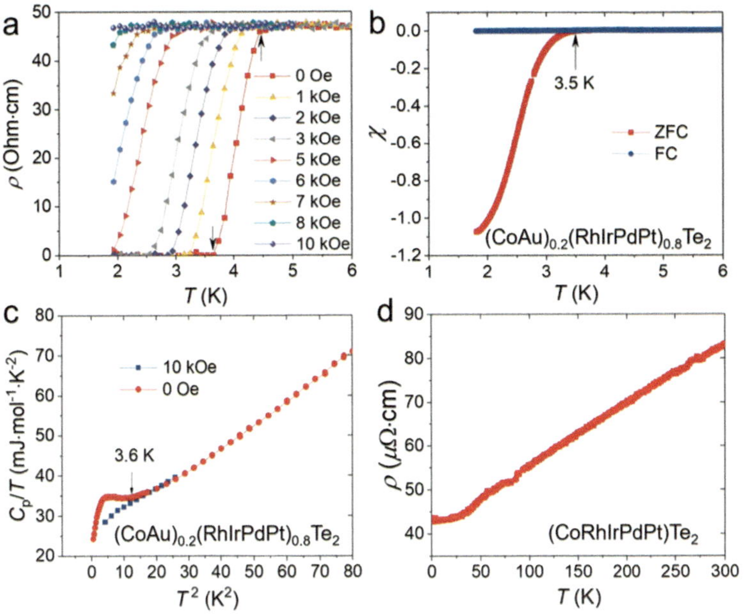

Fig. 9.2 **a–c** Resistivity, magnetization and heat capacity of $(CoAu)_{0.2}(RhIrPdPt)_{0.8}Te_2$, **d** resistivity of $(CoRhIrPdPt)Te_2$. (Reprinted with permission from ref 10. Copyright 2021 American Chemical Society)

that have earned MPX_3 the nickname of "magnetic graphene." While transition metals are the most common M ions, other options include alkaline earth metals like Mg in $MgPSe_3$, a combination of +1 (A^I) and +3 (M^{III}) cations in $Ag^IIn^{III}P_2Se_6$, and partially occupied M^{III} cations in $(Ga^{III})_{2/3}PSe_3$. The sum of the M valences must be +2 to maintain charge neutrality. Although numerous attempts have been carried out to modify the electrical transport and structure of these materials through field-gating, chemical doping, and molecular intercalation, the semiconducting or insulating behavior remains unchanged. One key structural feature of MPX_3 is the consistent distance between P-P dimers, which has been found to be approximately 2.3 Å regardless of the metal ions used.

Superconductivity has been long anticipated in MPX_3, but it has only been observed in $FePSe_3$ under external pressure of 8 GPa, with a transition temperature (T_c) of 4 K. The mechanism behind this observation is thought to involve a switch from high-spin to low-spin states in Fe $3d^6$ electrons, inducing a magnetic-nonmagnetic transition that is critical for superconductivity. However, recent neutron diffraction studies of pressurized $FePSe_3$ have revealed that the local magnetic moment persists at around 5 μ_B/Fe throughout the pressure range, casting doubt on the spin crossover explanation for superconductivity in $FePSe_3$.

Using iodine as the transport agent, the single crystalline $MEPSe_3$ can be successfully grown as shown in the inset of Fig. 9.4a. The nominal compositions of the synthesized samples were $FePSe_3$, $Fe_{0.8}Mn_{0.1}Cd_{0.05}In_{0.05}PSe_3$, $Fe_{0.7}(MnCd)_{0.1}In_{0.1}PSe_3$, and $(FeMnCd)_{0.25}In_{0.17}PSe_3$, which were referred to as $Fe_{1.0}$, $Fe_{0.8}$, $Fe_{0.7}$, and $Fe_{0.25}$, respectively. The ME atoms are octahedrally coordinated by six Se atoms, which are then covalently bonded to P-P dimers, forming an ethane-like $(P_2Se_6)^{4-}$ cluster. In the $Fe_{0.25}$ sample, the ME-ME distance is 3.69 Å, which is longer than the Fe–Fe distance of 3.44 Å and Mn–Mn distance of 3.50 Å in $FePSe_3$ and $MnPSe_3$, respectively. The $Fe_{0.25}$ sample exhibited good crystallinity with partial ME vacancies, as shown in Fig. 9.3c. The elemental distribution of the alloyed elements in the $Fe_{0.25}$ sample was homogeneous, as indicated by the EDS mapping shown in Fig. 9.3d. As the ionic radii of Mn, Cd, and In are larger than that of Fe, the diffraction peaks systematically shifted to lower angles as the content of dopants increased, indicating successful incorporation of these atoms at the Fe site.

The structural and physical properties of the four samples are presented in Fig. 9.4. The plot in Fig. 9.4a shows the evolution of lattice constants a and c with increasing dopant content. As the content of dopants increases, the a-axis expands from 6.26 to 6.38 Å, while the c-axis expands from 19.8 to 20.0 Å. The expansion of the c-axis is primarily due to the increase in P-P distances, which can be seen in Fig. 9.4b. In the $Fe_{0.25}$ sample, the P-P distance is 3.08 Å, which considerably exceeds the typical P-P dimer region of ~2.3 Å, indicating successful breaking of the P-P dimer. The electrical resistivity versus temperature (ρ-T) curves of the four samples suggest that they are semiconductors. The thermal activated gaps (E_a) obtained from the fitting of the ρ-T curves are plotted in Fig. 9.4b, and the E_a increases from 0.1 eV in $FePSe_3$ to 0.7 eV in $Fe_{0.25}$. In addition, the XPS of anion P is presented in Fig. 9.4c, which indicates a red-shift in P's $2p^{3/2}$ and $2p^{1/2}$ peaks, suggesting the formation of partial P^{3-} lone-pair electrons by breaking of the P-P dimers as the dopant content increases.

$Fe_{0.8}$ and $Fe_{0.25}$ samples were studied for pressure-induced structural and electrical transport property changes. Figure 9.5 shows the raw data of in-situ synchrotron XRD patterns of $Fe_{0.8}$ and $Fe_{0.25}$. The shifts of (113) and (300) peaks can be seen more clearly as the contour plot in Fig. 9.6a, b. Both angles shifted to higher values with increasing pressure, indicating unit cell shrinkage. At critical pressures of 12 GPa ($Fe_{0.8}$) and 28 GPa ($Fe_{0.25}$), a quasi-isostructural phase transition occurred with an abrupt decrement in lattice constants. The anisotropic compressibility of $FePX_3$ was analyzed using a trigonal space group R-3, and the normalized lattice constants were plotted. The total shrinkage ratios of a- and c-axis were 5.5 and 28.0% ($Fe_{0.8}$) and 13.0 and 36.0% ($Fe_{0.25}$). The giant shrinkage of 15% can lead to the recombination of P-P dimer.

The plot in Fig. 9.6e shows the unit-cell volume (V) of $Fe_{0.8}$ and $Fe_{0.25}$ samples under pressure. Both samples experience significant shrinkage under pressure, with shrinkage ratios of 38% and 52%, respectively, and collapse ratios at the phase transition pressure of 10.3% and 19.5%, respectively. By fitting of Birch-Murnaghan equation for the low/high-pressure phase of $Fe_{0.8}$ and $Fe_{0.25}$ samples, we estimate the low- (high-) pressure bulk modulus $B_{LP}(B_{HP})$ to be 55(3)/78(9) GPa and 27(2)/ 28(8) GPa, respectively. Notably, the B value of $Fe_{0.25}$ is the smallest ever reported

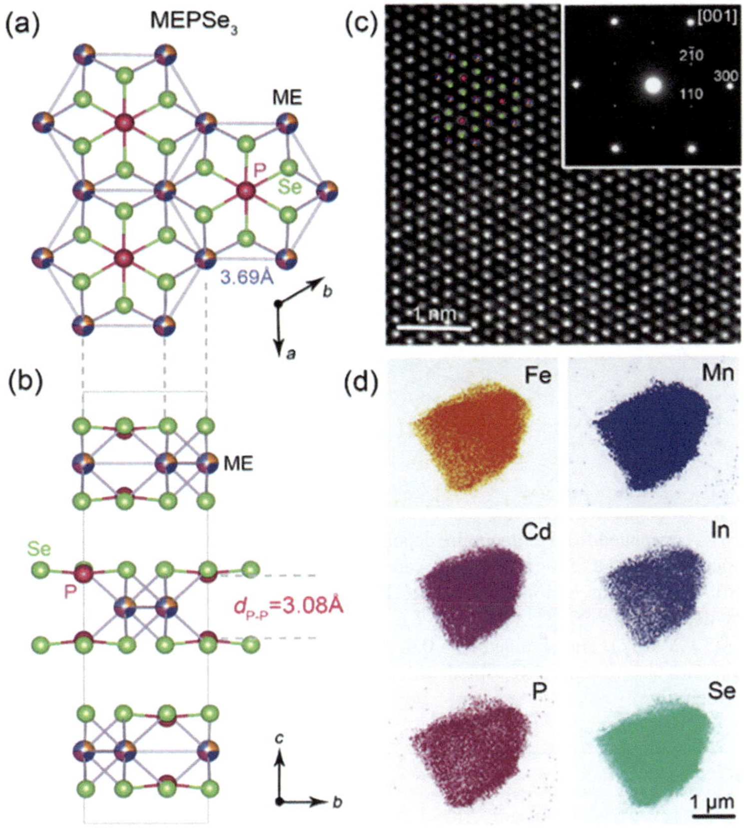

Fig. 9.3 **a, b** Top- and side-view of $(FeMnCd)_{0.25}In_{0.17}PSe_3$ along *c*-axis and *a*-axis, respectively. **c** HRTEM image of $(FeMnCd)_{0.25}In_{0.17}PSe_3$ taken along *c*-axis. Inset is indexed SAED (selected area electron diffraction) image. **d** EDS mapping of Fe (orange), Mn (Blue), Cd (magenta), In (purple), P (brown), and Se (green) of $(FeMnCd)_{0.25}In_{0.17}PSe_3$. The scale bar is 1 μm. (Reprinted with permission from ref 15. Copyright 2022 American Physical Society)

in MPX_3 compounds, and our results suggest that the existence of soft P-P covalent bonds greatly reduces B. These findings suggest that in layered compounds, B may primarily be determined by the strength and length of chemical bonds. Noticeably, the superconductivity in $HEPSe_3$ is accompanied by an out-of-plane collapse, different from the in-plane collapse observed in all the individual components, indicating a different superconducting mechanism from $FePSe_3$.

In Fig. 9.7, the semiconductor–metal transitions of $Fe_{0.8}$ and $Fe_{0.25}$ occur at 6 and 16 GPa, respectively. In both cases, the metallic state persists within a certain pressure range. The superconducting transitions in $Fe_{0.8}$ and $Fe_{0.25}$ are observed at $T_c = 4$ K and 3.2 K, respectively, at pressures above 18 and 32 GPa. The maximal T_c of the $Fe_{0.8}$ sample is 6.2 K and then gradually decreases, whereas for the $Fe_{0.25}$ sample, the T_c increases to 4.3 K at 46 GPa and then saturates to the highest pressure.

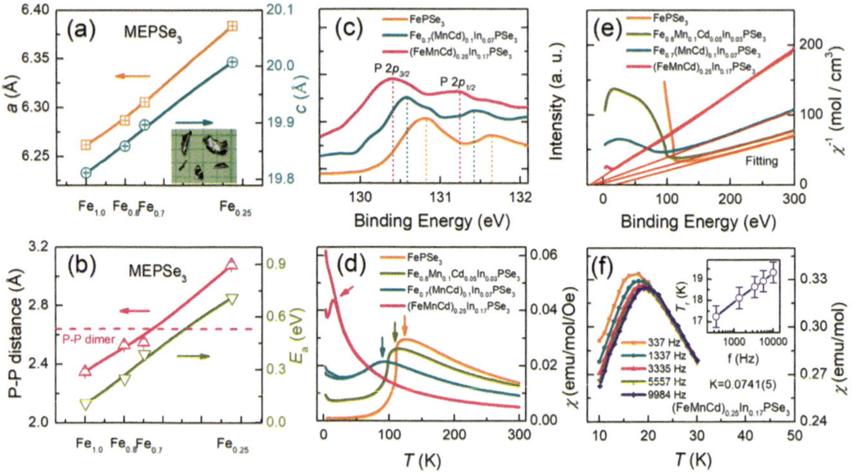

Fig. 9.4 **a** Lattice parameters a and c of $FePSe_3$ and three $MEPSe_3$ samples. **b** P-P distance and band gap (E_a) estimated from temperature dependence of conductivity in four samples. **c** P 2p XPS of three samples. **d**, **e** DC Susceptibility (χ) and $1/\chi$ of four samples as a function of temperature from 2 to 300 K. **f** AC susceptibility of $(FeMnCd)_{0.25}In_{0.17}PSe_3$ under different frequencies. Inset is frequency-dependent freezing temperature (T_f). Solid lines are fitting curves based on the equation, $K = \Delta T_f/(T_f\ \Delta \log f)$. The obtained $K = 0.0741(5)$ falls into the range of $0.0045 \leq K \leq 0.08$ for canonical spin-glass systems. (Reprinted with permission from ref 15. Copyright 2022 American Physical Society)

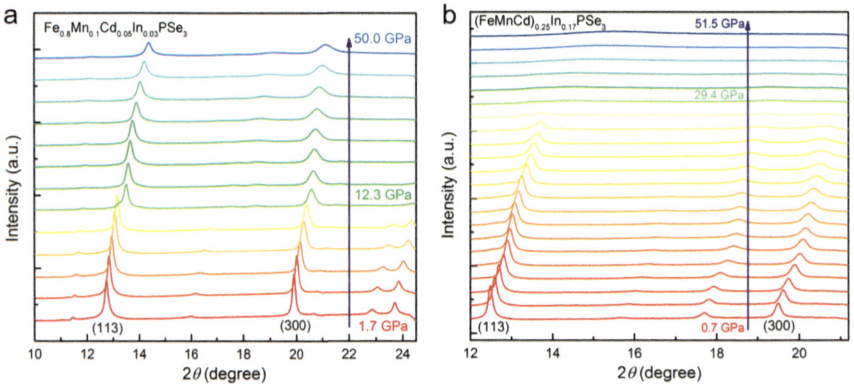

Fig. 9.5 In-situ synchrotron XRD patterns of $Fe_{0.8}Mn_{0.1}Cd_{0.05}In_{0.03}PSe_3$ (**a**) and $(FeMnCd)_{0.25}In_{0.17}PSe_3$ (**b**) under pressures. (Reprinted with permission from Ref. [15]. Copyright 2022 American Physical Society)

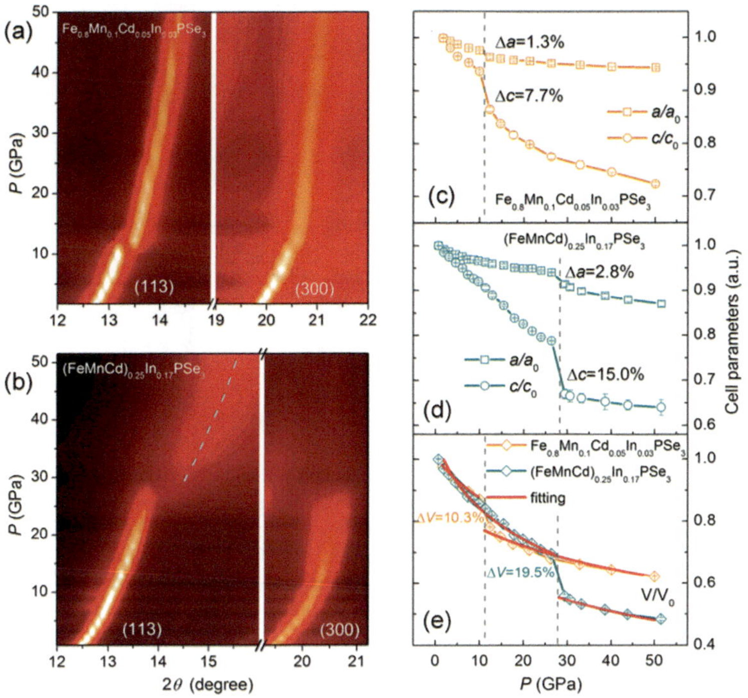

Fig. 9.6 **a, b** Color contour of the (113) and (300) diffraction peaks for $Fe_{0.8}Mn_{0.1}Cd_{0.05}In_{0.05}PSe_3$ ($Fe_{0.8}$) and $(FeMnCd)_{0.25}In_{0.17}PSe_3$ ($Fe_{0.25}$) under external physical pressure. **c–e** Pressure-dependent lattice parameters and volume of unit cell for both samples. (Reprinted with permission from Ref. [15]. Copyright 2022 American Physical Society)

Interestingly, once we fit all the metallic curves using the formula $R = R_0 + AT^n$ and extracted the coefficients in Fig. 9.8, peaks in A and dips in n at 12 and 27 GPa for $Fe_{0.8}$ and $Fe_{0.25}$ can be clearly seen, indicating the existence of quantum critical points on the verge of P-P dimer instabilities.

Under external pressure, $FePSe_3$ undergoes an in-plane collapse that results in the formation of a pseudo-trigonal monoclinic structure featuring intralayer Fe–Fe bonding. This behavior bears resemblance to the in-plane formation of Ir-Ir dimers in CdI_2-type $IrTe_2$, where charge redistribution between Ir-Ir and Te–Te regulates the CDW and superconductivity. The same type of behavior has been observed in the collapsed to uncollapsed transition in $LaFe_2As_2$. Furthermore, the breaking of the P-P dimer has been closely associated with a transition from magnetic to nonmagnetic in $Sr_{1-x}Ca_xCo_2P_2$, and a ferromagnetic quantum critical point has been identified on the verge of Ge(P)-Ge(P) dimer instability in $SrCo_2(Ge_{1-x}P_x)_2$. This study found that the collapse of the lone pair of P under high pressure triggers a transition from $(P_2)^{6-}$ to $(P\text{-}P)^{4-}$, releasing a certain number of electrons that significantly alter the electronic structure. This, in turn, increases the carrier concentration and density of states at the Fermi level in favor of superconductivity.

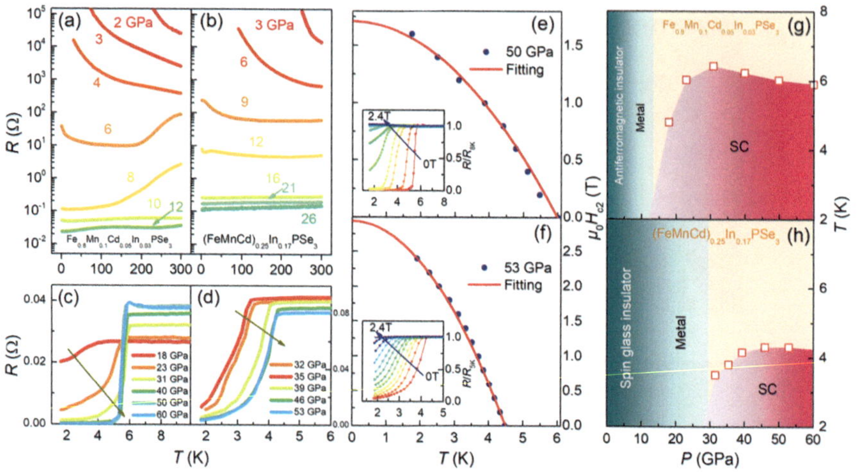

Fig. 9.7 **a**, **c** Temperature dependence of *R-T* curves of $Fe_{0.8}$ below and above 12 GPa. **b**, **d** Temperature dependence of *R-T* curves of $Fe_{0.25}$ below and above 26 GPa. Temperature dependent upper critical fields of **e** $Fe_{0.8}$ at 50 GPa and **f** $Fe_{0.25}$ at 53 GPa. Fitting curves are GL equation shown in main text. Insets are temperature dependence of the normalized resistances, respectively. **g**, **h** Electronic phase diagram of both samples versus pressure. (Reprinted with permission from Ref. [15]. Copyright 2022 American Physical Society)

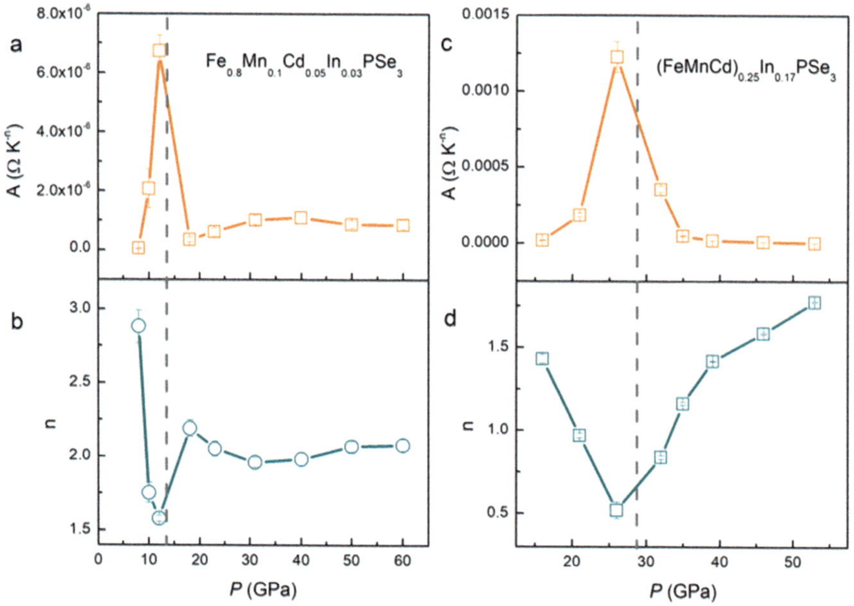

Fig. 9.8 Fitting of the pressure dependent resistance of Fe0.8 and Fe0.25 in the metallic region using the formula $R = R_0 + AT^n$. (Reprinted with permission from Ref. [15]. Copyright 2022 American Physical Society)

9.3 Forecast and Conclusion

Without doubt that the development of high-entropy van der Waals materials is still in its early stages. The incorporation of multiple elements into these materials has proven to be a highly promising approach to develop new superconducting materials. This method differs from conventional doping, as none of the incorporating elements can be viewed as constituting the parent compound, and instead, compositional fluctuations and electronic competitions are at play. This is in line with the spirit of condensed matter physics, where the idea that "more is different" holds true. In order to fully harness the potential of high-entropy van der Waals materials, it is crucial to continue the search for more distinct superconducting systems. Several promising aspects of this approach are worth mentioned:

1. Search unconventional superconductivity in magnetic matters. One promising approach for discovering unconventional superconductors is to search for materials with metastable magnetic ordering, which may serve as precursors to the superconducting state. This is evident in well-known examples such as the cuprates and iron-based superconductors. For canonical BCS superconductors, the introduction of magnetic elements was once thought to be detrimental to superconductivity. Recent research on high-entropy van der Waals materials, such as $(MnFeCoIn)PSe_3$, has shown that the introduction of magnetic elements like Mn and Co can actually break the d^6 electron rule and introduce superconductivity at high pressure. Exploring the intercalation of high-entropy van der Waals materials and studying their magnetic properties may lead to the discovery of new superconducting materials and further advance this field. In addition, introducing high entropy into 2D superconductivity can increase the tolerance of additional elements that may not conform to the same crystal structure as the alloying compounds. For example, $(TiVCrNb)_{0.8}(FeMn)_{0.2}Se_2$ and $(CoRhIrPdPt)Te_2$ exhibit dissimilar crystal structures in comparison to the rest of the compounds, yet their doping concentration can achieve levels as high as 20%. This high doping concentration is not attainable through doping monoelement TMD compounds. While these two compounds do not display superconductivity at ambient pressure, they offer a promising avenue for introducing desired species through the effect of high entropy. Imbuing the few existing high-entropy van der Waals superconductors with magnetic elements and corresponding order/disorder might induce the systems out of the BCS shackle, and exhibit competing or synergetic physical behavior.
2. Superconductivity through intercalation. The intercalation process involves the insertion of atomic or molecular species between the layers of a hexagonal lattice. This process can alter the electronic structure of the material, providing a pathway to modulate the electron–phonon coupling and enhance the superconducting properties of the material. The weak interlayer van der Waals interaction in HEX materials makes it easy to incorporate various species, allowing for a high degree of flexibility in controlling the material's electronic and crystalline structures. While the mixing of principal elements within the lattice

offers an intralayer method to modify the electronic properties of the material, intercalation provides a unique opportunity to investigate superconductivity triggered by layered compounds. However, the properties of intercalated HEX materials remain relatively unexplored. Prior research has demonstrated that the intercalation of $HEPS_3$ can effectively suppress magnetic order.

3. Novel phenomena approaching the monolayer limit. Another exciting aspect of high-entropy van der Waals superconductors is the potential to explore novel phenomena approaching the monolayer limit. While some HEX materials, such as $(TiVCrNiTa)S_2$, can be exfoliated into monolayer form, the superconducting HEX materials have not yet been realized in this form. Currently, the normal-pressure phase of high-entropy two-dimensional superconductors lacks single crystals, and the measurement of phase transitions, such as BKT and anisotropic superconductivity, has yet to be explored. However, with the intrinsic 2D forms of van der Waals materials, and the multiple degrees of freedom of their components, there is an exciting potential for new emergent phenomena. For instance, twisted angle phenomena, which have been studied extensively in graphene and other 2D materials and induced unconventional superconductivity, can also be explored in high-entropy van der Waals materials. Due to the complexity of their band structures, the emergent phenomena in these materials can be difficult to predict. Nonetheless, the unique properties of these materials provide an opportunity to explore new physics on the flatland.
4. Low temperature exploration of high-entropy superconductors. In previous studies, researchers have utilized a rapid quenching technique to obtain stable single phases and take advantage of the high entropy term-TdS. However, this technique has limited the available space groups for dozens of compounds, such as all chalcogenides, to the 1T (P-3m1) phase. This stands in contrast to TMD compounds, which have multiple potential phases, including 2H, 3R, and 1T′, each requiring different crystal growth temperatures. While it is possible to achieve other phases in high-entropy materials, doing so may require alternative synthesis methods, such as a solution growth process. Overall, exploring the low-temperature behavior of high-entropy superconductors presents a significant opportunity to uncover new and exciting phenomena in these materials, and further research is needed to fully realize their potential.

In this chapter, we have sketched the few recent progresses in high-entropy van der Waals superconductors, demonstrating their unique properties and potential for developing new superconducting systems. The incorporation of multiple elements into a single material not only provides a means to tune the electronic structure and enhance the superconductivity, but also offers a new avenue for exploring unconventional pairing mechanisms. However, it is important to note that this field is still in its nascent stage and there is a need for further research and experimentation to fully unlock the potential of high entropy van der Waals materials. As scientists continue to push the boundaries of these materials, we can anticipate that they will open up new pathways towards the development of superconducting technologies with unparalleled performance and functionality. With more innovation and exploration, we can expect these materials to mature from their incubating phase into something truly remarkable, leading to a new era in the field of superconductivity.

References

1. J.W. Yeh, S.K. Chen, S.J. Lin, J.Y. Gan, T.S. Chin, T.T. Shun, C.H. Tsau, S.Y. Chang, Adv. Eng. Mater. **6**, 299 (2004)
2. B. Cantor, I.T.H. Chang, P. Knight, A.J.B. Vincent, Mater. Sci. Eng. A **375–377**, 213 (2004)
3. P. Koželj, S. Vrtnik, A. Jelen, S. Jazbec, Z. Jagličić, S. Maiti, M. Feuerbacher, W. Steurer, J. Dolinšek, Phys. Rev. Lett. **113**, 1 (2014)
4. L. Sun, R. J. Cava **090301**, 1 (2019)
5. A.R. Mazza, X. Gao, D.J. Rossi, B.L. Musico, T.W. Valentine, Z. Kennedy, J. Zhang, J. Lapano, V. Keppens, R.G. Moore, M. Brahlek, C.M. Rost, T.Z. Ward, J. Vac. Sci. Technol. A **40**, 13404 (2021)
6. Y. Shukunami, A. Yamashita, Y. Goto, Y. Mizuguchi, Phys. C Supercond. Appl. **572**, 1353623 (2020)
7. R. Sogabe, Y. Goto, Y. Mizuguchi, Appl. Phys. Express **11**, 53102 (2018)
8. Y. Fujita, K. Kinami, Y. Hanada, M. Nagao, A. Miura, S. Hirai, Y. Maruyama, S. Watauchi, Y. Takano, I. Tanaka, ACS Omega **5**, 16819 (2020)
9. R. Sogabe, Y. Goto, T. Abe, C. Moriyoshi, Y. Kuroiwa, A. Miura, K. Tadanaga, Y. Mizuguchi, Solid State Commun. **295**, 43 (2019)
10. T. Ying, T. Yu, Y.S. Shiah, C. Li, J. Li, Y. Qi, H. Hosono, J. Am. Chem. Soc. **143**, 7042 (2021)
11. T. Ying, T. Yu, Y. Qi, X. Chen, H. Hosono, Adv. Sci. **9**, 2203219 (2022)
12. J. Cavin, A. Ahmadiparidari, L. Majidi, A.S. Thind, S.N. Misal, A. Prajapati, Z. Hemmat, S. Rastegar, A. Beukelman, M.R. Singh, K.A. Unocic, A. Salehi-Khojin, R. Mishra, Adv. Mater. **33** (2021). https://doi.org/10.1002/adma.202100347
13. Z. Du, C. Wu, Y. Chen, Z. Cao, R. Hu, Y. Zhang, J. Gu, Y. Cui, H. Chen, Y. Shi, J. Shang, B. Li, S. Yang, Adv. Mater. **2101473**, 1 (2021)
14. S.K. Nemani, B. Zhang, B.C. Wyatt, Z.D. Hood, S. Manna, R. Khaledialidusti, W. Hong, M.G. Sternberg, S.K.R.S. Sankaranarayanan, B. Anasori, ACS Nano **15**, 12815 (2021)
15. X. Chen, J. Wang, T. Ying, D. Huang, H. Gou, Q. Zhang, Y. Li, H. Hosono, J. Guo, X. Chen, Phys. Rev. B **106**, 184502 (2022)
16. K. Kudo, H. Ishii, M. Takasuga, K. Iba, S. Nakano, J. Kim, A. Fujiwara, M. Nohara, J. Phys. Soc. Jpn. **82**, 2 (2013)
17. J.J. Yang, Y.J. Choi, Y.S. Oh, A. Hogan, Y. Horibe, K. Kim, B.I. Min, S.-W. Cheong, Phys. Rev. Lett. **108**, 116402 (2012)

Chapter 10
Investigation of the High-Entropy Alloy Superconductor XZr_2 (X = Fe, Co, Ni, Rh, Ir) Using Muon Spin Spectroscopy (μSR)

Chennan Wang, Md. Riad Kasem, and Yoshikazu Mizuguchi

Abstract This chapter delves into the application of muon spin relaxation (μSR) for studying high entropy alloy (HEA) superconductors, with a specific focus on the XZr_2 family compounds. Initially, we introduce essential concepts of the μSR experiment, highlighting its strengths and limitations in exploring superconducting materials. The latter part of the chapter presents a detailed examination of the experimental results related to XZr_2, followed by an in-depth discussion of their significance.

10.1 The μSR Technique

The muon spin rotation and relaxation (μSR) technique [1–6] employs the elementary particle, the positive muon, as a probe to investigate the magnetic properties of materials. This method leverages the unique properties of muons, enabling detailed studies of magnetic and superconducting phenomena at the microscopic level.

10.1.1 Properties of the Muon

The positively charged muon (μ^+), a lepton, has a rest mass approximately 206.73 times that of an electron. It is produced during the decay of a positive pion (π^+):

$$\pi^+ \rightarrow \mu^+ + \nu_\mu. \tag{10.1}$$

C. Wang (✉)
Laboratory for Muon Spin Spectroscopy, Paul Scherrer Institute, Forschungsstrasse 111, 5232 Villigen, Switzerland

Department of Physics, University of Fribourg, Ch. du Musee 3, 1700 Fribourg, Switzerland
e-mail: chennan.wang@unifr.ch

Md. Riad Kasem · Y. Mizuguchi
Department of Physics, Tokyo Metropolitan University, Hachioji, Tokyo 192-0397, Japan
e-mail: mizugu@tmu.ac.jp

J. Kitagawa and Y. Mizuguchi (eds.), *High-Entropy Alloy Superconductors*, Springer Series in Solid-State Sciences 202, https://doi.org/10.1007/978-981-97-4129-8_10

This decay process, mediated by the weak interaction, links the momentum and spin direction of the muon. In the pion's rest frame, the muon emerges fully spin-polarized, with its spin oriented opposite to its momentum. A muon beam, created at a target surface and possessing a well defined momentum, is thus 100% spin polarized. These spin-polarized muons, when implanted in a sample, rapidly thermalize without significant loss of spin-polarization, typically stopping at interstitial lattice sites. The stopping power for a muon with a typical momentum of 27–30 MeV/c is around 150 mg/cm^2, which translates into a stopping distribution of about 100–200 μm for the example of cuprate superconductors.

Since muon has a spin 1/2 and a magnetic moment of 4.84×10^{-3} μB, in the presence of a local magnetic field, the muon precesses with the Larmor frequency

$$\omega_{\mathrm{L}} = \gamma_\mu B_{\mathrm{local}}, \tag{10.2}$$

where γ_μ is the gyromagnetic ratio $\gamma_\mu = 2\pi \times 135.53$ MHz/T.

The muon is an unstable particle, decaying into a positron (e^+) and two neutrinos (ν_e, ν_μ) with a mean lifetime of approximately $\tau_\mu = 2.2$ μs. The muon decay is also mediated by the weak interaction that violates the parity. Thus the positron emission is anisotropic, favoring the direction of the muon's spin. This provides an easy way of following the time-dependence of the spin-polarization of a muon ensemble, simply by monitoring the asymmetry of the positron emission,

$$\mu^+ \rightarrow e^+ + \nu_e + \bar{\nu}_\mu. \tag{10.3}$$

10.1.2 Principle of the μSR Experiment

10.1.2.1 Muon Polarization and Asymmetry of the Positron Emission

In a μSR experiment, the positron decay event is detected, which carries information about the muon's spin orientation at the time of decay. By detecting the time dependent change of these positrons, we gain insight into the spin dynamics of the muons which reflects the magnetic signature of the sample of interest.

The essence of a μSR experiment is tracking the evolution of the muon spin polarization, P(t), over time. An incoming muon will be detected firstly by a muon detector, which registers the arrival time of the muon and starts a clock. After the muon is decayed, one of the positron detectors surrounding the sample will detect the emitted positron and stop the clock. The polarization is determined from the spatial distribution asymmetry A(t) of the emitted positrons, which indicates the likelihood of a positron being emitted in respect with the muon spin orientation. The relationship between P(t) and A(t) is given by:

$$A(t) = A_0 \cdot P(t), \tag{10.4}$$

where A_0 represents the maximum observable asymmetry from the measurement. This relationship forms the basis of extracting meaningful information about the sample's magnetic properties from μSR data.

In the presence of a homogenous magnetic flux density B around a muon, the time evolution of the muon spin polarization, $P(t)$, is given as

$$P(t)/P(0) = \cos(\omega_L t + \phi) = \cos(\gamma_\mu B t + \phi), \tag{10.5}$$

where ϕ is the initial phase of the muon polarization. The histogram recorded by one positron detector can be expressed as

$$N(t) = N_0 e^{-t/\tau_\mu}[1 + A_0 P(t)] + B_0, \tag{10.6}$$

where N_0 is the initial positron rate of the positron detector at $t = 0$ (determined from the "prompt peak"), A_0 is the asymmetry of the positron emission and B_0 is a time-independent background events due to other charged particles or the dark current of the detector.

For a pair of positron detectors that are arranged in opposite directions, for example forward and backward detectors, denoted as f and b, the muon spin polarization direction is measured by detecting the asymmetry of positron emission:

$$A_{b,f}(t) = \frac{[N_b(t) - B_{0,b}] - [N_f(t) - B_{0,f}]}{[N_b(t) - B_{0,b}] + [N_f(t) - B_{0,f}]}. \tag{10.7}$$

The variation of the counter efficiency and the solid angles (due to a non-ideal sample and detector alignment) will cause a sensitivity deviation on each detector. In practice, calibration measurements are required to determine the values of N_0 and B_0. The muon spin polarization is eventually given by:

$$P(t) = \frac{A_{b,f}(t)}{A_{b,f}(0)}. \tag{10.8}$$

10.1.2.2 Muon Spin Depolarization

The spin polarization of the implanted muon ensemble typically decreases over time due to the internal magnetic distribution perceived by the muons [7]. Several factors may lead to a time-dependent depolarization or relaxation of the muon spin polarization, $P(t)$, which mainly contributes by spatial inhomogeneity (static relaxation) or fluctuations (dynamic relaxation). The former effect typically occurs in inhomogeneous magnetic materials and in the vortex state of a superconductor, while the latter arises from dynamical effects such as slow spin fluctuations that may occur above the transition temperature of a magnetic material.

The depolarization function in general can be represented in an integral form, where $f(B_\mu)$ is the probability distribution of the local magnetic fields probed by the muon ensemble

$$P(t) = \int_0^\infty f(B_\mu)\cos(\gamma_\mu B_\mu t + \phi)\,dB_\mu. \tag{10.9}$$

It is possible that the measurement of muon spin polarization along different crystallographic directions may exhibit distinct time dependence. In such cases, the local magnetic field anisotropy is manifested.

Several exemplary depolarization functions in analytical form will be presented in the next section.

10.1.2.3 μSR Measurement Configurations

In μSR experiments, three frequently utilized configurations are the zero-field (ZF), transverse-field (TF), and longitudinal-field (LF) modes, offering diverse types of information.

(i) ZF-μSR

In the ZF-μSR configuration, no external magnetic field is applied to the sample, such that a precession and relaxation of the muon spin is only caused by the internal magnetic properties of the sample. With a ZF-μSR measurement one can determine whether a sample is non-magnetic or if it contains certain magnetic ordering. Even short-range magnetic orders or a random spin arrangement, such as in spin glasses, can be identified.

Let us assume that the initial direction of the muon spin is along the z-direction, and a static magnetic order produces a field at the muon site, B_μ, that is tilted at an angle θ. The corresponding muon relaxation function along the initial muon spin direction is given as

$$P_z(t) = \cos^2(\theta) + \sin^2(\theta)\cos(\gamma_\mu B_\mu t + \phi). \tag{10.10}$$

If a sample comprises magnetic domains oriented in a random manner and each of these domains exhibits a identical type of long-range magnetic order, the local magnetic field B is averaged over all possible directions with respect to the initial direction of the muon spin, which yields

$$P_z(t) = \frac{1}{3} + \frac{2}{3}\cos(\gamma_\mu B_\mu t). \tag{10.11}$$

The 1/3 term comes from the muon spins parallel to the z-direction, hence called the longitudinal component of the muon depolarization. The 2/3 term with an oscillatory signal is generated from the muon spins perpendicular to the z-direction, which is known as the transverse component of the depolarization function. It is worth noting that the initial phase ϕ of the muon ensemble becomes nullified on average, resulting in its omission.

Additionally, the damping of the oscillatory signal is accounted by an additional Gaussian function with the damping rate $\sigma = \gamma_\mu \sqrt{\langle \Delta B_\mu^2 \rangle}$ which represents the inhomogeneity of the measured local magnetic field of the transverse component.

Another important relaxation function is for describing a sample with randomly oriented magnetic moments. This gives rise to a Gaussian frequency distribution that is centered around zero which is known as the Kubo-Toyabe function [8],

$$P_z(t) = \frac{1}{3} + \frac{2}{3} e^{-\sigma^2 t^2/2}(1 - \sigma^2 t^2). \tag{10.12}$$

(ii) TF-μSR

In the TF-μSR configuration, an external field is applied in the direction perpendicular to the initial muon spin polarization. Under this geometry, muons will precess in a local magnetic field which is composed by the internal and the applied field. A major usage of the TF-μSR mode is that at a sufficiently large external field the direction of the local magnetic field is almost parallel to the external field which is known. This allows one to reliably obtain information about the volume fractions of the magnetic and the non-magnetic regions in the sample. For a sample that consists of regions with different magnetic characters, the volume fraction for each magnetically distinct phase, f_i, is determined from

$$P_z(t) = \sum_i f_i \cos(\gamma_\mu B_{\mu,i} t + \phi) g_i(t), \tag{10.13}$$

where ϕ is the initial phase of the muon ensemble, and $g_i(t)$ is the depolarization function of each region.

(iii) LF-μSR

For LF-μSR an external magnetic field is applied parallel to the initial muon spin direction. If the applied field strength is significantly larger than the internal field, the local field at the muon sites will be almost parallel to the muon spin. As a result, for the case of static internal fields, there will be no muon precession and therefore no relaxation of the muon spin polarization. This so-called "decoupling experiment" can thus be used to differentiate between a static and a dynamic magnetic state. A small longitudinal field can also be used to remove the contribution from quasistatic nuclear moments from the muon spin relaxation. This allows one to reveal a specific contribution by a dynamic relaxation due to the low-energy spin fluctuations of electronic origin.

10.1.2.4 Field Distribution of the Superconducting Vortices

In type-II superconductors, quantized magnetic flux tubes known as "vortices" form spontaneously above H_{c1} [9]. Such effect leads to a variation in the magnetic field within the bulk material. The TF-μSR mode can be used to determine the internal magnetic field distribution of a superconductor in the vortex state. Since the distance between the vortex cores is orders of magnitude larger than the crystallographic unit cell, the muon sites are randomly distributed on the length scale of the vortex lattice such that a muon ensemble probes the inhomogeneity of the internal magnetic fields. A spread in the muon precession frequency leads to a characteristic relaxation of the μSR signal which depends on (but is not limited to) the magnetic penetration depth, λ.

For an ideal type-II superconductor, i.e. $\lambda \gg \xi$ (ξ is the coherence length of the Cooper pairs), the description of the field distribution of a single flux line is approximated as [10, 11]:

$$B(\mathbf{r}) = \frac{\Phi_0}{2\pi\lambda^2} K_0\left(\frac{\mathbf{r} - \mathbf{r}_0}{\lambda}\right), \tag{10.14}$$

where Φ_0 is the flux quantum and K_0 is the modified Hankel function of the zero-th order. Accordingly, the field distribution of a periodic flux line lattice (FLL) is formulated as:

$$B(\mathbf{r}) = \langle B \rangle \sum_{\mathbf{k}} \frac{\exp(i\mathbf{kr})}{1 + \lambda^2 |\mathbf{k}|^2}, \tag{10.15}$$

where $\langle B \rangle$ is the mean field within a superconductor and $\mathbf{k}$ are reciprocal vectors of the vortex lattice. The illustration in Fig. 10.1a displays the periodic modulation of the local magnetic field profile in a triangular lattice, with the magnetic field strength at the maximum, minimum and saddle positions highlighted.

The magnetic field variation in a perfect vortex lattice expressed in terms of the second moment of the field spatial distribution $B(\mathbf{r})$ can be derived as:

$$\langle \Delta B^2 \rangle_{\mathrm{FLL}} = \sum_{|\mathbf{k}|>0} \frac{B^2}{(1 + \lambda^2 |\mathbf{k}|^2)^2}. \tag{10.16}$$

For a triangularly arranged vortex lattice, $k = (16\pi^2/3a^2)(m^2 + mn + n^2)$ (m, n are integers); $a^2 = 2\phi_0/\sqrt{3}B$ (ϕ_0 is flux quantum), the approximated expression valid at an applied field below the upper critical field H_{c2} reads [11, 12]:

$$\langle \Delta B^2 \rangle_{\mathrm{FLL}} = \frac{0.172\phi_0}{2\pi}(1 - b)[1 + 1.21(1 - \sqrt{b})^3]\lambda^{-2}, \tag{10.17}$$

where $b \equiv \langle B \rangle / B_{c2}$. This formula can be further simplified under the condition $\langle B \rangle > 2B_{c1}$ [11, 12], such that

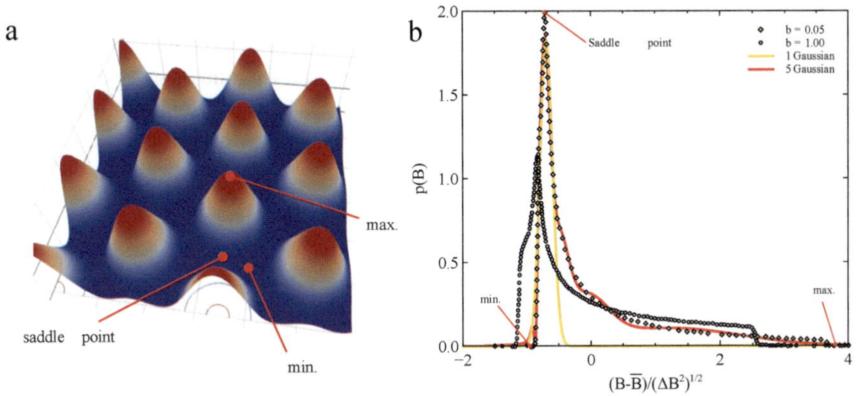

Fig. 10.1 **a** Sketch of the local magnetic field profile due to the FLL. **b** Neglecting the broadening by imperfections, the idealized p(B) from a solution to the G-L equations with b = B/B_{c2} values at 0.05 and 1.0 [12]. A combination of Gaussian functions was utilized to fit the b = 0.05 data, as demonstrated by the solid curves

$$\langle \Delta B^2 \rangle_{\text{FLL}} = 0.00371 \phi_0^2 \lambda^{-4}. \tag{10.18}$$

As the temperature rises, the superconductivity diminishes gradually. It can be inferred that $\langle \Delta B^2 \rangle_{\text{FLL}}$ will eventually reach zero at T_c. Assuming vortex dynamics is negligible, such as in a 3-dimensional bulk superconductor without anisotropy, the estimated magnetic penetration depth derived from Eq. 10.18 remains a valid approximation throughout the entire temperature range below T_c. In contrast, in anisotropic 2-dimensional superconductors where strong vortex dynamics exist, the aforementioned mathematical relation significantly undervalues the magnetic penetration depth, particularly in the vicinity of T_c [13–18].

In summary, Eq. 10.18 is a useful approximated relation only describing the second moment of the field distribution of a perfectly arranged triangular FLL in an isotropic strong type-II superconductor.

10.1.2.5 The Link Between $\langle \Delta B^2 \rangle_{\text{FLL}}$ and σ_μ

To determine the internal magnetic field distribution of a superconductor in the vortex state, the TF-μSR mode is used and the superconducting sample should be field-cooled (FC) below the superconducting transition temperature. Because the distance between the vortex cores is orders of magnitude larger than the crystallographic unit cell, the muon sites are thus randomly distributed on the length scale of the vortex lattice such that an ensemble of muon spins probes the distribution of the magnetic fields and results in the spread of the muon precession frequency leading to a characteristic relaxation, σ_μ, of the μSR signal.

As depicted in Fig. 10.1b, the field distribution of an FLL is distinct from a Gaussian distribution, which is the typical internal field distribution of magnetic samples with long-range magnetic ordering [7]. In common practice, one or several Gaussian relaxation functions with variance σ, are used to evaluate the second moment of the local field of the FLL in an approximated way by defining $\gamma_\mu\sqrt{\langle\Delta B^2\rangle_{\text{FLL}}} \equiv \sigma_\mu$, and

$$\sigma_\mu^2/\gamma_\mu^2 = \frac{1}{F}\sum_i f_i[\sigma_i^2/\gamma_\mu^2 + (B_i - \langle B\rangle)^2], \tag{10.19}$$

where f_i is the weighting factor for each component, $F = \sum f_i$ and $\langle B\rangle$ is the first moment of the field distribution.

An alternative way to analyze $\langle\Delta B^2\rangle_{\text{FLL}}$ is through approximation using the asymmetrical Gaussian distribution (also known as the skewed Gaussian function). The widths on the left and right sides of $\langle B\rangle$ are denoted by two independent variances, namely σ_- and σ_+, respectively. The variance σ_μ is effectively given by [19]

$$\sigma_\mu^2 = \frac{1}{\pi}[(\pi - 2)\sigma_-^2 - (\pi - 4)\sigma_-\sigma_+ + (\pi - 2)\sigma_+^2]. \tag{10.20}$$

In fact, acquiring the second moment of the internal FLL field profile can be accomplished more directly through numerical analysis of a Fourier transformed μSR time-spectra [20]. The real part of this transformed spectra, denoted as $p(B)$, is the probability density of the internal field distribution.

There is no general superiority of a particular methodology over another as each technique presents its distinct merits and drawbacks. The first two techniques directly engage with the measured muon time spectra, but they are more vulnerable to fitting instability when increasing the number of fitting parameters. Furthermore, these methods display limited accuracy in the description of the internal field profile. While the last method circumvents potential biases associated with selecting a modeling function, it necessitates measurement data with greater statistics to produce a accurate Fourier transformed spectra [15].

10.1.2.6 Uncertainty of the Determined λ_{sc} by μSR

The μSR technique is capable of microscopically detecting magnetic field variations with high sensitivity, without the need for demanding experimental preparations. Although measuring the superconducting vortex state is a straightforward task and the measurement is uncomplicated to perform, obtaining credible λ_{sc} values is not always trivial [14]. It is essential to be aware of the relevant sources of systematic errors elaborated below:

(a) As elucidated in the previous section, the determination of $\langle\Delta B^2\rangle_{\text{FLL}}$ based on σ_μ depends on the analytical method chosen [12].

(b) The simple relevance between λ_{sc} and $\langle\Delta B^2\rangle_{\text{FLL}}$ is based on the assumption for a specific vortex configuration, which typically assumes a perfectly arranged triangular lattice [11]. However, it can already been observed in actual measurements that the field distribution of an FLL is commonly smeared out due to departing from an idealized condition within the examined volume of a bulk sample [13, 21]. A variety of intrinsic and extrinsic causes can be responsible for such a deviation, e.g. vortex lattice disorder and defect pinning, deformation of the triangular lattice, and sample inhomogeneity [12, 14–16, 18], etc. Unfortunately, a number of these factors are often unspecified due to lacking of information from the actual sample.

(c) The bulk μSR experiment can result in significant inaccuracies when determining λ_{sc} for 2D superconductors. Due to the presence of a sheet superfluid density, $\rho_{sc,2D}$, the interlayer coupling strength is reduced, thus resulting in discontinuity of the columnar vortices along the field direction. This leads to the formation of pancake vortices that are confined within the superconducting plane with a larger characteristic lateral scale known as the Pearl length, $\lambda_P = 2\lambda_L^2/d$, where λ_L is the London penetration depth and d is the interlayer spacing [22]. When this occurs, the pancake vortices lose their coherent arrangement across different sheets and create a disordered state that hinders accurate determination of $\langle \Delta B^2 \rangle_{FLL}$ through bulk μSR experiments [13–16].

Overall, the bulk μSR method is primarily susceptible to the disorderedness in the FLL. It is imperative to consider the validity of the underlying assumptions employed when deducing λ_{sc} from the muon spin depolarization rate, particularly for 2D superconductors.

10.2 Applying μSR to Investigate Zirconium-Based High-Entropy Alloy Superconductors

In chapter 4, a series of prototypical zirconium-based high-entropy alloy (HEA) superconductors XZr_2 (X=Fe, Co, Ni, Rh, Ir), were introduced. This chapter will present additional complementary details about their magnetism and superconductivity based on the μSR technique.

10.2.1 Probing Magnetism at Microscopic Scale

Although magnetic elements are present in the chemical formula, the results from macroscopic measurements, specifically SQUID magnetometry, do not indicate the existence of static ferromagnetic or antiferromagnetic orders in XZr_2. It remains unclear whether other forms of magnetic correlations, such as spin fluctuations or short-range magnetic order, are presented. The μSR experiment may offer complementary insights in this regard.

It is convenient to apply a weak-transverse field (wTF) μSR experiment to evaluate both the magnetic moment strength and magnetic volume fraction of a sample. In

Table 10.1 Normal state muon depolarization rate in XZr_2

X=	Cu	Co	Rh	Fe, Co, Ni	Fe, Co, Ni, Rh, Ir
T (K)	6.0(2)	10.0(2)	12.0(2)	8.0(2)	10.0(2)
A_0	0.281(2)	0.227(2)	0.266(2)	0.230(2)	0.273(2)
σ (MHz)[a]	0.05(1)	0.05(1)	0.05(1)	0.05(1)	0.05(1)
λ (MHz)	0.04(1)	0.16(1)	0.00(1)	0.18(1)	0.00(1)

[a]Value determined from global fitting of the depolarization function at all temperatures

our measurement, the arc sintered ingot samples were grinded and suspended in the muon beam with a 30 Oe magnetic field applied transversely to the muon spin direction. The muon time spectra are analyzed using the following function:

$$A(t) = A_0\cos(\gamma_\mu B_\mu t + \phi)g(t)f(t), \tag{10.21}$$

where A_0 is the experimentally determined amplitude of the muon asymmetry, $g(t) = e^{-\sigma^2 t^2/2}$ is the Gaussian depolarization function, and $f(t) = e^{-\lambda t}$ is the exponential depolarization function. Here, σ represents the muon spin depolarization rate caused by the static nuclear moments and λ represents the transverse muon spin depolarization rate caused either by static or dynamical electronic moment contributions other than the nuclear contributions.

As summarized in Table 10.1, the measurements were performed in a temperature range between 6–12 K, which is just above the superconducting transition temperature for each compound. Although powder X-ray diffraction has confirmed that our samples are single phase, for (Fe,Co,Ni)Zr_2 and (Fe,Co,Ni,Rh,Ir)Zr_2, their A_0 values are observably lower than the anticipated full asymmetry value of 0.282 for a paramagnetic sample (as shown in Fig. 10.2). A reduction in overall asymmetry for metallic samples suggests the presence of a significant depolarization field within the sample, resulting from its sizeable magnetic moment that is not uniformly distributed throughout the volume of the sample. This observation commonly serves as an indicator of magnetic phase separation, either chemically or electronically. The latter scenario may arise from the formation of dilute magnetic moments or small magnetic clusters.

In the remaining portion of the sample volume where muon depolarization is not prominently manifested, it can be attributed primarily to two mechanisms. The first contribution stems from the quasi-static nuclear magnetic moment, leading to a Gaussian-shaped depolarization function in the wTF condition with a rate of $\sigma = 0.05(1)$ MHz, which is a relatively minor value in its magnitude. The second contribution is derived from the presence of electronic magnetic moments that give rise to a unique exponential function, whose magnitude is denoted by λ. The $CoZr_2$ and (Fe,Co,Ni)Zr_2 specimens exhibit relatively higher values of λ, accompanied by smaller A_0. Additionally, the temperature dependence of λ (as shown in Fig. 10.3) does not exhibit a distinct magnetic transition; instead, it overall demonstrates a

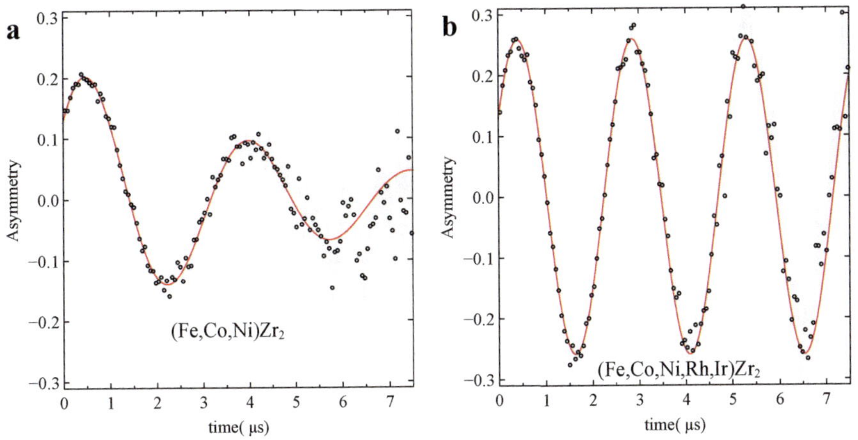

Fig. 10.2 Muon time spectra from the weak transverse field experiments on the polycrystalline samples **a** $(Fe,Co,Ni)Zr_2$, at 20 Oe, and **b** $(Fe,Co,Ni,Rh,Ir)Zr_2$, at 30 Oe

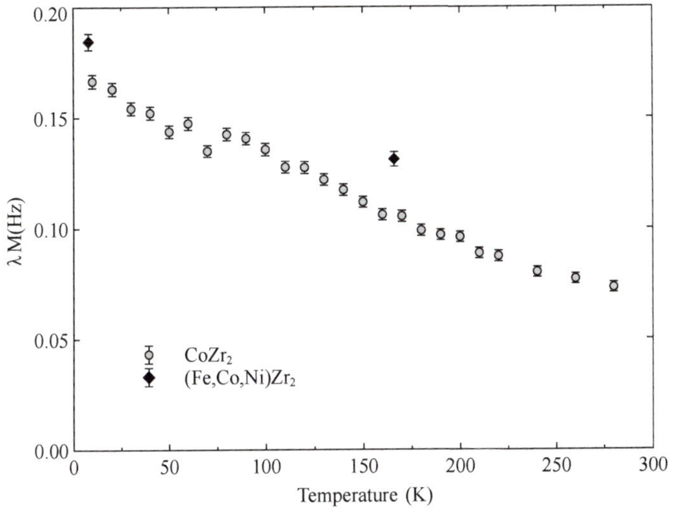

Fig. 10.3 λ as a function of temperature measured using the TF-μSR method in an applied weak magnetic field

monotonic increase as the temperature decreases, which resembles a crossover behavior.

This finding implies that the origin of the detected muon depolarization in the normal state of the XZr_2 samples can be predominantly attributed to the existence of a minority magnetic phase. Within the time scale of μSR it does not show convincing presence of magnetism microscopically in the majority phase of the sample, although further evidence is needed to ascertain whether the minor magnetic phase is intrinsic or extrinsic.

10.2.2 Investigation of Superconductivity in XZr_2

The influence of compositional mixing on the superconducting pairing nature in high-entropy alloy superconductors (HEAs), particularly with respect to site entropy, requires further investigation. To address this issue, we conducted a μSR study involving several zirconium-based HEA compounds (XZr_2 with X=Fe, Co, Ni, Rh and Ir) and analyzed their respective microscopic indicators for superconductivity.

As introduced in an earlier section, the characteristic length scale λ_{sc} can be determined by μSR through the measurement of the FLL field profile occurring in the bulk. In our study, polycrystalline samples were field cooled (FC) below the superconducting transition temperature to achieve a vortex state. However, as our samples are provided in polycrystalline form, it is reasonable to anticipate a modification of the FLL linewidth that will be used to determine λ_{sc}. There are primarily two factors that can be attributed to the randomly orientated grains: (i) destruction of the ideally structured Abrikosov lattice and triggering of quasiperiodic vortex arrangement which leads to additional local field broadening; (ii) mixing of the magnetic penetration depth across three principal axis. In this case, an effective magnetic penetration depth is defined as [16]

$$\lambda_{\text{eff}}^{-2} = \frac{1}{\lambda_{ab}\lambda_c}, \tag{10.22}$$

where λ_{ab} and λ_c are magnetic penetration along the crystallographic *ab*- and *c*-directions, respectively. Thus, it is essential to note that the determined λ^*_{sc} values from our experiment are not entirely inherent and influenced by the aforementioned constraints.

Table 10.2 provides a summary of the superconducting characteristics of the samples used in our μSR investigations [23]. The typical muon time spectra of the sample measured below and above the superconducting transition temperature are presented in Fig. 10.4. Below the superconducting transition temperature, the formation of an FLL is evident in the time trace of the μSR signal, as manifested by noticeable damping. In the normal state, the μSR spectra exhibit full asymmetry and negligible depolarization rates, indicating an absence of magnetism in all measured compounds.

The magnetic field penetration depth of a superconductor and the generation of its vortices will be considerably impacted by the intensity of an applied external field. This interrelation was quantified in Eq. 10.16. For our HEA superconductor samples, such behavior was measured by FC in a series of applied magnetic fields and the results are presented in Fig. 10.5. The data demonstrate a typical trend in which σ^*_{sc} decreases as the magnetic field increases. The maximum value is attained at $H = 2H_{c1}$, which reaches approximately twice its first critical field value.

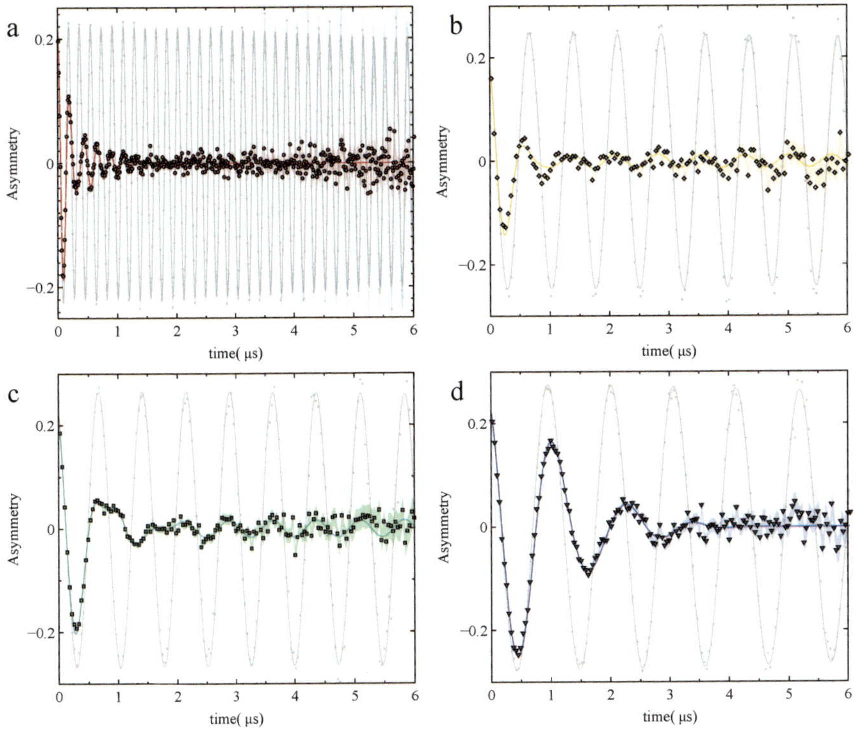

Fig. 10.4 Transverse field μSR spectra acquired in both the field-cooled state at a temperature of 1.5 K, and the normal state immediately above the superconducting transition temperature (in grey color) for **a** $CoZr_2$, **b** $RhZr_2$, **c** $(Co,Rh,Ir)Zr_2$ and **d** $(Fe,Co,Ni,Rh,Ir)Zr_2$

For the measurements conducted on single crystals, the field dependence of σ^*_{sc} can be used to identify the existence of multiple superconducting gaps. However, determining the number of these gaps in polycrystalline samples becomes impractical due to ambiguity resulting from measuring a combination of both σ^*_{ab} and σ^*_{c} when there is microscopic anisotropy present in the magnetic penetration depth, i.e. $\lambda_{ab} \neq \lambda_c$.

In Fig. 10.5, the fitting results for both single and two magnetic penetration depth scenarios are presented. Overall, the experimental values of σ^*_{sc} must be taken into

Table 10.2 List of the basic superconducting parameters of the sample

$X=$	Co	(Co,Rh,Ir)	(Fe,Co,Ni,Rh,Ir)	Rh
T_c(K)	5.5(2)	8.6(2)	5.2(2)	11.0(2)
$H_{c2}(0)$(T)	1.2(1)	9.3(1)	5.5(1)	9.1(1)
ΔS/R	0	1.45	1.65	0

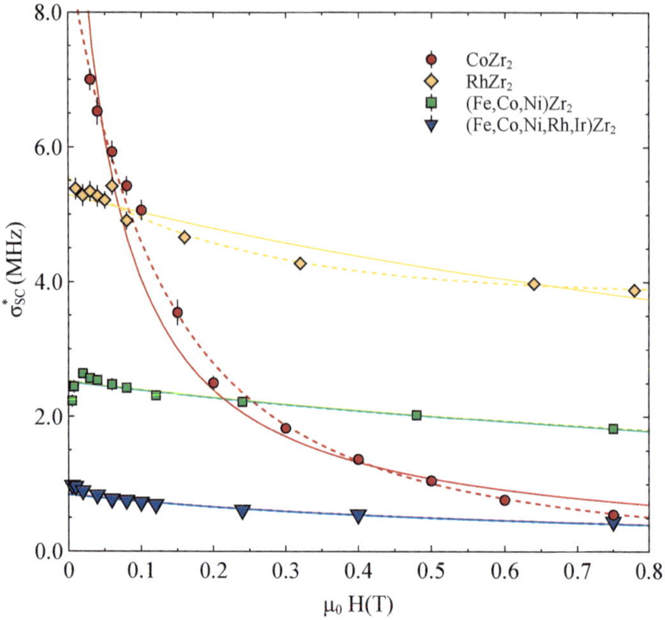

Fig. 10.5 Nuclear contribution corrected muon depolarization acquired at 1.5 K as a function of different applied external magnetic fields after FC. The quantity σ^*_{sc} serves as a means of assessing the effective magnetic penetration depth, despite being influenced by vortex disorder. The solid and dash lines represent the fitted curves that assume the presence of a single or two distinct superconducting gaps/ magnetic penetration depths, respectively

account to any potential systematic uncertainties as explained in the previous section. There is no obvious qualitative significance between these two scenarios that can be concluded at this stage. On the other hand, it can be observed that $CoZr_2$ displays a more rapid decrease in σ^*_{sc} as the applied field increases in comparison with its counterparts within the XZr_2 superconductor category. This suggests possible very different microscopic characteristics for $CoZr_2$.

Further insights regarding the pairing symmetry can be obtained via the temperature dependence of σ^*_{sc}. As elaborated in Sect. 10.1.2.4, it is imperative to first verify the validity of employing Eq. 10.18. Based on the field dependence results, it can be inferred that the superconductivity in XZr_2 compounds exhibits isotropic behavior. Moreover, it is reasonable to conjecture that the vortex lattice maintains its rigidity until reaching T_c due to its 3-dimensional tetragonal $CuAl_2$-type structure (I4/mcm).

The inverse squared effective magnetic penetration depth $1/{\lambda^*_{eff}}^2$ as a function of temperature is presented in Fig. 10.6. At low temperatures, the superfluid density, $\rho_{sc} \propto 1/{\lambda^*_{eff}}^2$, of all compounds exhibits a notable saturation, which implies the absence of nodal features in the superconducting gap. To quantify the microscopic parameter, the data are fitted with a superconducting gap function. Assuming an isotropic s-wave gap, the temperature-dependent relation between the penetration

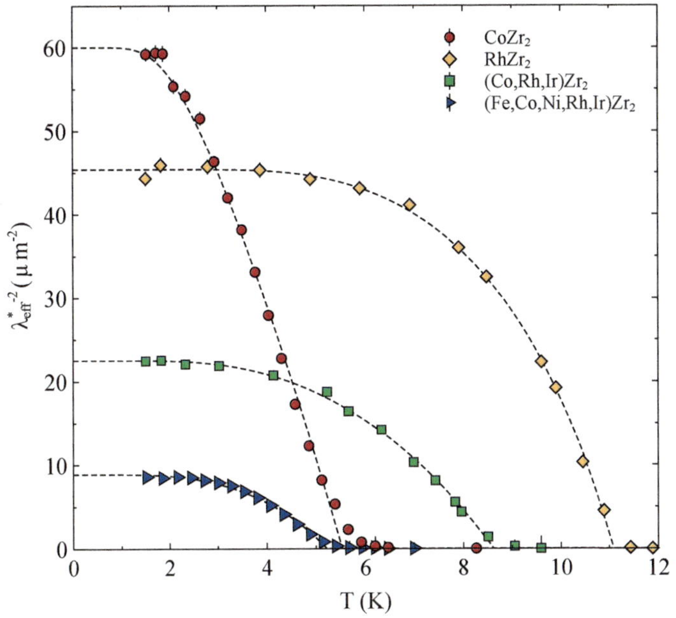

Fig. 10.6 Effective superfluid density represented in magnetic penetration depth as a temperature-dependent superconducting order parameter. The magnetic penetration depth is approximated using σ_{sc}^* (dashed lines). The superconducting gap functions that assume a single s-wave gap within the BCS regime

Table 10.3 Superconducting parameters extracted from the temperature dependent penetration depth

$X=$	Co	(Co,Rh,Ir)	(Fe,Co,Ni,Rh,Ir)	Rh
T_c(K)	5.5(2)	8.6(2)	5.2(2)	11.0(2)
$1/\lambda_{0,\mathrm{eff}}^{*\,2}$ $(1/\mu\mathrm{m}^2)$	60.1(1)	22.4(1)	8.8(2)	45.4(1)
Δ_0 (meV)	0.80(1)	1.58(1)	1.05(1)	2.58(1)

depth and gap size is described as [10]

$$\frac{\lambda(0)^2}{\lambda(T)^2} = \frac{\Delta(T)}{\Delta_0} \tanh\left[\frac{\Delta(T)}{2k_B T}\right], \tag{10.23}$$

which relation is derived within the Bardeen-Cooper-Schrieffer (BCS) scheme, i.e. $\Delta_0 = 1.76\, k_B T_c$ and in the dirty limit ($\xi \gg l$).

The extracted parameters from the self-consistent gap function by fitting the temperature-dependent $1/\lambda_{\mathrm{eff}}^{*\,2}$, are summarized in Table 10.3. The $CoZr_2$ exhibits a superconducting coupling strength denoted by α, which is measured by the superconducting gap amplitude ($2\Delta_0$) divided by the product of Boltzmann's constant (k_B) and the critical temperature for superconductivity (T_c). The calculated value of α for $CoZr_2$ is found to be 3.38(3), closely resembling the BCS ratio of 3.53 that represents the weak-coupling regime in conventional phonon-mediated pairing superconductors. This result confirms that the applied s-wave gap function is appropriate for describing $CoZr_2$.

However, despite the chemical resemblance between $RhZr_2$ and $CoZr_2$, the BCS ratio obtained for $RhZr_2$ is $\alpha = 5.38(3)$, which has a strikingly larger value than $CoZr_2$. This significant deviation observed in the BCS gap ratio implies that the assumed weak-coupling s-wave model lacks adequacy quickly and that there could be a higher possibility of systematic error in the determined gap value for $RhZr_2$.

As presented in Fig. 10.7a, the obtained BCS ratios for the high-entropy counterparts $(Co,Rh,Ir)Zr_2$ and $(Fe,Co,Ni,Rh,Ir)Zr_2$ are lower than the value for $RhZr_2$ but remain higher than the value for $CoZr_2$. This finding may imply that there exist limits to the coupling strength in HEA superconductors, which is dictated by both the upper and lower bounds of the corresponding non-HEA type compounds with a binary composition. Nevertheless, further theoretical study could help to comprehend the lattice vibrational modes mediating Cooper pairing within high-entropy superconductors.

In addition, we conducted a comparative analysis of other significant variables that contribute to comprehending the superconductive properties exhibited by HEA superconductors. These factors are listed in Fig. 10.7b–d.

The Pauli paramagnetic limit for a weak-coupling superconductor is commonly defined as $\mu_0 H_{\text{Pauli}} = \Delta_0/\sqrt{2}\mu_B \simeq 1.86T_c$. As illustrated in Fig. 10.7b, $CoZr_2$ exhibits of approximately 10% in relative to its Pauli limit, in contrast with other compounds that exhibit values exceeding 40%. Specifically, the value for the HEA superconductors is approximately 60% which surpasses the levels attained by the relevant binary counterparts. A high ratio of H_{c2}/T_c is commonly associated with a reduced Cooper pair coherent length, which may imply unconventional mechanisms responsible for pairing due to factors such as increased spin-orbit interactions and stronger superconducting fluctuations.

To avoid biased results caused by measurement inaccuracies or systematic errors inherent in the chosen methodology, we conducted additional evaluations using alternative criteria to corroborate the nature of the superconductivity in our HEA compounds via different perspectives. As presented in Fig. 10.7c, it shows that the superconducting transition temperature (T_c) to Fermi temperature (T_F) ratio is higher in HEA compounds, indicating larger superconducting pairing interactions and the superconductivity is less susceptible to external perturbations such as thermal excitations. Here, the Fermi temperature T_F is proportional to ${\sigma_{sc}}^{3/4}/\gamma^{1/4}$, where σ_{sc} is the muon depolarization rate due to the FLL at zero temperature and γ is the Sommerfeld coefficient, whose values are obtained from the heat capacity measurements [24]. Meanwhile, it can be observed from Fig. 10.7d that the variation in the Ginzburg-Landau parameter κ across multiple samples indicates a more robust type-II superconductivity ($\kappa \gg 1/\sqrt{2}$) in HEA superconductors with an increase in

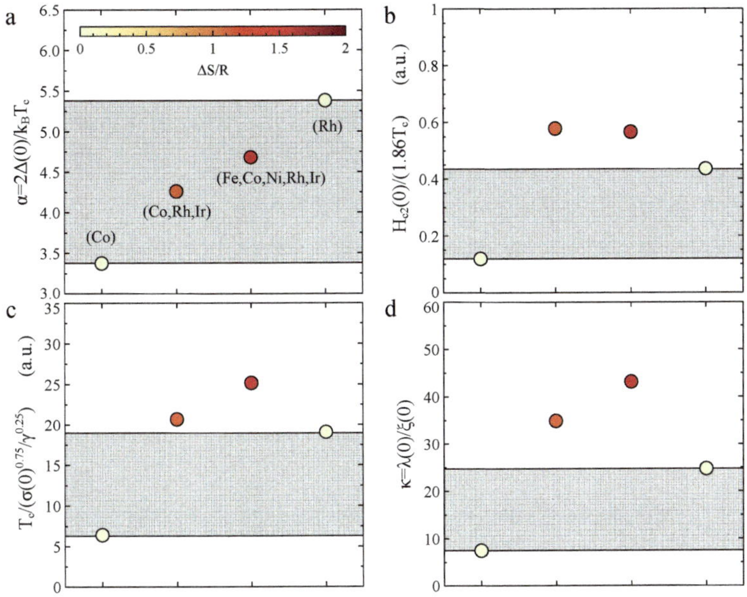

Fig. 10.7 Comparison between microscopic parameters evaluated for different compounds: **a** BCS ratio between the zero-temperature gap value $\Delta(0)$ and T_c, **b** ratio between the estimated zero-temperature upper critical field $H_{c2}(0)$ and superconducting transition temperature T_c, **c** ratio between T_c and the Fermi temperature $T_F \propto \sigma_{sc}(0)^{3/4}/\gamma^{1/4}$, **d** Ginzburg-Landau parameter in terms of the estimated zero-temperature effective magnetic penetration depth $\lambda(0)$ and coherence length $\xi(0)$.The mixing entropy ΔS normalized by the gas constant R is indicated by color scale

the κ value of five times compared to $CoZr_2$, which could not exclude conventional pairing.

Finally, we summarize the measurement results on a scaling plot, where the bulk condensate density (in case of 2D superconductors, the sheet superfluid density shall be used instead of the bulk value) is plotted versus T_c. It has been shown that the corresponding superconducting condensation density of the high-T_c superconductors is correlated with the superconducting critical temperature, and is used for categorizing different types of superconductors. According to the diagram depicted in Fig. 10.8, it is evident that $CoZr_2$ falls within the unshaded region, indicating its association with conventional superconducting pairing. Conversely, (Fe,Co,Ni,Rh,Ir)Zr_2, which exhibits the most extensive site mixing among the compounds mentioned, is situated within the shaded region and aligns similarly with heavy fermion superconductors, thereby signifying possible unconventional pairing characteristics for HEA XZr_2 superconductors.

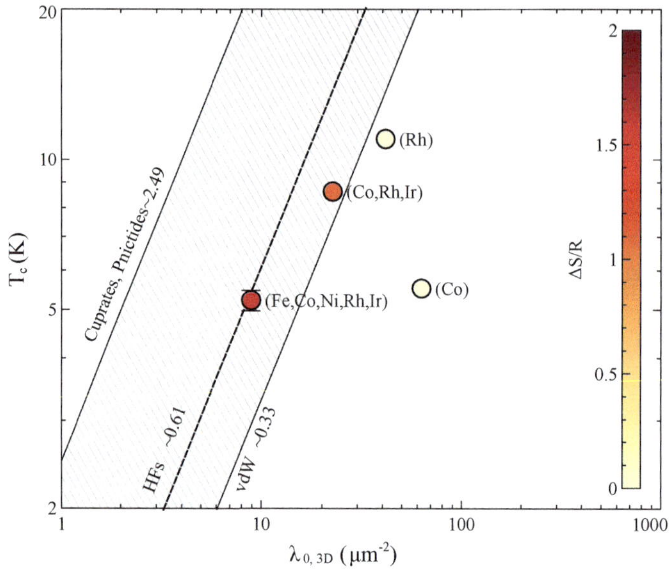

Fig. 10.8 The scaling relation between T_c and the effective bulk superfluid density of different XZr_2 superconductors. The regime associated with unconventional pairing properties is highlighted by the shaded area, where the dashed lines are characteristic scalings generalized from experimental values [25] (HF: Heavy fermions, vdW: van der Waals materials with superconductivity). The linear relation applies for weak coupling BCS superconductors in the dirty limit [26]. The level of the mixing entropy ΔS normalized by the gas constant R is depicted using color scale

10.3 Conclusion and Outlook

In this chapter, we present investigations of the normal and superconducting states of several zirconium-based (XZr_2) high-entropy alloy (HEA) superconductors with different metal compositions, including Fe, Co, Ni, Rh and Ir. In their normal state, although magnetic ions are present in most compunds, no clear evidence for inherent magnetism such as spin fluctuations or long-range static magnetic ordering can be detected by μSR. In their superconducting state, μSR results suggest that HEA superconductors possess distinct superconducting states that are different from the amalgamations of chemically similar alloy superconductors. Based on a consistent trend observed in a range of microscopic indicators, we discovered a correlation between the increase in site entropy and proximity to unconventional superconducting pairing regimes for XZr_2 HEA superconductors.

Further clarification regarding the influence of site mixing entropy on the electronic band structure and electronic correlations is essential to understand its impact on the superconducting pairing nature. Improvement in single crystal synthesis would also be beneficial, e.g., in investigating the properties of superconductivity concerning anisotropy in superfluid density. Our findings indicate that there are unique physical characteristics present in HEA superconductors, which warrant additional exploration.

Acknowledgements Muon measurements were performed at the Swiss Muon Source, Paul Scherrer Institute and we are grateful for the provision of beamtime.

References

1. J.H. Brewer, Problems and opportunities: the early history of μsr. Phys. Procedia **30**, 2–6 (2012). 12th International Conference on Muon Spin Rotation, Relaxation and Resonance (μSR2011). https://doi.org/10.1016/j.phpro.2012.04.027
2. A. Schenck, Muon Spin Rotation Spectroscopy: Principles and Applications in Solid State Physics. A. Hilger, Bristol, Boston (1985)
3. J.H. Brewer (ed.), Muon spin rotation/relaxation/resonance. Wiley-VCH Verlag GmbH & Co KGaA (2003). https://doi.org/10.1002/3527600434.eap258
4. R. De Renzi, Introduction to μSR, pp. 149–177. Springer, Milano (2007). https://doi.org/10.1007/978-88-470-0532-7_6
5. S.J. Blundell, R.D. Renzi, T. Lancaster, F.L. Pratt (eds.), *Muon Spectroscopy—An Introduction* (Oxford University Press, New York, 2021)
6. A. Amato (ed.), Physics with Muons: From Atomic Physics to Solid State Physics. Lecture Notes, https://www.psi.ch/de/lmu/lectures (2022)
7. A. Yaouanc, P.D.d. Réotier, Muon Spin Rotation, Relaxation, and Resonance: Applications to Condensed Matter. Oxford University Press, Oxford (2011)
8. R.S. Hayano, Y.J. Uemura, J. Imazato, N. Nishida, T. Yamazaki, R. Kubo, Zero-and low-field spin relaxation studied by positive muons. Phys. Rev. B **20**(3), 850–859 (1979). https://doi.org/10.1103/PhysRevB.20.850
9. A.A. Abrikosov, The magnetic properties of superconducting alloys. J. Phys. Chem. Solids **2**(3), 199–208 (1957). https://doi.org/10.1016/0022-3697(57)90083-5
10. M. Tinkham, *Introduction to Superconductivity, 2*, ed. (Dover books on physics. Dover Publ, Mineola, NY, 2015)
11. E.H. Brandt, Properties of the ideal Ginzburg-Landau vortex lattice. Phys. Rev. B **68**(5), 054506 (2003). https://doi.org/10.1103/PhysRevB.68.054506
12. E. Brandt, Flux distribution and penetration depth measured by muon spin rotation in high-Tc superconductors. Phys. Rev. B **37**(4), 2349–2352 (1988). https://doi.org/10.1103/PhysRevB.37.2349
13. J.R. Clem, Two-dimensional vortices in a stack of thin superconducting films: A model for high-temperature superconducting multilayers. Phys. Rev. B **43**, 7837–7846 (1991). https://doi.org/10.1103/PhysRevB.43.7837
14. E.M. Forgan, M.T. Wylie, S. Lloyd, M.P. Nutley, S.L. Lee, R. Cubitt, C. Aegerter, H. Keller, T.W. Li, μSR studies of the vortex lattice in high-Tc and other superconductors. Hyperfine Interact. **105**(1/4), 61–69 (1997). https://doi.org/10.1023/A:1012654027039
15. C.M. Aegerter, S.L. Lee, Using μSR to investigate the vortex lattice in high-temperature superconductors. Appl. Magn. Reson. **13**(1–2), 75–93 (1997). https://doi.org/10.1007/BF03161972
16. J.E. Sonier, J.H. Brewer, R.F. Kiefl, μSR studies of the vortex state in type-II superconductors. Rev. Modern Phys. **72**(3), 769–811 (2000). https://doi.org/10.1103/RevModPhys.72.769
17. E.H. Brandt, The vortex lattice in conventional and high-Tc superconductors. Braz. J. Phys. **32**(3) (2002). https://doi.org/10.1590/S0103-97332002000400002
18. J.E. Sonier, μSR studies of Cuprate superconductors. J. Phys. Soc. Japan **85**(9), 091005 (2016). https://doi.org/10.7566/JPSJ.85.091005

19. A. Suter, B.M. Wojek, Skewed gaussian—musrfit memorandum, musrfit: a free platform-independent framework for μSR data analysis. Phys. Procedia **30**, 69–73 (2012). https://doi.org/10.1016/j.phpro.2012.04.042. 12th International Conference on Muon Spin Rotation, Relaxation and Resonance (μSR2011)
20. B.D. Rainford, G.J. Daniell, μSR frequency spectra using the maximum entropy method. Hyperfine Interact **87**(1), 1129–1134 (1994). https://doi.org/10.1007/BF02068515
21. D. Charalambous, E.M. Forgan, S. Ramos, S.P. Brown, R.J. Lycett, D.H. Ucko, A.J. Drew, S.L. Lee, D. Fort, A. Amato, U. Zimmerman, Driven vortices in type-II superconductors: a muon spin rotation study. Phys. Rev. B **73**(10) (2006). https://doi.org/10.1103/PhysRevB.73.104514. Accessed 2020-05-23
22. J. Pearl, Current distribution in superconducting films carrying quantized fluxoids. Appl. Phys. Lett. **5**(4), 65–66 (1964). https://doi.org/10.1063/1.1754056
23. Y. Mizuguchi, M.R. Kasem, T.D. Matsuda, Superconductivity in CuAl2 -type Co0.2Ni0.1Cu0.1Rh0.3Ir0.3Zr2 with a high-entropy-alloy transition metal site. Mater. Res. Lett. **9**(3), 141–147 (2021). https://doi.org/10.1080/21663831.2020.1860147
24. M.R. Kasem, A. Yamashita, T. Hatano, K. Sakurai, N. Oono-Hori, Y. Goto, O. Miura, Y. Mizuguchi, Anomalous broadening of specific heat jump at T_c in high-entropy-alloy-type superconductor $TrZr_2$. Supercond. Sci. Technol. **34**(12), 125001 (2021). https://doi.org/10.1088/1361-6668/ac2554
25. Y. Fang, W.-L. You, M. Li, Unconventional superconductivity in CuxBi2Se3 from magnetic susceptibility and electrical transport. New J. Phys. **22**(5), 053026 (2020). https://doi.org/10.1088/1367-2630/ab7fca
26. V.G. Kogan, Homes scaling and BCS. Phys. Rev. B **87**(22), 220507 (2013). https://doi.org/10.1103/PhysRevB.87.220507

Chapter 11
HEA Effects in Layered Unconventional Superconductors: Cuprates and BiS_2-based Superconductors

Masanori Nagao

Abstract The high-entropy alloy (HEA) concept was introduced to the *RE* site (*RE*: rare earth element) for $REBa_2Cu_3O_x$ cuprate and $REOBiS_2$ BiS_2-based layered superconductors. $REBa_2Cu_3O_x$ single-crystalline whiskers and $REOBiS_2$ single crystals with HEAs were grown using the flux method. The superconducting properties (superconducting transition temperature: T_c and superconducting anisotropy: γ_s) were little affected by the mixed entropy (ΔS_{mix}) at the *RE* site. These results were observed using single-crystalline samples. From another perspective, the HEAs concept can be applied to $REBa_2Cu_3O_x$ and $REOBiS_2$ superconductors without degrading their superconducting properties. However, for the $REOBiS_2$ superconductor, the atomic configuration of the *RE* site (equivalent or nonequivalent) affected the superconducting properties.

11.1 The Concept of High-Entropy Alloys (HEAs) in Cuprate and BiS_2-Based Superconductors

This chapter presents the cuprate and BiS_2-based superconductors that introduced the HEA concept. These superconductors are typical layered superconductors that form analogs and show interesting other properties. Many compounds of cuprate superconductors, such as $(RE, A)_2CuO_4$ ((*RE*, *A*)-214 phase) (*RE*: rare earth element, *A*:Ba, Sr, Ce) [1], $REBa_2Cu_3O_x$ (*RE*-123 phase) [2], and $Bi_2Sr_2Ca_{n-1}Cu_nO_x$ ($n = 1$: Bi2201, $n = 2$: Bi2212, $n = 3$: Bi2223 phase) [3], have been reported. In BiS_2-based superconductors, several compounds, such as $Bi_4O_4S_3$ [4], $REO_{1-x}F_xBiS_2$ [5, 6], and $Eu_3F_4Bi_2S_4$ [7], have been reported. This chapter focused on $REBa_2Cu_3O_x$ and $REOBiS_2$, which have *RE* sites. The *RE* site in these superconductors was subsequently substituted with various rare earth elements to introduce the concept of HEAs. The crystal structure of $REBa_2Cu_3O_x$ has stacked superconducting and blocking

M. Nagao (✉)
Center for Crystal Science and Technology, University of Yamanashi, Miyamae 7-32, Kofu 400-0021, Yamanashi, Japan
e-mail: mnagao@yamanashi.ac.jp

J. Kitagawa and Y. Mizuguchi (eds.), *High-Entropy Alloy Superconductors*, Springer Series in Solid-State Sciences 202, https://doi.org/10.1007/978-981-97-4129-8_11

layers, which are CuO_2–*RE*–CuO_2 and BaO–CuO–BaO, respectively. On the other hand, *RE*OBiS_2 is composed of alternating stacking sequences of BiS_2 and *RE*O layers, which are superconducting and blocking layers, respectively. Figure 11.1 shows the crystal structures of (a) $ErBa_2Cu_3O_{6.91}$ (RE = Er, x = 6.91) [8] and (b) $CeOBiS_2$ (RE = Ce, x = 0) [9], which are typical $REBa_2Cu_3O_x$ and $REOBiS_2$ superconductors, respectively. The crystal structure was drawn by VESTA computer software [10]. The superconducting properties and the high-entropy effect at the *RE* site were investigated. The mixed entropy (ΔS_{mix}) at the *RE* site was introduced to evaluate the high-entropy effect. The ΔS_{mix} was calculated from the analytical contents of the *RE* site using the following equation: [11]

$$\Delta S_{\text{mix}} = -R \sum (C_{RE} \ln C_{RE})\ (RE : \text{rare earth element}) \tag{11.1}$$

Here, R = 8.314 J/mol·K (gas constant), and the analytical content of each rare earth element was defined as the C_{RE}. To reveal the intrinsic properties of these materials, single crystals were necessary. Single-crystal growth of those layered superconductors introduced the HEA concept at the *RE* site. The relationship between ΔS_{mix} and superconducting properties was subsequently investigated in terms of the superconducting transition temperature (T_c) and superconducting anisotropy (γ_s) using single crystals.

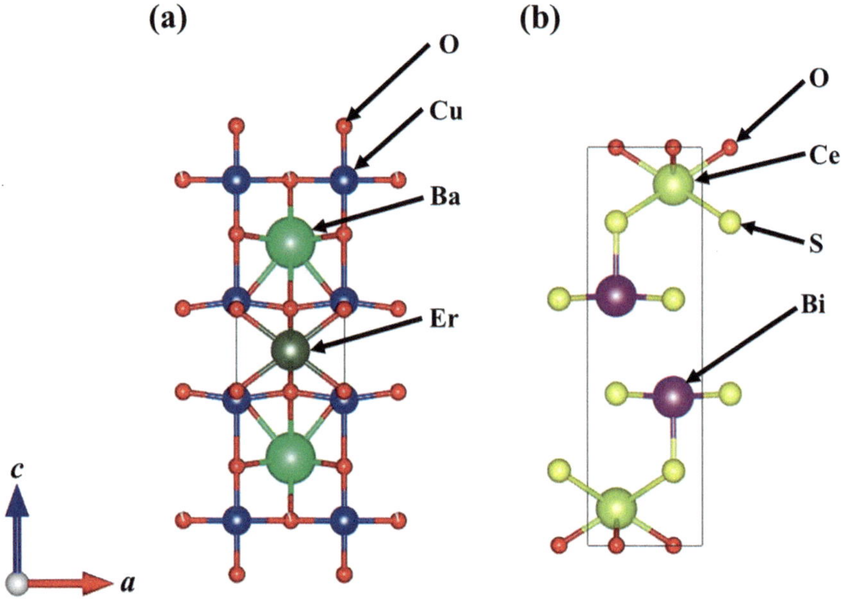

Fig. 11.1 Schematic image of the crystal structure of **a** $ErBa_2Cu_3O_{6.91}$ and **b** $CeOBiS_2$ with the HEA concept at the *RE* site. (Drawn by VESTA computer software [10])

11.2 *RE*-123 Cuprate Superconductor with HEAs Concept

11.2.1 Poly-Crystalline Samples

Polycrystalline bulk samples of *RE*-123 with the HEA concept introduced in the *RE* site were synthesized by a solid-state reaction method. The *RE* site was substituted by six kinds of rare earth elements, such as Y, La, Nd, Sm, Eu, and Gd. Moreover, ΔS_{mix} did not affect the superconducting transition temperature (T_c) or critical current density (J_c) of the *RE*-123 samples. The effect of orthorhombicity on the *RE*-123 phase was greater than that on the ΔS_{mix} phase [12]. However, an improvement in the intragrain critical current density ($J^{local}{}_c$) in the high ΔS_{mix} region (more than 1.3 *R*) was reported [13]. In another report, for four rare earth element-substituted *RE*-123 polycrystalline bulk samples, the high-entropy effect did not affect the irreversible field (H_{irr}), superconducting transition temperature (T_c), or effective magnetic moment [14]. On the other hand, *RE*-123 thin films with the HEA concept introduced at the *RE* site were obtained. *RE*-123 thin films with three to five rare earth element substitutions at the *RE* site were grown by the pulsed laser deposition (PLD) method. No correlation between the rare earth element number and superconducting properties was observed, and T_c and J_c did not deteriorate with increasing ΔS_{mix} [15]. According to other reports, a *RE*-123 (*RE* = Y + Dy + Gd + Sm + Eu) thin film was grown using the trifluoroacetate-metal organic deposition (TFA-MOD) method, and the superconducting properties improved under magnetic fields, which increased the maximum pinning force density [16]. Moreover, the RE_2CuO_4 Ruddlesden–Popper phase is the parent compound of ((*RE*, *A*)-214 cuprate superconductors. Using the PLD method, the growth of RE_2CuO_4 thin films with the introduction of the HEA concept has been reported [17].

11.2.2 Single-Crystalline Whiskers [18]

Single-crystalline whiskers are suitable for investigating intrinsic properties because of their useful characteristics, such as perfect crystallinity, predictable growth directions, and intrinsic superconducting properties. Single-crystalline whiskers of *RE*-123 with the HEAs concept introduced in the *RE* site were grown using the Te-doping method [19–22]. The Ba site substitution of rare earth elements should be suppressed when investigating the high-entropy effect at the *RE* site. The superconducting properties of *RE*-123 are affected by Ba site substitution, which depends on the ionic radius of the *RE* elements. Therefore, Dy, Ho, Er, Tm, Yb, and Lu, which can be barely substituted at the Ba site, were chosen as the rare earth elements. Moreover, because Gd has a lower substitution at the Ba site than does other rare earth elements (La–Eu), Gd was employed to achieve a higher ΔS_{mix}.

RE-123 superconducting single-crystalline whiskers with the HEAs concept were grown from Te-doped precursors. The precursor powders were synthesized using a solid-state reaction, in which $RE_{1.5}Ba_3Cu_3Te_{0.5}O_x$ was weighed and RE = $Gd_aDy_bHo_cEr_dTm_eYb_fLu_g$ ($a + b + c + d + e + f + g = 1.5$) was used for the nominal compositions. The starting materials are RE_2O_3, $BaCO_3$, CuO, and TeO_2. The additional RE_2O_3, $BaCO_3$, and TeO_2 from the $REBa_2Cu_3O_x$ stoichiometric composition play a role in the flux. The starting materials were ground in a mortar and calcined in air at 750, 780, and 800 °C for 10 h (the grinding and calcination steps were repeated three times.). The obtained precursor powders (0.5 g) were pressed into a pellet 10 mm in diameter and approximately 3 mm in thickness. For single-crystalline whisker growth, the precursor pellets were subjected to heat treatment in dry air at T_{max} for 10 h, followed by slow cooling to 900 °C at a rate of 0.5 °C/h. The T_{max} values were optimized for the growth length of the whiskers between 940 and 990 °C. After the temperature was slowly cooled to 900 °C, the mixture was cooled to approximately 300 °C at 8 h and then spontaneously cooled to room temperature (30 °C). (Note: The superconducting properties of *RE*-123 superconductors affect the oxygen amount, which is changed by annealing at 300–500 °C. In this section, the temperature of the furnace controller was set to cool to room temperature (30 °C) after 8 h. However, the actual furnace temperature was not set below 300 °C. Therefore, temperatures below 300 °C resulted in spontaneous cooling.)

Figure 11.2 shows a typical scanning electron microscopy (SEM) image of the grown *RE*-123 superconductors with the HEA concept of a single-crystalline whisker. The lengths of the grown whiskers were 0.5–1.0 mm. The flat surface was *ab*-plane and 10–50 μm wide. The thickness was approximately 10 μm, which was the *c*-axis direction. Evidently, the width of the *ab*-plane was greater than the thickness of the *c*-axis. Table 11.1 summarizes the optimized heat treatment temperature (T_{max}) and chemical compositions of the obtained single-crystalline whisker samples: analytical content (normalized using $\Sigma(C_{RE}) = 1.00$), average ionic radii of *RE* sites, and ΔS_{mix} at the *RE* site. For *RE* elements with small ionic radii (Tm, Yb, and Lu), the values of the analytes were significantly lower than those of the nominal compositions. This result suggested that *RE* elements with small ionic radii are difficult to substitute for *RE*-123 single-crystalline whiskers. The average Ba and Cu contents normalized using the relation $\Sigma(C_{RE}) = 1.00$ were 2.1 ± 0.1 and 3.1 ± 0.2, respectively. Accordingly, the substitution of the Ba site by *RE* elements was expected to be low. The superconducting properties of *RE*-123 superconductors are affected by the average ionic radius of the *RE* sites and the Ba site substitution. Consequently, whiskers with similar average ionic radii for the *RE* sites and Ba contents were selected because the effect of ΔS_{mix} at the *RE* site was investigated. Incidentally, the *c*-axis lattice constants of those *RE*-123 single-crystalline whiskers were estimated to be in the range of 11.63–11.66 Å, which are similar values.

Figure 11.3a shows the resistivity–temperature (ρ–T) characteristics along the *ab*-plane in the temperature range of 60–300 K for the *RE*-123 single-crystalline whiskers, which are shown in Table 11.1. The dashed lines for each sample are linear extrapolations of the resistivity from the normal state. A superconducting transition with zero resistivity (T_c^{zero}) was observed in the 88.8–92.7 K range, for which the criterion was defined as 10 μΩcm in the ρ–T characteristics. Metallic behavior was observed in the normal state (higher than 100 K). The T_c^{zero} and normal-state

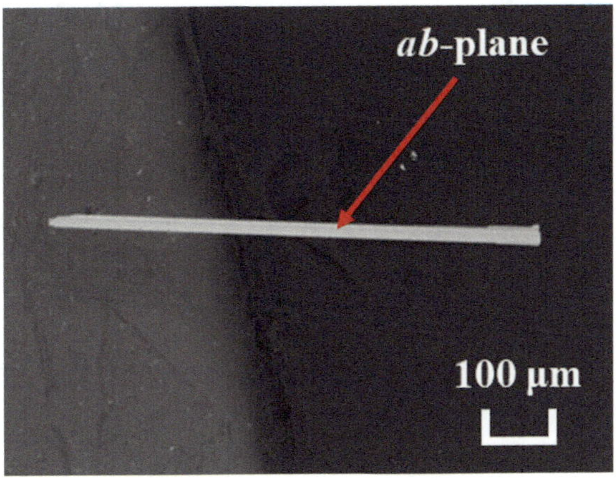

Fig. 11.2 Scanning electron microscopy (SEM) image of the as-grown *RE*-123 single-crystalline whisker with the HEAs concept [39]

resistivity slopes were observed to be uncorrelated with ΔS_{mix}, as shown in Fig. 11.3b, c, respectively.

The superconducting anisotropies (γ_s) of those *RE*-123 single-crystalline whiskers were estimated in the vortex flux liquid state using an effective mass model [23]. The angular (θ_a) dependence of the resistivity (ρ) in the vortex flux liquid state under various magnetic fields (H) was measured. The reduced field (H_{red}) was subsequently calculated using the following equation for an effective mass model:

$$H_{\mathrm{red}} = H\left(\sin^2\theta_a + \gamma_s^{-2}\cos^2\theta_a\right)^{1/2} \tag{11.2}$$

Here, θ_a is the angle between the *ab*-plane and the magnetic field [23]. γ_s was estimated from the best scaling of the ρ–H_{red} relationship. The details of the estimation method have been described previously [24, 25]. Figure 11.4 shows that the estimated values of γ_s have no clear relationship with ΔS_{mix}. γ_s was only slightly affected by ΔS_{mix} in the *RE*-123 superconductor.

The ΔS_{mix} at the *RE* site of *RE*-123 superconductors was uncorrelated with the physical properties, which were the T_c^{zero}, γ_s, and normal-state resistivity slopes. In particular, the ΔS_{mix} at the *RE* site for the *RE*-123 superconductor hardly affects the superconducting properties compared to the oxygen content in the CuO chain layer [26]. These results indicated that both the superconducting properties and normal-state transport properties were insensitive to ΔS_{mix}, which affects local disorders. From another perspective, the high-entropy alloy (HEA) concept can be introduced in *RE*-123 superconductors without degrading the superconducting properties.

Table 11.1 Nominal compositions, analytical contents, T_{max}, average ionic radius of *RE* sites, and ΔS_{mix} at the *RE* site in the grown *RE*-123 single-crystalline whiskers Reproduced with permission from [18]

Sample	Nominal	Gd:*a*	Dy:*b*	Ho:*c*	Er:*d*	Tm:*e*	Yb:*f*	Lu:*g*	Ba	Cu	T_{max} (°C)	Average ionic radius of *RE* sites (Å)	ΔS_{mix} (*R*: Gas constant)
	Analytical	C_{Gd}	C_{Dy}	C_{Ho}	C_{Er}	C_{Tm}	C_{Yb}	C_{Lu}	C_{Ba}	C_{Cu}			
RE1	Nominal				1.5				3.0	3.0	965	1.00	0
	Analytical				1.00				1.95(2)	2.95(1)			
RE2	Nominal			0.78	0.72				3.0	3.0	970	1.01	0.67*R*
	Analytical			0.59(1)	0.41(1)				2.08(3)	3.17(3)			
RE3	Nominal			0.40	0.50	0.60			3.0	3.0	965	1.01	1.01*R*
	Analytical			0.44(1)	0.34(1)	0.19(1)			2.18(6)	3.2(1)			
RE4	Nominal		0.25	0.30	0.40	0.55			3.0	3.0	970	1.01	1.35*R*
	Analytical		0.26(1)	0.34(0)	0.29(1)	0.15(0)			1.97(1)	2.87(1)			
RE5	Nominal		0.1	0.2	0.2	0.4	0.6		3.0	3.0	960	1.00	1.60*R*
	Analytical		0.16(4)	0.17(5)	0.20(2)	0.22(7)	0.25(5)		1.9(2)	2.9(3)			
RE6	Nominal	0.15	0.13	0.17	0.20	0.35	0.50		3.0	3.0	990	1.01	1.78*R*
	Analytical	0.23(1)	0.17(1)	0.15(1)	0.15(2)	0.19(1)	0.20(1)		2.14(8)	3.34(7)			
RE7	Nominal	0.15	0.10	0.10	0.15	0.20	0.30	0.50	3.0	3.0	970	1.01	1.91*R*
	Analytical	0.20(1)	0.13(1)	0.10(1)	0.15(1)	0.09(1)	0.20(1)	0.15(1)	2.0(1)	3.0(2)			

Nominal: Nominal composition ($a + b + c + d + e + f + g = 1.5$), Analytical: Analytical content (normalized using $\Sigma(C_{RE}) = 1.00$)

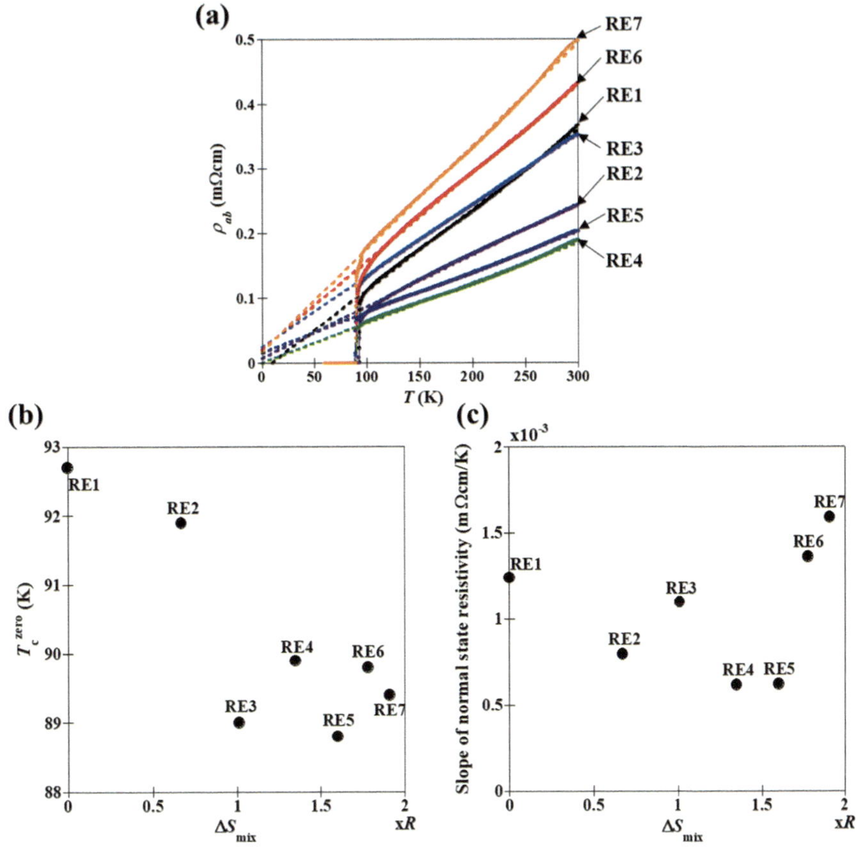

Fig. 11.3 **a** ρ–T characteristics along the *ab*-plane of the *RE*-123 single-crystal whiskers. The dashed lines are linear extrapolations of the resistivity from the normal state. **b** T_c^{zero} and **c** slopes of the ρ–T curves in the normal state as a function of ΔS_{mix}. Reproduced with permission from [18]

11.3 *RE*OBiS$_2$ Superconductor with HEAs Concept

11.3.1 Poly-Crystalline Samples

To introduce a high-entropy effect at the *RE* site, polycrystalline bulk samples of *RE*(O, F)BiS$_2$ with five rare earth elements (La, Ce, Pr, Nd, and Sm) substituted at the *RE* site were synthesized by a solid-state reaction method [27]. The anisotropic displacement parameters of the in-plane S-site in *RE*(O, F)BiS$_2$ decreased with increasing ΔS_{mix} at the *RE* site. Thus, the in-plane disorder on the BiS$_2$ layer was suppressed [28]. In addition, when the value of ΔS_{mix} was approximately 1 *R*, the local maximum value of the Grüneisen parameter was reported [29].

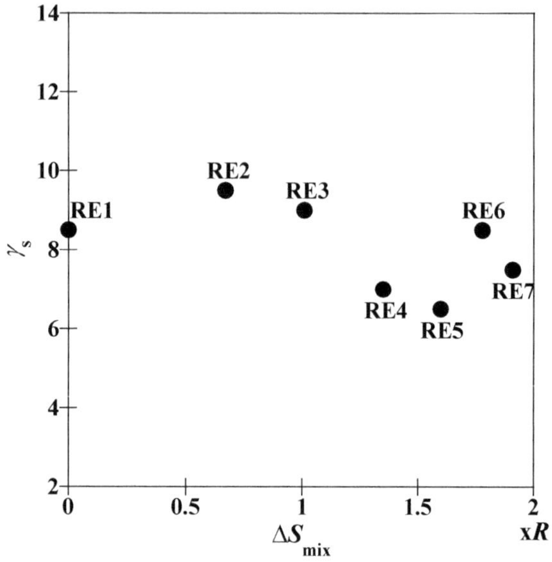

Fig. 11.4 ΔS_{mix} dependence of γ_s for the RE1-7 samples in the *RE*-123 single-crystal whiskers

11.3.2 Single-Crystalline Samples [30, 31]

Single-crystalline samples are required for investigating intrinsic properties. *RE*$OBiS_2$ single crystals with the HEAs concept introduced in the *RE* site were grown using the alkali metal chloride flux [6, 32], which was CsCl (melting temperature is 645 °C.) or a mixture of CsCl and KCl (molar ratio of CsCl:KCl = 5:3, eutectic temperature = 616 °C [33]). The raw materials used were La_2S_3, Ce_2S_3, Pr_2S_3, Nd_2S_3, Sm_2S_3, Gd_2S_3, Bi_2S_3, and Bi_2O_3, which were weighed to obtain a nominal composition $(La_ACe_BPr_CNd_DSm_EGd_F)OBiS_2$ $(A + B + C + D + E + F = 1.0)$. A mixture of raw materials (0.8 g) and alkali metal chloride flux (5.0 g) was ground using a mortar and pestle and then sealed in an evacuated (~10 Pa) quartz tube. The quartz tube was heated at T_{max} for 10 h, followed by cooling to T_{end} at a rate of 1 °C/h. The sample names are listed in Table 11.2. CsCl was employed as the flux, except for samples #RE-2 and #RE-4-e, for which the T_{max} and T_{end} were 950 and 650 °C, respectively. The CsCl/KCl mixture was employed as the flux for samples #RE-2 and #RE-4-e, with T_{max} and T_{end} values of 850 and 600 °C, respectively. Then, the quartz tube was spontaneously cooled to room temperature (10–30 °C). The cooled quartz tube was opened in air, and the obtained materials were washed and filtered with distilled water to remove the alkali metal chloride flux.

Figure 11.5 shows a typical SEM image of the *RE*$OBiS_2$ single crystal with the HEA concept at the *RE* site. The single crystals have plate-like shapes with dimensions of 0.5 × 0.5–1 × 1 mm^2 and a thickness of approximately 100–300 μm. According to the $2\theta/\theta$ scan XRD measurements, the *c*-plane was a flat surface and a well-developed plane of the grown single crystals.

Table 11.2 Nominal compositions and analytical contents, ΔS_{mix} of the *RE* site, and *c*-axis lattice constants of the grown $REOBiS_2$ single crystals

Sample	Nominal	La: *A*	Ce: *B*	Pr: *C*	Nd: *D*	Sm: *E*	Gd: *F*	Bi	S	ΔS_{mix} (*R*: Gas constant)	*c*-axis lattice constant (Å)
	Analytical	C'_{La}	C'_{Ce}	C'_{Pr}	C'_{Nd}	C'_{Sm}	C'_{Gd}	C'_{Bi}	C'_{S}		
#RE-1 [41]	Nominal		1.00					1.00	2.00	0	13.57
	Analytical		1.0(1)					1.01(5)	2.00		
#RE-2	Nominal		0.40		0.60			1.00	2.00	0.6687*R*	13.62
	Analytical		0.39(1)		0.61(1)			0.98(1)	2.00(5)		
#RE-2-e [42–44]	Nominal		0.50		0.50			1.00	2.00	0.6931*R*	13.60
	Analytical		0.50(1)		0.50(1)			0.98(1)	2.00(5)		
#RE-3	Nominal		0.30	0.30	0.40			1.00	2.00	1.091*R*	13.62
	Analytical		0.32(1)	0.29(1)	0.39(1)			0.99(2)	1.99(3)		
#RE-3-e	Nominal		0.33	0.33	0.33			1.00	2.00	1.098*R*	13.61
	Analytical		0.33(1)	0.33(1)	0.33(1)			1.00(6)	2.12(7)		
#RE-4	Nominal		0.30	0.40	0.20	0.10		1.00	2.00	1.302*R*	13.63
	Analytical		0.27(1)	0.40(1)	0.21(1)	0.12(2)		1.08(3)	2.16(3)		
#RE-4-e	Nominal	0.25	0.25		0.25	0.25		1.00	2.00	1.386*R*	13.63
	Analytical	0.26(1)	0.24(1)		0.25(1)	0.25(1)		0.97(4)	1.97(9)		
#RE-5	Nominal	0.10	0.30	0.30	0.10	0.20		1.00	2.00	1.481*R*	13.62
	Analytical	0.09(1)	0.30(2)	0.32(1)	0.09(1)	0.20(1)		1.01(5)	2.10(8)		
#RE-5(Gd)	Nominal	0.20	0.30	0.30		0.10	0.10	1.00	2.00	1.493*R*	13.61
	Analytical	0.20(1)	0.31(1)	0.30(1)		0.09(1)	0.10(1)	0.88(1)	2.00(3)		
#RE-5-e [30]	Nominal	0.20	0.20	0.20	0.20	0.20		1.00	2.00	1.598*R*	13.63
	Analytical	0.23(1)	0.21(1)	0.19(2)	0.19(1)	0.17(1)		0.96(3)	2.17(3)		

(continued)

Table 11.2 (continued)

Sample	Nominal	La: *A*	Ce: *B*	Pr: *C*	Nd: *D*	Sm: *E*	Gd: *F*	Bi	S	ΔS_{mix} (*R*: Gas constant)	*c*-axis lattice constant (Å)
	Analytical	C'_{La}	C'_{Ce}	C'_{Pr}	C'_{Nd}	C'_{Sm}	C'_{Gd}	C'_{Bi}	C'_{S}		
#RE-6(Gd)	Nominal	0.20	0.30	0.20	0.10	0.10	0.10	1.00	2.00	1.665*R*	13.62
	Analytical	0.21(1)	0.31(1)	0.21(1)	0.08(1)	0.08(1)	0.11(1)	0.95(2)	2.07(5)		

Nominal: Nominal composition ($A + B + C + D + E + F = 1.0$), Analytical: Analytical content (normalized using $\Sigma(C'_{RE}) = 1.00$)

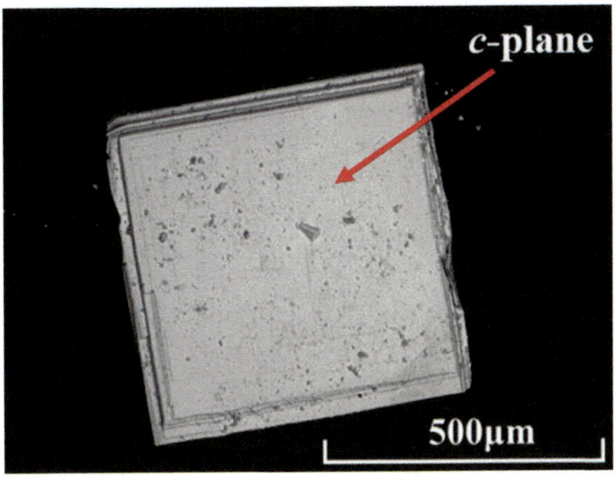

Fig. 11.5 SEM image of the as-grown *RE*OBiS_2 single crystal with the HEA concept [40]

Table 11.2 shows the nominal compositions and analytical contents of the *RE*OBiS_2 single crystals obtained with the HEA at the *RE* site. The analytical content of each element was defined as C'_{XX} (*XX*: element symbol) in this section. The obtained values were normalized using the atomic content obeying $\Sigma(C'_{RE}) = 1.00$. The analytical contents almost corresponded to the nominal compositions. *RE*OBiS_2 single crystals with more than 15 at% Gd substituted into the *RE* site have never been grown, which suggested that the *RE*OBiS_2 structure becomes unstable from small ionic radius element substitution. The ΔS_{mix} at the *RE* site was calculated from the analytical content (C'_{RE}) using Eq. 11.1, and the results are summarized in Table 11.2. The analytical contents of Bi and S were normalized using the relation $\Sigma(C'_{RE}) = 1.00$ and were 0.98 ± 0.03 and 2.07 ± 0.10, respectively, which almost corresponded to the stoicheiometric values. The *c*-axis lattice constants of those *RE*OBiS_2 single crystals were within 13.60–13.63 Å, except for that of CeOBiS_2 (#RE-1, 13.57 Å). The average ionic radius of the *RE* sites in *RE*OBiS_2 superconductors affects the superconducting properties. Only Ce has a mixed valence state configuration in which the electronic states are trivalent (Ce^{3+}) or tetravalent (Ce^{4+}) [30]. The ratio of the Ce valence states (Ce^{3+} and Ce^{4+}) at the *RE* site in those *RE*OBiS_2 single crystals was estimated by X-ray absorption fine structure (XAFS) spectroscopy. Consequently, the average ionic radius [34] of the *RE* sites considering the valence state (Ce^{3+} and Ce^{4+}) was evaluated for those samples. Table 11.3 summarizes the ratios of the Ce valence states and the average ionic radii of the *RE* sites. The Ce^{4+} concentrations were approximately 10 at% for all the samples. The average ionic radii of the *RE* sites were within a similar range of 1.101–1.109 Å, except for CeOBiS_2 (#RE-1, 1.127 Å). Therefore, except for #RE-1, the effect of the HEA concept on *RE*OBiS_2 superconductors could be evaluated using *RE*OBiS_2 single crystals without the influence of the *RE* site ionic radius or Ce^{4+} concentration.

Figure 11.6 shows the ΔS_{mix} dependence of T_c^{zero}, which was defined as 60 μΩcm in the ρ–T measurement. The large increase in T_c^{zero} from samples #RE-1 to #RE-2 is due to the chemical pressure effect at the *RE* site [35]. On the basis

Table 11.3 Ce^{3+} and Ce^{4+} contents at the *RE* site and the average ionic radius of the *RE* site in the grown *RE*OBiS$_2$ single crystals

Sample	Ce^{3+} and Ce^{4+} contents at *RE*-site		Average *RE*-site ionic radius (Å) (Calculated from the ratio of Ce^{3+} and Ce^{4+})
	Ce^{3+}	Ce^{4+}	
#RE-1	0.91	0.090	1.127
#RE-2	0.30	0.088	1.105
#RE-2-e [44]	0.37	0.13	1.104
#RE-3	0.23	0.090	1.109
#RE-3-e	0.23	0.097	1.101
#RE-4	0.18	0.093	1.109
#RE-4-e	0.13	0.11	1.104
#RE-5	0.18	0.12	1.102
#RE-5(Gd)	0.18	0.13	1.104
#RE-5-e [30]	0.12	0.095	1.104
#RE-6(Gd)	0.19	0.12	1.105

of these findings, a simple correlation between ΔS_{mix} and T_c^{zero} was not observed. However, except for #RE-2-e, the sample with an equivalent *RE* element composition (i.e., samples #RE-3-e, #RE-4-e, and #RE-5-e) at the *RE* site exhibited lower T_c^{zero} values than did similar ΔS_{mix} samples. Therefore, the equivalent and nonequivalent composition samples are plotted in blue and red, respectively, in Fig. 11.6. Different trends of T_c^{zero} behavior between the equivalent- and nonequivalent-composition samples were observed.

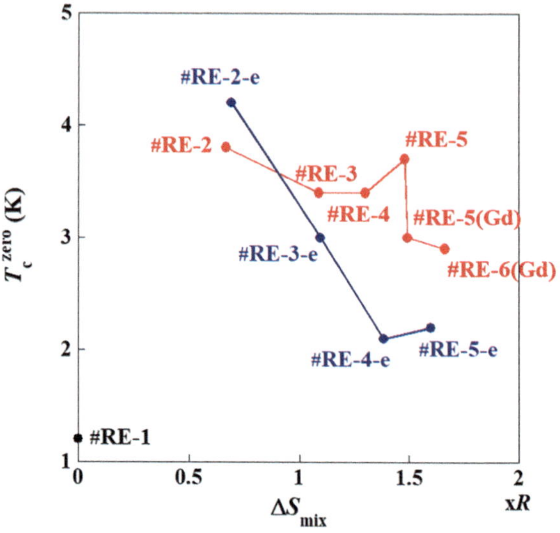

Fig. 11.6 ΔS_{mix} dependence of T_c^{zero} for the *RE*OBiS$_2$ single crystals. Reproduced with permission from [31]

Then, the superconducting anisotropies (γ_s) of those *RE*OBiS$_2$ single crystals were estimated using Eq. 11.2, which is the same as that in Sect. 2.2. Figure 11.7 shows the ΔS_{mix} dependence of γ_s. The γ_s values of the equivalent *RE* elemental composition samples were lower than those of the nonequivalent *RE* elemental composition samples. The γ_s value showed a different trend between the equivalent- and nonequivalent-composition samples, which also exhibited a similar trend in T_c^{zero}. These results indicate that the ΔS_{mix} at the *RE* site does not directly affect the superconducting properties (T_c^{zero} and γ_s). Instead, they are affected by an equivalent or nonequivalent *RE* element composition at the *RE* site. These results show that, compared to those of ΔS_{mix}, the superconducting properties (T_c^{zero} and γ_s) of these materials are significantly affected by the equivalent or nonequivalent *RE* element composition. A possible reason for this difference is the difference in atom configurations, which are similar to the ordering of *RE* elements. However, in HEA materials, the nonequivalent component $Ta_{34}Nb_{33}Hf_8Zr_{14}Ti_{11}$ becomes superconductivity [36], but superconductivity in the equivalent-composition Ta–Nb–Hf–Zr–Ti is not observed [37]. The *RE* element distribution in *RE*OBiS$_2$ was postulated as follows: The equivalent-composition and non-equivalent-composition samples were homogeneous and inhomogeneous at the *RE* site, respectively. In the nonequivalent-composition samples, specific *RE* elements become localized (disordered) and introduce a local and/or correlated disorder [38], which may affect the superconducting properties (T_c^{zero} and γ_s). Therefore, for the *RE*OBiS$_2$ superconductors, the superconducting transition temperature of the disordered (nonequivalent composition at the *RE* site) samples was higher than that of the ordered (equivalent composition at the *RE* site) samples.

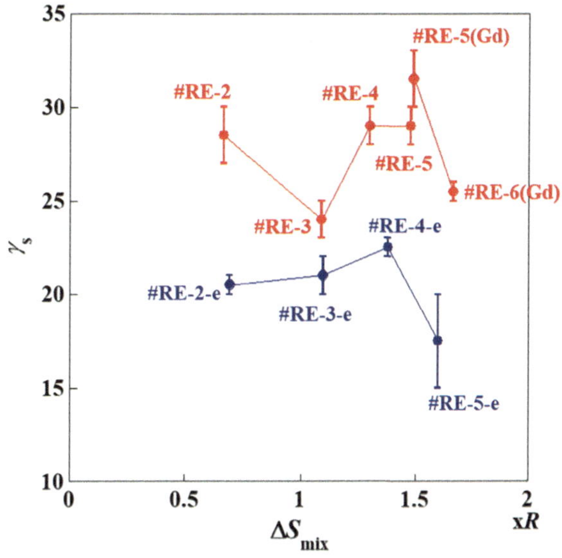

Fig. 11.7 ΔS_{mix} dependence of γ_s for the *RE*OBiS$_2$ single crystals. Reproduced with permission from [31]

11.4 Summary

RE-123 single-crystalline whiskers and *RE*OBiS_2 single crystals with the HEAs concept introduced in the *RE* site were grown using the Te-doping method and the alkali metal chloride flux method, respectively. No correlation or little effect between the superconducting properties (superconducting transition temperature: T_c and superconducting anisotropy: γ_s) and the mixed entropy (ΔS_{mix}) at the *RE* site were revealed using single crystals in those superconductors. In other words, the *RE*-123 and *RE*OBiS_2 superconductors can introduce the HEA concept without changing the superconducting properties. However, the superconducting properties of *RE*OBiS_2 were observed to be affected by equivalent and nonequivalent compositions at the *RE* site.

The authors would like to thank Mr. Y. Fujita and Mr. Y. Suzuki (University of Yamanashi) for their experimental execution and useful discussion.

Acknowledgements This work was partially supported by JSPS KAKENHI (Grant-in-Aid for Scientific Research (B): Grant Number 21H02022; Grants-in-Aid for Scientific Research (C): Grant Number JP19K05248; and Grant-in-Aid for Challenging Exploratory Research: Grant Number 21K18834).

References

1. J.G. Bednorz, K.A. Müller, Possible high-T_c superconductivity in the Ba–La–Cu–O system. Z. Phys. B Condens. Matter **64**, 189 (1986). https://doi.org/10.1007/BF01303701
2. M.K. Wu, J.R. Ashburn, C.J. Thorng, P.H. Hor, R.L. Meng, L. Gao, Z.J. Huang, Y.Q. Wang, C.W. Chu, Superconductivity at 93 K in a new mixed-phase Y–Ba–Cu–O compound system at ambient pressure. Phys. Rev. Lett. **58**, 908 (1987). https://doi.org/10.1103/PhysRevLett.58.908
3. H. Maeda, Y. Tanaka, M. Fukutomi, T. Asano, A new high-T_c oxide superconductor without a rare earth element. Jpn. J. Appl. Phys. **27**, L209 (1988). https://doi.org/10.1143/JJAP.27.L209
4. Y. Mizuguchi, H. Fujihisa, Y. Gotoh, K. Suzuki, H. Usui, K. Kuroki, S. Demura, Y. Takano, H. Izawa, O. Miura, BiS_2-based layered superconductor $Bi_4O_4S_3$. Phys. Rev. B **86**, 220510 (2012). https://doi.org/10.1103/PhysRevB.86.220510
5. Y. Mizuguchi, S. Demura, K. Deguchi, Y. Takano, H. Fujihisa, Y. Gotoh, H. Izawa, O. Miura, Superconductivity in novel BiS_2-based layered superconductor $LaO_{1-x}F_xBiS_2$. J. Phys. Soc. Jpn. **81**, 114725 (2012). https://doi.org/10.1143/JPSJ.81.114725
6. M. Nagao, A. Miura, I. Ueta, S. Watauchi, I. Tanaka, Superconductivity in $CeOBiS_2$ with cerium valence fluctuation. Solid State Commun. **245**, 11–14 (2016). https://doi.org/10.1016/j.ssc.2016.07.017
7. H.F. Zhai, P. Zhang, S.Q. Wu, C.Y. He, Z.T. Tang, H. Jiang, Y.L. Sun, J.K. Bao, I. Nowik, I. Felner, Y.W. Zeng, Y.K. Li, X.F. Xu, Q. Tao, Z.A. Xu, G.H. Cao, Anomalous Eu valence state and superconductivity in Undoped $Eu_3Bi_2S_4F_4$. J. Am. Chem. Soc. **136**, 15386 (2014). https://doi.org/10.1021/ja508564s
8. J. Mesot, P. Allenspach, U. Staub, A. Furrer, H. Mutka, R. Osborn, A. Taylor, Neutron-spectroscopic studies of the crystal field in $ErBa_2Cu_3O_x$ ($6<x<7$). Phys. Rev. B **47**, 6027–6036 (1993). https://doi.org/10.1103/PhysRevB.47.6027
9. R. Sagayama, H. Sagayama, R. Kumai, Y. Murakami, T. Asano, J. Kajitani, R. Higashinaka, T.D. Matsuda, Y. Aoki, Symmetry lowering in $LaOBiS_2$: a mother material for BiS_2-based

layered superconductors. J. Phys. Soc. Jpn. **84**, 123703 (2015). https://doi.org/10.7566/JPSJ.84.123703
10. K. Momma, F. Izumi, VESTA: a three-dimensional visualization system for electronic and structural analysis. J. Appl. Crystallogr. **41**, 653–658 (2008). https://doi.org/10.1107/S0021889808012016
11. J.W. Yeh, S.K. Chen, S.J. Lin, J.Y. Gan, T.S. Chin, T.T. Shun, C.H. Tsau, S.Y. Chang, Nanostructured high-entropy alloys with multiple principal elements: novel alloy design concepts and outcomes. Adv. Eng. Mater. **6**, 299–303 (2004). https://doi.org/10.1002/adem.200300567
12. Y. Shukunami, A. Yamashita, Y. Goto, Y. Mizuguchi, Synthesis of RE123 high-T_c superconductors with a high-entropy-alloy-type RE site. Physica C **572**, 1353623 (2020). https://doi.org/10.1016/j.physc.2020.1353623
13. A. Yamashita, Y. Shukunami, Y. Mizuguchi, Improvement of critical current density of $RE\mathrm{Ba_2Cu_3O_{7-\delta}}$ by increase in configurational entropy of mixing. R. Soc. Open Sci. **9**, 211874 (2022). https://doi.org/10.6084/m9.figshare.c.5899201
14. K. Wang, Q. Hou, A. Pal, H. Wu, J. Si, J. Chen, S. Yu, Y. Chen, W. Lv, J.-Y. Ge, S. Cao, J. Zhang, Z. Feng, Structural and physical properties of high-entropy $RE\mathrm{Ba_2Cu_3O_{7-\delta}}$ Oxide superconductors. J. Supercond. Nov. Magn. **34**, 1379–1385 (2021). https://doi.org/10.1007/s10948-021-05855-5
15. A. Yamashita, K. Hashimoto, S. Suzuki, Y. Nakanishi, Y. Miyata, T. Maeda, Y. Mizuguchi, Fabrication of high-entropy $\mathrm{REBa_2Cu_3O_{7-\delta}}$ thin films by pulsed laser deposition. Jpn. J. Appl. Phys. **61**, 050905 (2022). https://doi.org/10.35848/1347-4065/ac5b39
16. J. Chen, R. Huang, X. Zhou, D. Zhou, M. Li, C. Bai, Z. Liu, C. Cai, Nucleation and epitaxy growth of high-entropy $\mathrm{REBa_2Cu_3O_{7-\delta}}$ (RE= Y, Dy, Gd, Sm, Eu) thin films by metal organic deposition. J. Rare Earths **41**, 1091–1098 (2022). https://doi.org/10.1016/j.jre.2022.06.006
17. A.R. Mazza, X. Gao, D.J. Rossi, B.L. Musico, T.W. Valentine, Z. Kennedy, J. Zhang, J. Lapano, V. Keppens, R.G. Moore, M. Brahlek, C.M. Rost, T.Z. Ward, Searching for superconductivity in high entropy oxide Ruddlesden-popper cuprate films. J. Vac. Sci. Technol. A **40**, 013404 (2022). https://doi.org/10.1116/6.0001441
18. Y. Suzuki, M. Nagao, Y. Maruyama, S. Watauchi, I. Tanaka, Growth of $RE\mathrm{Ba_2Cu_3O}_x$ single-crystal whiskers utilizing the concept of high-entropy alloys. Jpn. J. Appl. Phys. **62**, 033001 (2023). https://doi.org/10.35848/1347-4065/acbf91
19. M. Nagao, M. Sato, H. Maeda, S.-J. Kim, T. Yamashita, Growth and superconducting properties of $\mathrm{Bi_2Sr_2CaCu_2O_{8+\delta}}$ single-crystal whiskers using tellurium-doped precursors. Appl. Phys. Lett. **79**, 2612–2614 (2001). https://doi.org/10.1063/1.1401784
20. M. Nagao, M. Sato, H. Maeda, S.-J. Kim, T. Yamashita, Growth and superconductivity of $\mathrm{(BiPb)_2Sr_2Ca_2Cu_3O_{10+\delta}}$ single-crystal whiskers. Jpn. J. Appl. Phys. **41**, L43–L45 (2002). https://doi.org/10.1143/JJAP.41.L43
21. M. Nagao, T. Kawae, K.S. Yun, H. Wang, Y. Takano, T. Hatano, T. Yamashita, M. Tachiki, H. Maeda, Intrinsic Josephson junctions in $\mathrm{Y_1Ba_2Cu_3O}_x$ single-crystal whiskers grown using Te-doped precursors. J. Appl. Phys. **98**, 073903 (2005). https://doi.org/10.1063/1.2064312
22. M. Nagao, S. Watauchi, I. Tanaka, T. Okutsu, Y. Takano, T. Hatano, H. Maeda, Growth and anisotropic properties of $\mathrm{RBa_2Cu_3O}_x$ single-crystal whiskers. Jpn. J. Appl. Phys. **49**, 033101 (2010). https://doi.org/10.1143/JJAP.49.033101
23. G. Blatter, V.B. Geshkenbein, A.I. Larkin, From isotropic to anisotropic superconductors: a scaling approach. Phys. Rev. Lett. **68**, 875–878 (1992). https://doi.org/10.1103/PhysRevLett.68.875
24. Y. Iye, I. Oguro, T. Tamegai, W.R. Datars, N. Motohira, K. Kitazawa, Anisotropy of $\mathrm{Bi_2Sr_2CaCu_2O}_{8+y}$ revisited. Physica C **199**, 154–160 (1992). https://doi.org/10.1016/0921-4534(92)90554-P
25. H. Iwasaki, O. Taniguchi, S. Kenmochi, N. Kobayashi, Anisotropic behavior of superconductivity in $\mathrm{Y}_{1-x}\mathrm{Pr}_x\mathrm{Ba_2Cu_3O}_y$ films. Physica C **244**, 71–77 (1995). https://doi.org/10.1016/0921-4534(95)00034-8
26. T. Katayama, J. Shimoyama, H. Ogino, S. Horii, K. Kishio, Evaluation of J_c anisotropy in Nd123 single crystals. J. Phys. Conf. Ser. **97**, 12144 (2008). https://doi.org/10.1088/1742-6596/97/1/012144

27. R. Sogabe, Y. Goto, Y. Mizuguchi, Superconductivity in $REO_{0.5}F_{0.5}BiS_2$ with high-entropy-alloy-type blocking layers. Appl. Phys. Express **11**, 053102 (2018). https://doi.org/10.7567/APEX.11.053102
28. R. Sogabe, Y. Goto, T. Abe, C. Moriyoshi, Y. Kuroiwa, A. Miura, K. Tadanaga, Y. Mizuguchi, Improvement of superconducting properties by high mixing entropy at blocking layers in BiS2-based superconductor $REO_{0.5}F_{0.5}BiS_2$. Solid State Commun. **295**, 43–49 (2019). https://doi.org/10.1016/j.ssc.2019.04.001
29. F.I. Abbas, Y. Nakahira, A. Yamashita, Md. R. Kasem, M. Yoshida, Y. Goto, A. Miura, K. Terashima, R. Matsumoto, Y. Takano, C. Moriyoshi, Y., Mizuguchi, estimation of the Grüneisen parameter of high-entropy alloy-type functional materials: the cases of $REO_{0.7}F_{0.3}BiS_2$ and MTe. Condens. Matter **7**, 34 (2022). https://doi.org/10.3390/condmat7020034
30. Y. Fujita, K. Kinami, Y. Hanada, M. Nagao, A. Miura, S. Hirai, Y. Maruyama, S. Watauchi, Y. Takano, I. Tanaka, Growth and characterization of $ROBiS_2$ high-entropy superconducting single crystals. ACS Omega **5**, 16819–16825 (2020). https://doi.org/10.1021/acsomega.0c01969
31. Y. Fujita, M. Nagao, A. Miura, D. Urushihara, Y. Mizuguchi, Y. Maruyama, S. Watauchi, Y. Takano, I. Tanaka, Effects of equivalent composition on superconducting properties of high-entropy $REOBiS_2$ (RE = La, Ce, Pr, Nd, Sm, Gd) single crystals. Physica C **608**, 1354254 (2023). https://doi.org/10.1016/j.physc.2023.1354254
32. M. Nagao, S. Demura, K. Deguchi, A. Miura, S. Watauchi, T. Takei, Y. Takano, N. Kumada, I. Tanaka, Structural analysis and superconducting properties of F-substituted $NdOBiS_2$ single crystals. J. Phys. Soc. Jpn. **82**, 113701 (2013). https://doi.org/10.7566/JPSJ.82.113701
33. FT-salt-FACT salt-phase diagrams (CsCl-KCl). https://www.crct.polymtl.ca/fact/phase_diagram.php?file=CsCl-KCl.jpg&dir=FTsalt
34. R.D. Shannon, Revised effective ionic radii and systematic studies of interatomic distances in halides and chalcogenides, Acta Crystallogr. Sect. A Found Adv. A**32**, 751–767 (1976). https://doi.org/10.1107/S0567739476001551
35. Y. Mizuguchi, A. Miura, J. Kajitani, T. Hiroi, O. Miura, K. Tadanaga, N. Kumada, E. Magome, C. Moriyoshi, Y. Kuroiwa, In-plane chemical pressure essential for superconductivity in $BiCh_2$-based (Ch: S, Se) layered structure. Sci. Rep. **5**, 14968 (2015). https://doi.org/10.1038/srep14968
36. P. Koželj, S. Vrtnik, A. Jelen, S. Jazbec, Z. Jagliči, S. Maiti, M. Feuerbacher, W. Steurer, J. Dolinšek, Discovery of a superconducting high-entropy alloy. Phys. Rev. Lett. **113**, 107001 (2014). https://doi.org/10.1103/PhysRevLett.113.107001
37. S. Marik, M. Varghese, K.P. Sajilesh, D. Singh, R.P. Singh, Superconductivity in equimolar Nb-Re-Hf-Zr-Ti high entropy alloy. J. Alloys Compd. **769**, 1059–1063 (2018). https://doi.org/10.1016/j.jallcom.2018.08.039
38. D.A. Keen, A.L. Goodwin, The crystallography of correlated disorder. Nature **521**, 303–309 (2015). https://doi.org/10.1038/nature14453
39. Y. Suzuki, Personal communication
40. Y. Fujita, Graduation Thesis (2020)
41. M. Tanaka, M. Nagao, R. Matsumoto, N. Kataoka, I. Ueta, H. Tanaka, S. Watauchi, I. Tanaka, Y. Takano, Superconductivity and its enhancement under high pressure in "F-free" single crystals of $CeOBiS_2$. J. Alloy. Compd. **722**, 467–473 (2017). https://doi.org/10.1016/j.jallcom.2017.06.125
42. M. Nagao, A. Miura, D. Urushihara, Y. Maruyama, Y. Goto, Y. Mizuguchi, C. Moriyoshi, Y. Kuroiwa, Y. Wang, S. Watauchi, T. Asaka, Y. Takano, K. Tadanaga, I. Tanaka, Flux growth and superconducting properties of (Ce, Pr)OBi_S2 single crystals. Front. Chem. **8**, 44 (2020). https://doi.org/10.3389/fchem.2020.00044
43. N. Kase, M. Matsumoto, K. Kondo, J. Gouchi, Y. Uwatoko, T. Sakakibara, N. Miyakawa, Superconductivity of electron-doped $NdOBiS_2$ by substitution of mixed-valence Ce Ions. J. Phys. Soc. Jpn. **88**, 103703 (2019). https://doi.org/10.7566/JPSJ.88.103703
44. Y. Hanada, Master's Thesis (2020)

Correction to: High-Entropy Alloy Superconductors

Jiro Kitagawa and Yoshikazu Mizuguchi

Correction to:
J. Kitagawa and Y. Mizuguchi (eds.), *High-Entropy Alloy Superconductors*, Springer Series in Solid-State Sciences 202, https://doi.org/10.1007/978-981-97-4129-8

The original version of the book was inadvertently published with an incorrect affiliation of the second Editor (Yoshikazu Mizuguchi), which has now been changed from "Research Fellow at, UEH University Ho Chi Minh City Vietnam, Ho Chi Minh, Vietnam" to "Department of Physics, Tokyo Metropolitan University, Tokyo, Japan". The book has been updated with the change.

The updated version of the book can be found at
https://doi.org/10.1007/978-981-97-4129-8

J. Kitagawa and Y. Mizuguchi (eds.), *High-Entropy Alloy Superconductors*, Springer Series in Solid-State Sciences 202, https://doi.org/10.1007/978-981-97-4129-8_12

Index

J. Kitagawa and Y. Mizuguchi (eds.), *High-Entropy Alloy Superconductors*, Springer Series in Solid-State Sciences 202, https://doi.org/10.1007/978-981-97-4129-8

GPSR Compliance

The European Union's (EU) General Product Safety Regulation (GPSR) is a set of rules that requires consumer products to be safe and our obligations to ensure this.

If you have any concerns about our products, you can contact us on ProductSafety@springernature.com

In case Publisher is established outside the EU, the EU authorized representative is:

Springer Nature Customer Service Center GmbH
Europaplatz 3
69115 Heidelberg, Germany

Batch number: 10703855

Printed by Printforce, the Netherlands